stereochemistry of organometallic and inorganic compounds 4

Stereochemical Control, Bonding and Steric Rearrangements

stereochemistry of organometallic and inorganic compounds 4

Stereochemical Control, Bonding and Steric Rearrangements

Edited by

IVAN BERNAL

Department of Chemistry, University of Houston, Houston, TX 77004, U.S.A.

ELSEVIER

Amsterdam — Oxford — New York — Tokyo   1990

ELSEVIER SCIENCE PUBLISHERS B.V.
Sara Burgerhartstraat 25
P.O. Box 211, 1000 AE Amsterdam, The Netherlands

Distributors for the United States and Canada:

ELSEVIER SCIENCE PUBLISHING COMPANY INC.
655, Avenue of the Americas
New York, NY 10010, U.S.A.

ISBN 0-444-88841-1

Printed in The Netherlands

Preface

In the introduction to volume 3, I suggested that we were witnessing the start of a new era in synthetic chemistry inasmuch as we were, by natural progression of events in chemical sophistication, by new pressures brought about by environmental concerns, larger demands from the medical and pharmaceutical sectors and the economics of the market place, being forced into demanding larger and larger stereochemical control and better and better product yields from chemical reactions capable of producing useful products.

Probably no better illustration of the pressures to sharpen our stereochemical control tools can be cited than that generated by the drug industry. Of 528 chiral compounds used in the treatment of medical problems in humans, 467 of them are administered in the form of racemates as a result of the difficulties associated with either the synthesis of stereochemically pure product or the separation of racemates obtained via uncontrolled reactions (source: Chemical and Engineering News, March 18, 1990, p. 38). In that regard, the increased interest in the methods of conglomerate crystallization witness the fact that solution of these problems are sought from any methodology which may provide aid in the elimination of valid concerns associated with the use of these mixtures.

In view of the well-known tragedy of thalidomide, which was administered as a racemate, and whose medically non-useful enantiomer was the cause of the teratogenic deformations, there is no doubt that the manufacturers and marketers of other such racemates are taking a chance that currently unforeseen, undesirable long term effects will not be found among other such racemic drugs. Thus, both new approaches to tried and true chemical methodology and radically new methods of approaching chemistry and stereochemistry will be useful in attempts to remedy the, apparently unavoidable, current need to dispense chemicals of the type under discussion.

Thus, it is timely to have a review of newer discoveries in such tried and true methods of C-C bond formation as alkylations and aldol reactions of metal enolates. Professor Yamamoto has contributed, in collaboration with Dr. Sasaki, such a review -- the first chapter of this volume. The senior author is no stranger to this series, having contributed a chapter to volume 3.

Dr. S. Sakaki gauges for us the ability of ab-initio methods to justify the results of empirical observations in the field of transition metal derivatives of small molecules such as N_2, CO_2, etc... Moreover, once having established the strengths and weaknesses of the various approaches to such theoretical calculations, he proceeds to a more interesting evaluation of these methods; e.g., their ability to predict, in those areas in which they are particularly strong and reliable, chemical and stereochemical events and/or results in advance of experiments, later carried out in the laboratory.

It is precisely such a balanced evaluation of strengths and weaknesses of the various methodologies of molecular orbital theory that is needed for us to make progress in the valuable arena of theoretical understanding vs. experimental fact. It is hoped that this review help the non-participant of theoretical research (the user of results) to obtain a better understanding and appreciation of the value and power of such theories. Hopefully, as a result, more experiments will be designed to expose more clearly the fine line between success and failure of numerical predictions. Better yet, to conceive untried empirical methods whose early seeds appeared first as suggestions derived from a series of numbers in some of the tabular results of a theorist. We have already seen some examples of that wish-come-true, but we need more; hopefully, soon. And, even more hopefully, Dr. Sakaki's review may be the catalyst needed by a careful and hopeful reader to take a bold step in this direction.

The awesome thoroughness and clarity with which Professor Zanello has reviewed the stereochemical results of electron transfer reactions in mono-nuclear copper compounds is remarkable. Even more remarkable is that 322 references were found in the literature despite the fact that the discipline reviewed was limited to discussion of those mono-nuclear Cu compounds for which stereochemical changes were determined or inferred in the course of electron transfer chemistry. At a time when such compounds are being implicated in a bewildering array of chemical and biochemical processes, it is imperative that order be introduced into the vast literature that has appeared, and he has done so in a magisterial way.

The enzymatic aspects of this field are well known thanks to such recent reviews as those cited by Professor Zanello in the introduction to his chapter. Many other aspects of this field are emerging in areas of equal importance in chemistry and biochemistry. For example, the recent implication of redox $Cu(I)<-->Cu(II)$ transformations in compounds found useful in the mitigation of radiation damage, and the stereochemical transformations they undergo during such valuable redox processes, are the crux of Professor Zanello's survey-- steric changes accompanying redox transformations. He has done a great service to the chemical community by his magnificent effort.

Finally, I would like to bring to the attention of the readers of this volume the fact that during the past five years we have produced four volumes of this series, and that the fifth one is well on its way towards completion. That is not only a rapid pace under the most ideal circumstances but, in fact, it is quite remarkable given the personal and professional problems faced by some contributors and editor alike.   Thus, it is with gratitude for those who have made this series possible that I face the task of writing an introduction to another volume of this series. My sincere thanks go equally to the authors contributing to the four volumes and   to the staff of Elsevier for their co-operation and patience.

Without going into details here, I specially thank the contributors to this volume for their forbearance in seeing this project completed. A number of my colleagues at the University of Houston and I had the misfortune of having our offices and laboratories seriously damaged in the early hours of the 24th of December, 1989. Such an event is difficult to recover from, specially when not only hardware, but files (computer as well as paper) are seriously damaged. The delays in production of the volume caused by this event does not affect the quality of the work done by the contributors, but will, unavoidably, date the coverage of their topics by a few months. No one could regret this incident more than I.

Ivan Bernal

Houston, Texas
June 5th, 1990.

VIII

List of Contributors

Nobuki Sasaki   Department of Chemistry, Faculty of Science, Tohoku University, Sendai 980, Japan

Shigeyoshi Sakaki   Department of Applied Chemisitry, Faculty of Engineering, Kumamoto University, Kurokami, Kumamoto 860, Japan

Yoshinori Yamamoto   Department of Chemistry, Faculty of Science, Tohoku University, Sendai 980, Japan

Piero Zanello   Dipartimento di Chimica, Università di Siena, Piano dei Mantellini 44, I-53100 Siena, Italy

Table of Contents

Chapter 1
THE STEREOCHEMISTRY OF C-C BOND FORMATION VIA METAL ENOLATES.
ALKYLATION AND HETEROATOM INTRODUCTION

Y. Yamamoto and N. Sasaki

Chapter 2
TRANSITION-METAL COMPLEXES OF N_2, CO_2, AND SIMILAR SMALL MOLECULES. AB-
INITIO MO STUDIES OF THEIR STEREOCHEMISTRY AND COORDINATE BONDING NATURE

S.Sakaki

Chapter 3
ELECTROCHEMISTRY OF MONONUCLEAR COPPER COMPLEXES.
STRUCTURAL REORGANIZATIONS ACCOMPANYING REDOX CHANGES

P. Zanello

THE STEREOCHEMISTRY OF C-C BOND FORMATION VIA METAL ENOLATES. ALKYLATION AND HETEROATOM INTRODUCTION

Y. Yamamoto and N. Sasaki

ABBREVIATIONS

BMDA	bromomagnesium diisopropylamine
BMHMDS	bromomagnesium hexamethyldisilazide
CP	η^5 - C_5H_5
CP*	η^5 - C_5Me_5
DME	dimethoxyethane
DMF	dimethylformamide
DTBAD	di-tert-butyl azodicarboxylate
(DBAD)	
HMDS	hexamethyldisilazide
HMPA	hexamethylphosphoric triamide
LICA	lithium cyclohexylisopropylamide
LDA	lithium diisopropylamide
LDEA	lithium diethylamide
LHMDS	lithium hexamethyldisilazide
LiTMP	lithium 2, 2, 6, 6 - tetramethylpiperidide
(LTMP)	
LOBA	lithium t-octyl-t-butylamide
TASCN	tris (diethylamino) sulfonium cyanide
TBDMS(Cl)	t-butyldimethylsilyl (chloride)
TMEDA	N, N, N', N'-tetramethylethylenediamine
TMS(Cl)	trimethylsilyl (chloride)

THE STEREOCHEMISTRY OF C-C BOND FORMATION VIA METAL ENOLATES. ALKYLATION AND HETEROATOM INTRODUCTION.

Yoshinori Yamamoto* and Nobuki Sasaki

Comprehensive reviews on the alkylation reactions and aldol reactions of metal enolates were published by Evans in 1984 (alkylation)[1] and in 1982 (aldol)[2], by Heathcock in 1984 (aldol)[3,4], by Mukaiyama in 1982 (aldol)[5], by Masamune in 1985 (aldol)[6], and by many other active workers in this field (acyclic stereocontrol with organometallic reagents)[7]. The research in this field has increased enormously during the past decade. According to the CA file for the period 1980-1986, the number of papers on metal enolates and aldol reactions is over 650. Earlier reviews which include structure and reactivity of alkali metal enolates, as well as other aspects of the alkylation and acylation reactions, have been published by Gompper[8], Sheveiev[9], House[10], d'Angelo[11], Hajos[12], Caine[13], and Jackman[14].

The reactions of metal enolates are roughly divided into two classes; (i) alkylation and (ii) aldol reactions. This review article is therefore concentrated on the alkylation and related reactions of metal enolates. The aldol and related reactions will be treated elsewhere. In addition to the alkylation, we mention on the introduction of the heteroatoms such as O and N at the α-position of carbonyl groups via the metal enolates.

R=C $\lesssim$ alkylation   aldol   X=O,NR'

R=O   or   N (heteroatom introduction)

1. STEREOCHEMISTRY OF ALKYLATION OF ACYCLIC SYSTEMS

1. 1 BASIC PRINCIPLES

The stereocontrol of cyclic metal enolates, like **1**, has been achieved much more easily than control over acyclic systems (**2**) since the conformation of **1** is fixed and, thus, electrophiles may predominantly attack either of the enolate faces from the more stereoelectronically favorable direction. On the other hand, the conformation of the acyclic enolate is flexible, as shown in **2**, and the reaction takes place with several possible conformations, resulting in low stereoselectivity. If there is a heteroatom such as O, N, S..., at an appropriate position of acyclic chain, the conformation of the acyclic system is fixed via metal chelation as shown in **3**. In this case, very high stereocontrol has been realized (chelation control); electrophiles attack from the less hindered side of **3** (from the α-face).

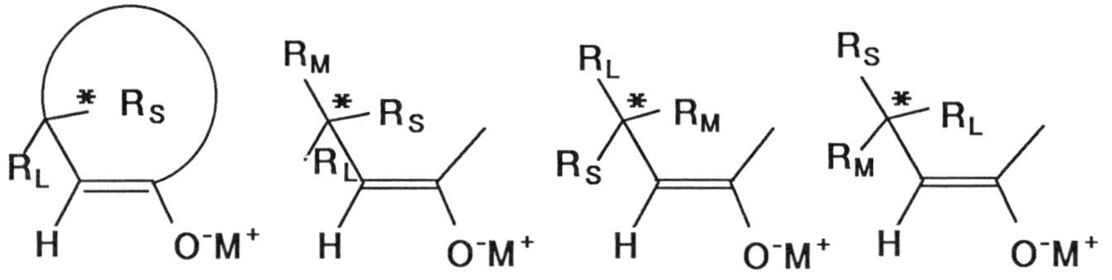

* ,asymmetric center

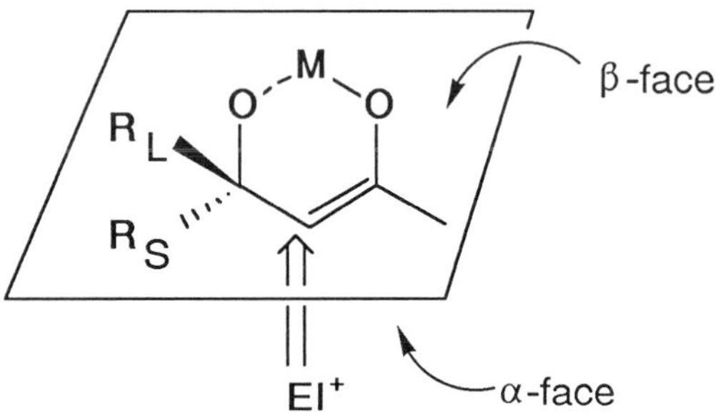

The allylic strain concept is useful for predicting the stereochemical outcome of electrophilic reactions of ordinary enolates which have no heteroatoms at the carbon chain. Let's consider the lithium ester enolate **4**, which possesses a chiral center at the β position of the corresponding ester (that is, the allylic position of **4**). Three possible conformations are conceivable (**4a**, **4b** and **4c**). It is obvious that there is a severe steric repulsion between R_L and OLi in **4c** and a similar repulsion between R_S and OLi in **4b** ($A^{(1,3)}$-strain). Therefore, **4a** is the most stable conformation and the reaction takes place mostly through this conformation. An electrophile attacks from the less hindered face (α face) of **4a**, producing the syn isomer **5** predominantly. Consequently, the allylic strain ($A^{(1,3)}$-strain), which stems from the steric congestion between C-1 and C-3 substituents of allylic systems (e.g.R_L and OLi in **4c**), forces the enolate to take the conformation **4a** (the eclipsed model)(Houk model)[15].

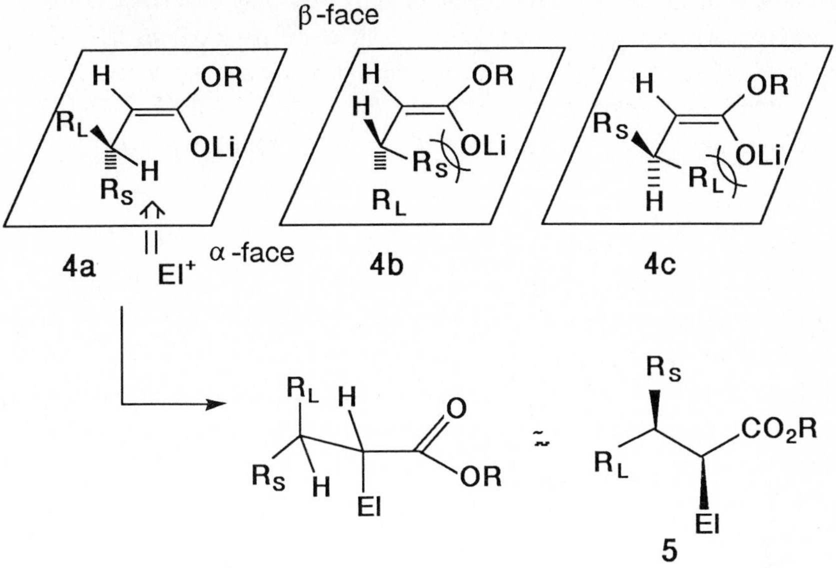

β-face

4a   El⁺   α-face   4b   4c

5

The conjugate addition of BuCu·BF₃ to **6** followed by hydrolysis produced a mixture of syn-**7** and anti-**8** in the ratio of 30:70. Treatment of **9** with LDA(lithium diisopropylamide) followed by addition of MeI gave a mixture of syn -**7** and anti-**8** in the ratio of 65:35. These diastereoselectivities can be well explained by the eclipsed model(**10** and **11**). The enolate derived from the conjugate addition undergoes protonation to **10** from the less hindered side, giving **8** predominantly. The lithium enolate derived from **9** is attacked by MeI from the less hindered side of **11**, producing **7** preferentially.[18]

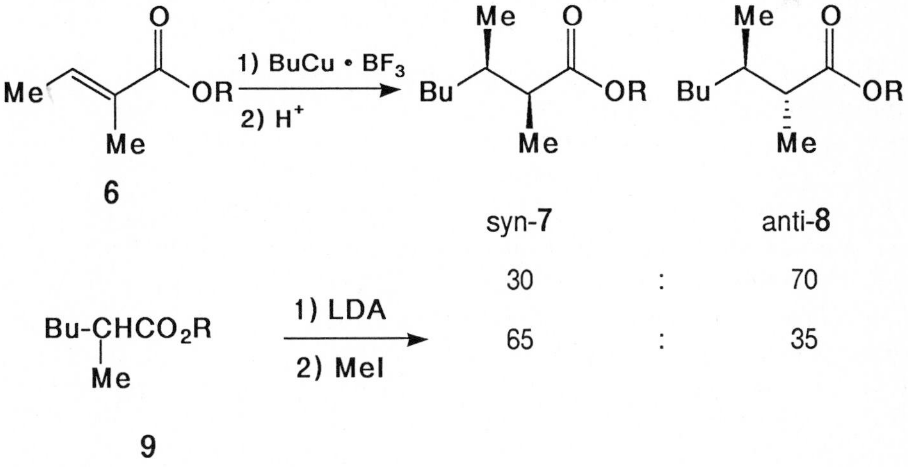

	syn-**7**		anti-**8**
	30	:	70
	65	:	35

6

9

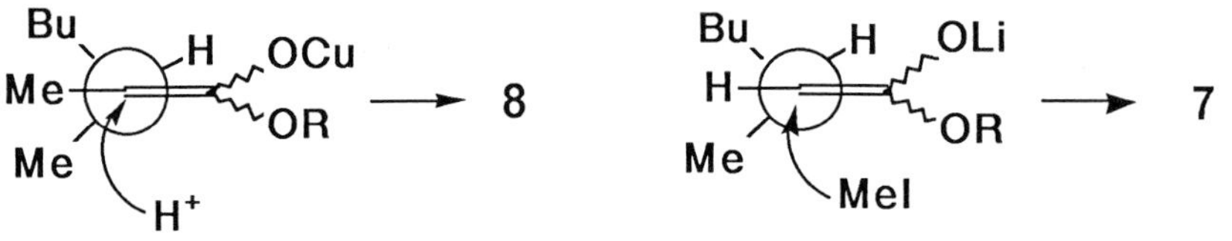

8 **7**

10 **11**

Theoretical calculations have been carried out to provide a model for the electrophilic attack to double bonds[15]. The rule is summarized in **12**, which is quite similar to **4a**. The preferred conformation possesses the small group (R_S) eclipsing the double bond. The electrophile attacks from the less hindered β face to give **13** preferentially. The diastereoselectivity of nucleophilic attack to aldehydes and ketones is reasonably explained by Cram's rule[16] and/or Felkin-Anh[17] model which is summarized in **14**. The nucleophile attacks again from the face opposite to the large group (R_L), giving **15** predominantly. The R_S occupies the inside position in **12**, while the R_S occupies the outside position in **14**. Therefore, electrophilic attack must occur from the opposite side to that of the corresponding nucleophilic attack on carbonyl groups. The following examples clearly demonstrate that the electrophilic rule is prone to be the opposite of Cram's rule (Felkin-Anh rule)[19].

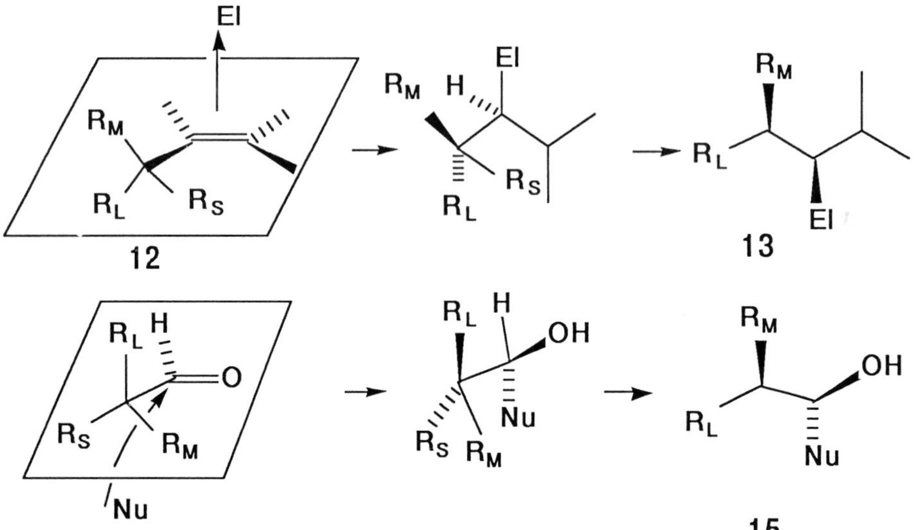

12 **13**

14 **15**

The methylation of lithium enolate (16) gave a 60:40 mixture of 17 and 18. Here again, the syn-isomer (17) was produced predominantly. On the other hand, the protonation of 19 afforded the anti-isomer (18) preferentially; 17:18=14:86. This trend is in good agreement with the above observation (see 6 and 9). The methylation of aldehyde (20) with methyl Grignard reagent produced the anti-isomer (22) predominantly and the hydride reduction of ketone (23) with LiAlH₄ gave the syn-isomer (21) preferentially. The diastereoselectivity of the electrophilic methylation is opposite to that of the nucleophilic methylation. Further, the diastereoselectivity of the electrophilic protonation is opposite to that of the nucleophilic hydride addition.

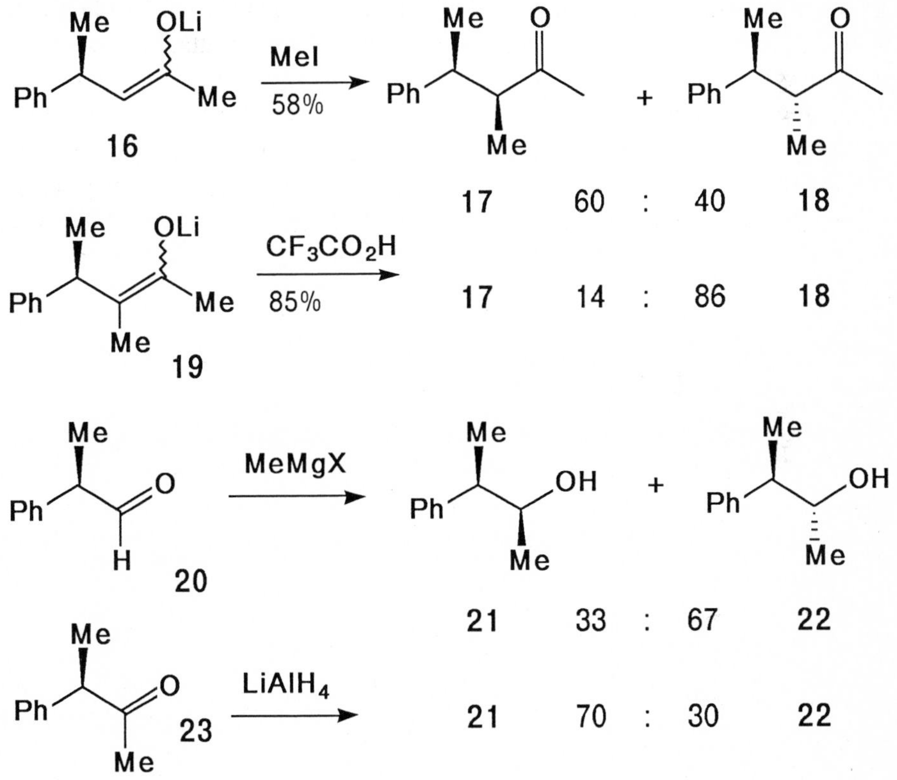

1.2 HOW TO ENHANCE THE DIASTEREOSELECTIVITY

The conjugate addition of the silyl-cuprate reagent to **24** followed by methylation gave the syn-isomer (**25**) with very high diastereoselectivity (syn:anti=97:3).[20] This selectivity is very high in comparison with the diastereoselectivity of the other methylation (**9** or **16**). The conjugate addition to **26** followed by protonation produced the anti-isomer (**27**) in the ratio of 85:15. The protonation selectivity is not so high as the methylation diastereoselectivity, but still higher than the ordinary protonation reaction (**6**). The methylation reaction proceeds through the eclipsed model (**28**). The electrophile attacks anti to the silyl group for steric or for electronic reasons, or for both combined. Quite interestingly, the conjugate addition of the silyl cuprate to **29**, followed by methylation, gave a mixture of adducts in the ratio of 54:46[21]. The low diastereoselectivity is due to lack of the allylic strain in the intermediate (**30**).

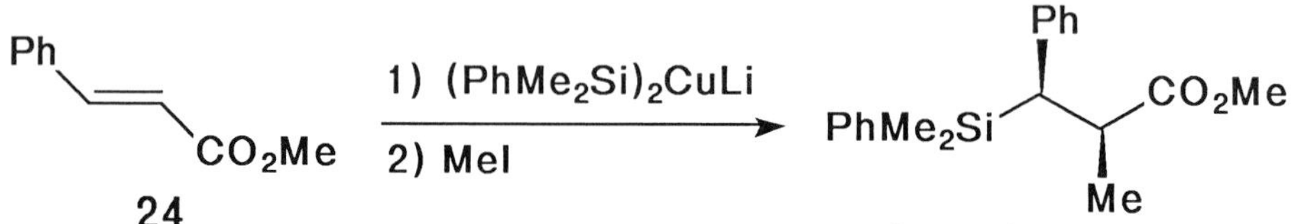

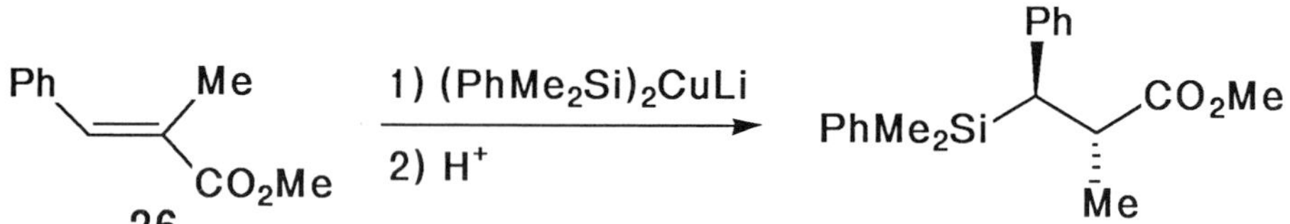

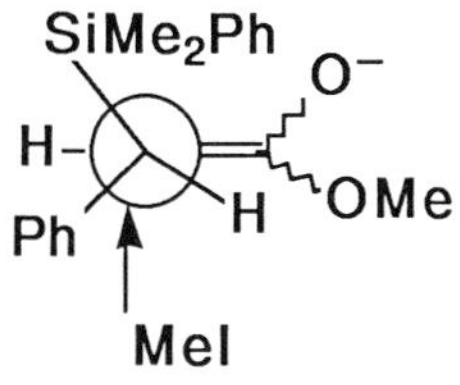

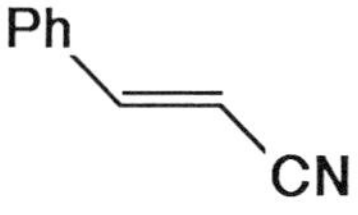

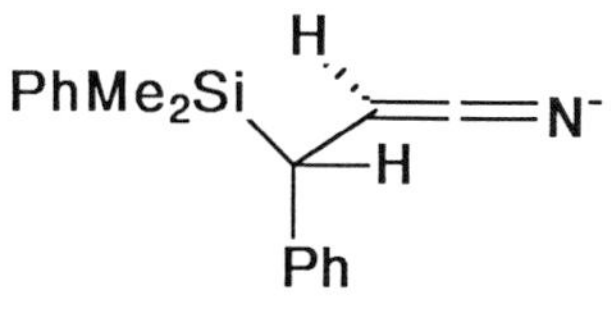

The lowest-energy conformation (**28**) of the enolate is attacked anti to the silyl group. In the allylic silane derivatives, electrophiles also attack the double bond anti to the silyl group (anti-S_E') owing to σ-π conjugation.[22] The difference in the two series is that the electrophile attacks on C-3 of the allylsilanes while it attacks on C-2 of the enolates. Instead of methyl cinnamate (24), the R' substituted esters (R'CH=CHCO$_2$ Me) were also treated with the silyl cuprate followed by methyl iodide. The degree of diastereoselectivity depended upon R' and decreased in the series Ph(97:3) > Me(91:9) > i-Pr(85:15) > t-Bu(66:34), but retained the same sense. Since the effective size of t-Bu is larger than that of a trimethylsilyl group,[23] a substantial portion of the diastereoselectivity is electronic in origin.

This is supported by the results of conjugate addition of an α trialkyltin group, instead of the silyl.[21,24] Ethyl crotonate gave 31 in the ratio of 98:2 upon treatment with n-Bu$_3$SnLi - MeI. Based upon available A values[23], the steric requirements of the asymmetric substituents apparently follow the order n-Bu$_3$Sn < Me < Me$_2$PhSi.[25] If the diastereoselectivity is dictated only by the steric effect of the eclipsed model, such a high selectivity must not be realized or even the opposite diastereoselectivity (anti) may be produced, owing to the order n-Bu$_3$Sn < Me. Consequently, the high diastereoselectivity appears to be substantially electronic in origin (σ-π conjugation). In the conjugate addition of n-Bu$_3$SnLi , the perpendicular model (32) is proposed instead of the eclipsed model. Regardless of 32 or 28 , the electrophilic attack takes place in the same sense.

In conclusion, one way to enhance the diastereoselectivity is to introduce the Si and Sn groups at the asymmetric α-position of enolates. The stereoelectronic effect of these asymmetric substituents forces the enolate to take either the eclipsed or perpendicular conformation, making the electrophilic attack anti to this substituent.

The steric effect of the asymmetric substituent at the α-position is also important in enhancing the diastereoselectivity.[26] The conjugate addition of dithioacetal anion (34) to 35 produced the lithium enolate (36). Addition of MeI to 36 took place anti to the Nu substituent and gave the syn-isomer (37) exclusively. Further, treatment of 38 with 34, followed by protonolysis, produced the anti-isomer (**40**) either exclusively or predominantly. Here again, the protonolysis takes place anti to the sterically bulky Nu substituent of **39**. However, there is a possibility that the intermediate enolate takes the chelate form (**41**) owing to the interaction between S and Li. In **41**, R^1 group is bigger than H; and, thus, methylation occurs also from the α face. In order to distinguish whether the high diastereoselectivity is due to the chelation (**41**) or to the steric effect of the allylic strain (**36**), the trans enolate (**42**) was prepared in which Li metal could not chelate to Nu substituent. Methylation of **42** produced a similar selectivity as **36**. Consequently, the steric effect at the γ-position, as well as the stereoelectronic effect, plays an important role to enhance the diastereoselectivity.

NuLi(34)
THF
MeI
35
36
37 >99 : 1
34
AcOH
38
39
40 >99 : 1 ~ 76 : 24
R1=Me,Ph; R2=Me,t-Bu.
NuLi(34); Ph(MeS)2CLi
Me3Si(MeS)2CLi
41
42

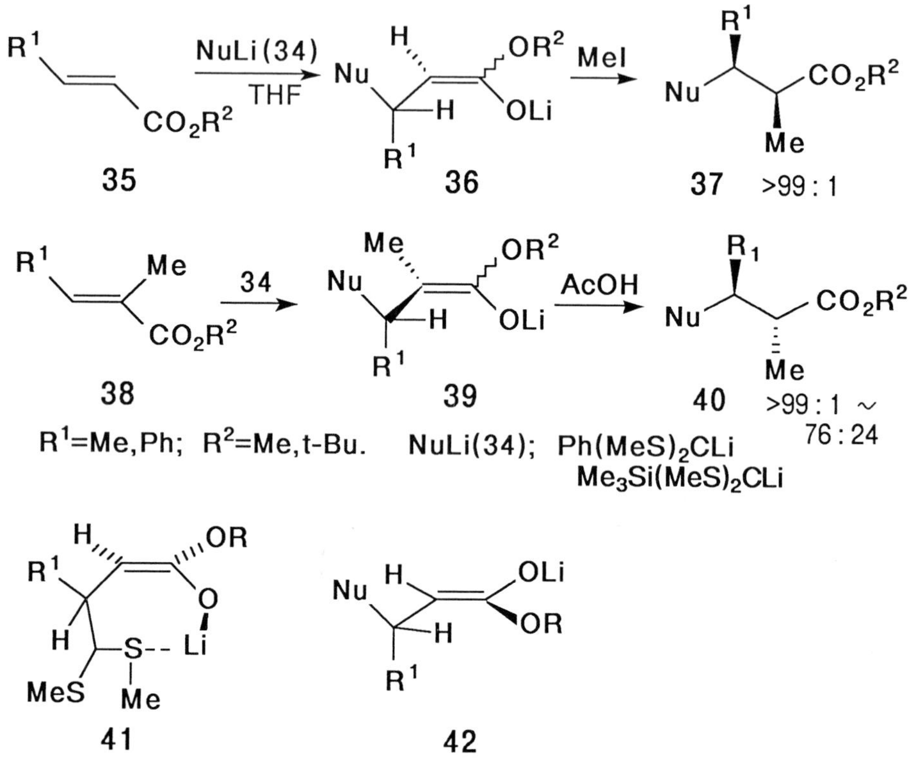

Treatment of **43** with LDA induced the intramolecular alkylation to give **45** with very high diastereoselectivity (98:2)[27]. The enolate intermediate **44** takes the eclipsed conformation and the intramolecular alkylation gives **45**. Therefore, it is now clear that the eclipsed model can be applicable to the intramolecular delivery of an alkyl group. Interestingly, the cyclic enolate derived from **46** produced **47** predominantly upon intermolecular methylation.

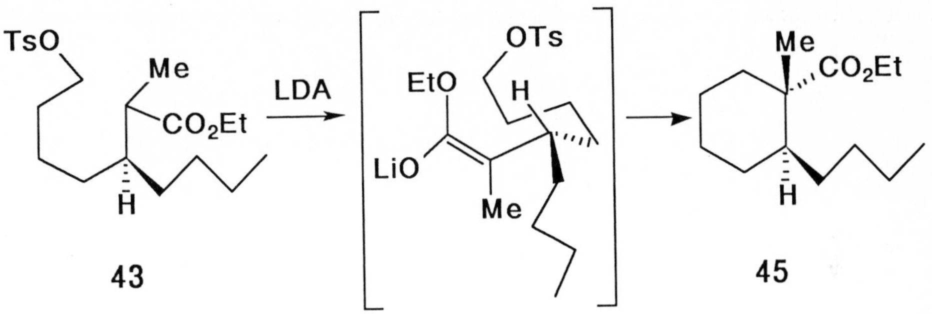

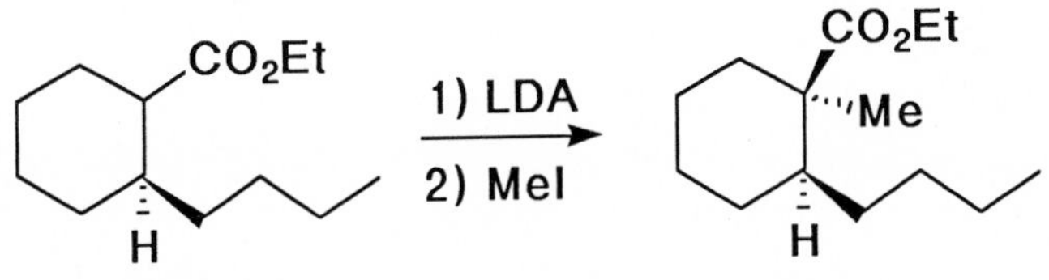

The formation of 3-alkoxy ester enolates has been known for a number of years.[28] Methylation of the dianion (**48**) gave the anti-isomer (**49**) with very high diastereoselectivity (98:2).[29] This high selectivity is ascribed to the chelation effect (as shown in **48**) which results in the shielding of the β face of the enolate by the R group. The chelation effect also works well for the methylation of δ-hydroxy carboxylic ester (**50**).[30] Treatment of **50** with 2 equiv. of LiNEt₂ at -100 -78°C in THF-HMPA, followed by methylation with Me_2SO_4, gave the syn-isomer (52) very predominantly (92:8). This remote stereocontrol, explained by the chelate formation (51), resulted from lithium-olefin π-coordination, instead of the ordinary O-Li...O chelation, as in 53, and of the α-carbon-Li...O chelation, like 54. The reactions of dimetalated succinamides (55) with electrophiles (RX) gave 2,3-dialkylated adducts with high diastereoselectivity.[31] Here again, the high stereoselectivity may be explained by reference to a chelation model of the monoalkylated enolate.

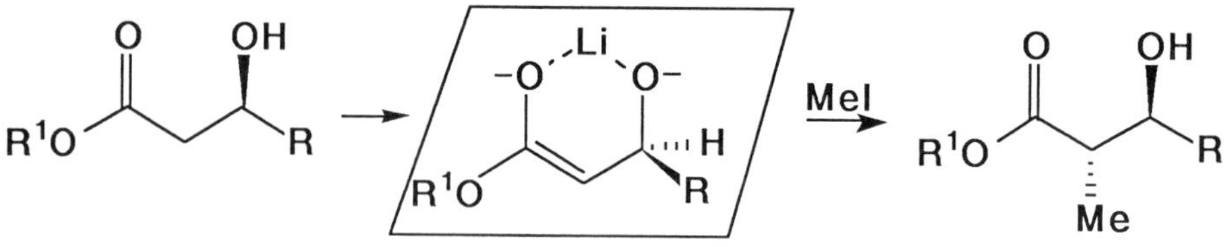

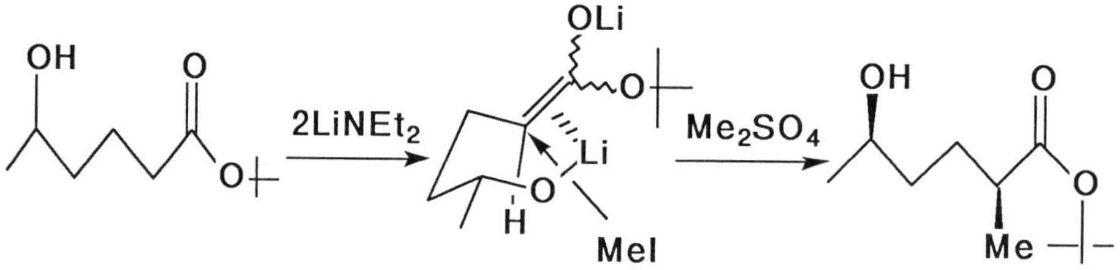

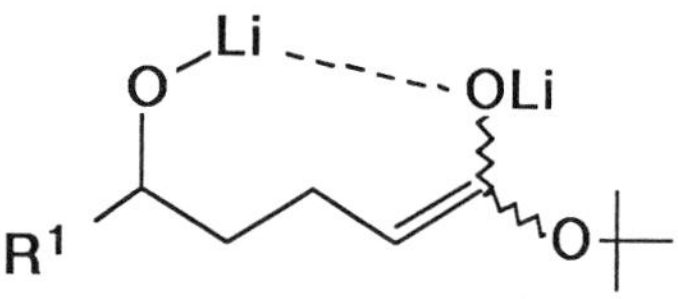

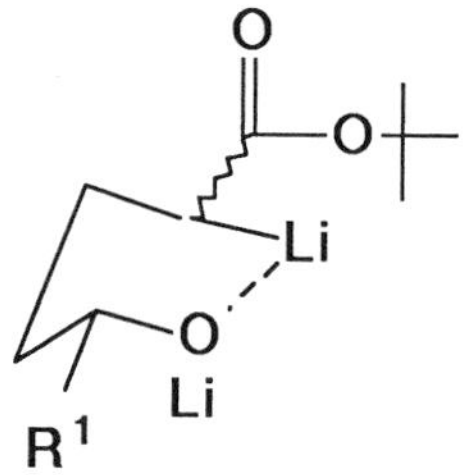

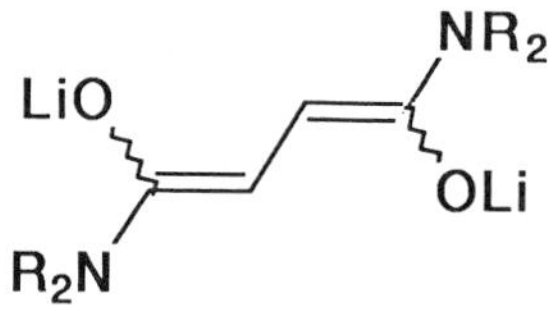

As mentioned above, protonation of diastereotopic enolates gives the anti-product predominantly, in agreement with the eclipsed model (**19, 26,** and **39**). However, the syn-stereoselectivity is reported in a related protonation, in which the chelation effect causes this unusual selectivity.[32] The 1,4-addition of lithiated methyl dithioacetate with α,β-disubstituted enones afforded the syn-isomer (**56**) with moderate to high (>95:5) stereoselectivity. This anti-Houk selectivity arises from an auto-protonation, involving transfer of the hydrogen α to the thiocarbonyl group towards the enolate moiety (**57**). The stereoselectivity is thus controlled at this intramolecular proton delivery step. An intermolecular protonation would probably lead to the anti-isomer (Houk product). A pseudo-cyclic transition state leading to the syn product is proposed (**58**). Here again, the chelation between O-Li and S controls the diastereoselectivity.

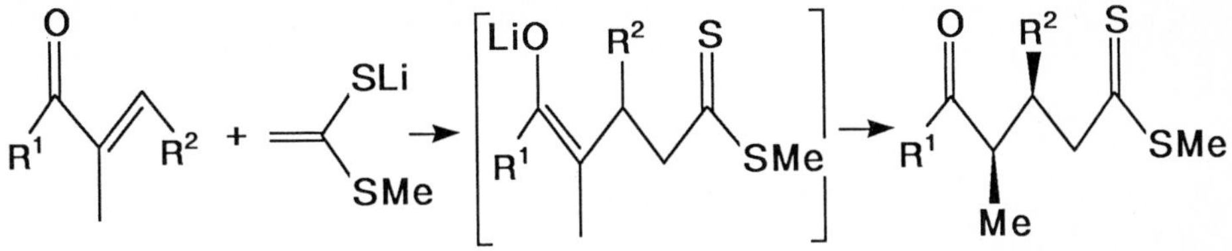

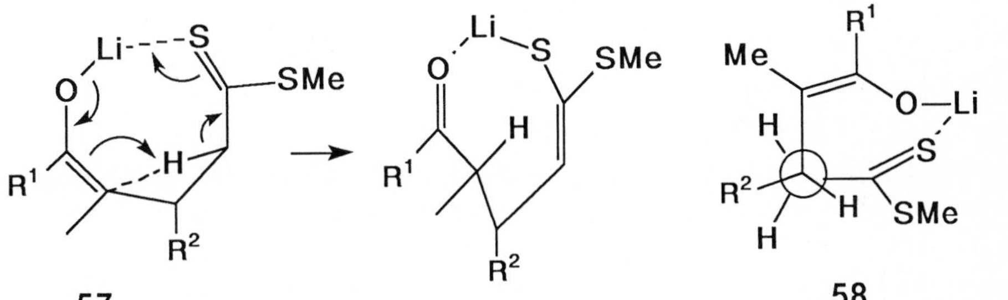

In summary, the following factors are important to control and to enhance the diastereoselectivity of alkylation reactions in ordinary acyclic enolates; (i) chelation control, (ii) allylic strain-stereoelectronic control, and (iii) allylic strain-steric control.

A different approach to control the stereochemistry of acyclic alkylation involves diastereoselective α-allylation of tertiary thioamides.[32a] Deprotonation of the thioamide produced a Z-thioamide anion[69], which reacted with stereodefined crotyl tosylate to give the stereodefined S-allylated adduct. Subsequent thio-Claisen rearrangement gave the anti-adduct from trans-crotyl tosylate and the syn-adduct from cis-crotyl tosylate. By using related intramolecular rearrangements, such as enolate-Claisen and [2,3]-Wittig sigmatropic reactions, the stereochemical control of acyclic molecules has been achieved.[32b] However, such rearrangement reactions are beyond the boundary of this review.

2. STEREOCHEMISTRY OF ALKYLATION OF CYCLIC SYSTEMS
2.1 BASIC PRINCIPLES

In cyclic enolates, the conformational ambiguity observed in the acyclic systems does not exist, and the stereoselectivity is largely determined by steric effects. For example, let's consider the exocyclic enolate (59) which is conformationally locked by the t-butyl group. The equatorial attack of electrophiles on 59 gives 60, while the axial attack produces 61. The axial attack would be hindered by the axial hydrogens at C-3 and C-5 but the equatorial attack is relatively unhindered. In fact, equatorial attack is preferentially observed;[33] 60/61=85/15(X=Me,El=MeI), 84/16(OMe,MeI), 87/13(OMe, nBuBr), 59/41(OLi, MeI). The less nucleophilic enolate 59(X=Me) exhibits greater diastereoselectivity than the more nucleophilic enolate 59(X=OLi) toward methylation. In conclusion, the steric bias present in this system dictates largely the diastereoselectivity.

The stereoelectronic effect should also be considered in accounting for control of the stereoselectivity. Alkylation of the endocyclic enolate (62), derived from 4-t-butylcyclohexanone, demonstrates the significance of stereoelectronic control.[34] Axial alkylation of 62 gives 63 via a chairlike transition state, while equatorial alkylation produces 64 via a boatlike transition state. Since the chair conformation is more stable than the twist-boat conformation in a six membered ring, axial attack may be preferable in comparison with equatorial attack. Actually, 63 is always produced predominantly; 63/64=54/46 (R=H, El=EtI), 55/45 (H, MeI), 70/30 (Me, CD3I), 77/23 (CN, MeI). This result suggests that the energetic bias of axial alkylation, via a chairlike transition state, is relatively small. Taken together, the enolate-alkyl halide transition state is mostly reactantlike, and the diastereoselectivity is dictated largely by steric factors. Further precise data and discussions are presented in ref 1 and 13. More recent results are presented in this review. Needless to say, chelation control is also a very important concept to control stereochemistry in cyclic systems.

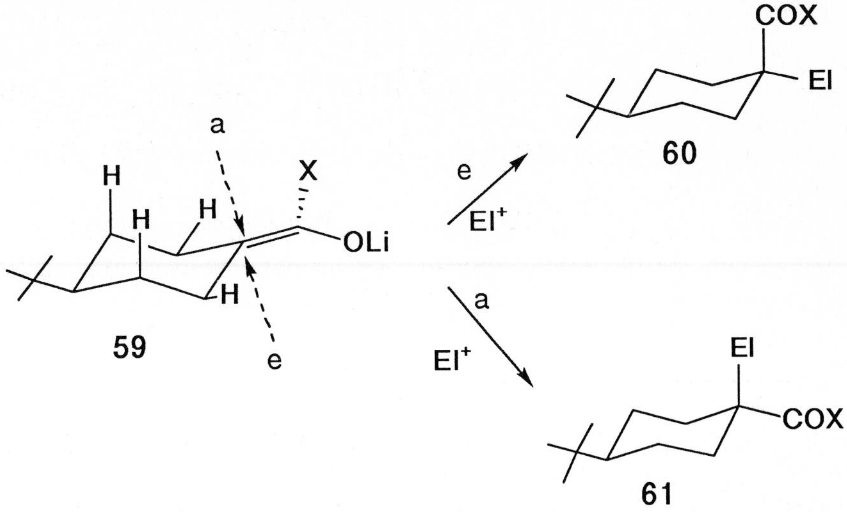

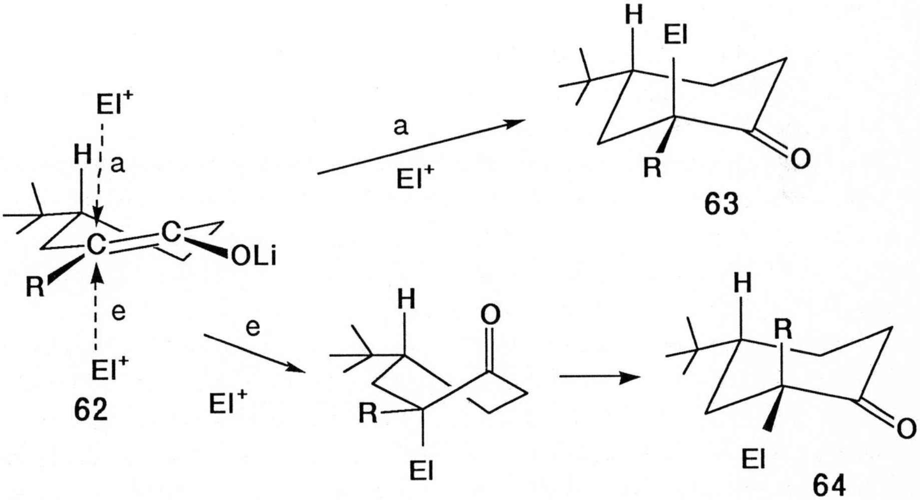

In summary, most experimental results suggest that the stereochemistry of cyclic enolate alkylation is controlled by the steric approach factor (less hindered side attack): the stereochemical consequence of alkylation is usually explained by two factors,[34d] steric approach control (early transition state) and product development control (late transition state).

2. 2 HOW TO ENHANCE THE DIASTEREOSELECTIVITY

The conjugate addition of the sterically bulky NuLi (**34**) to **65** gave the enolate **66** in which the conformation must be fixed (as shown) owing to the allylic strain. Alkylation with allyl bromide produced **67** with very high diastereoselectivity; the isomer ratio was 18:1 at -78°C and 26:1 at -100°C.[35] Here again, attack by RX takes place anti to the bulky dithioacetal group for steric reasons. Therefore, the origin of the very high stereoselectivity attained here is ascribed to the allylic strain and the very bulky asymmetric substituent at the α-position, which also cause excellent stereocontrol in acyclic systems. However, the conjugate addition of **34** to **68** followed by alkylation resulted in low diastereoselectivity.[36] The conjugate addition presumably takes place from the α-face owing to the presence of a very bulky OCPh₃ substituent in the β-face. The resulting enolate takes the conformation shown in **69**, owing to the allylic strain. The attack of electrophiles either from the β-face, or from the α-face, is hindered by the very bulky groups (OCPh₃ and Nu); thus, there is no big difference in steric factors between the β-face attack and α-face attack. On the other hand, **70** gave very high diastereoselectivity. In this case, the α-face becomes a sterically unhindered side, since the Nu is in the β-face and the Me is in the α-face owing to the allylic strain.

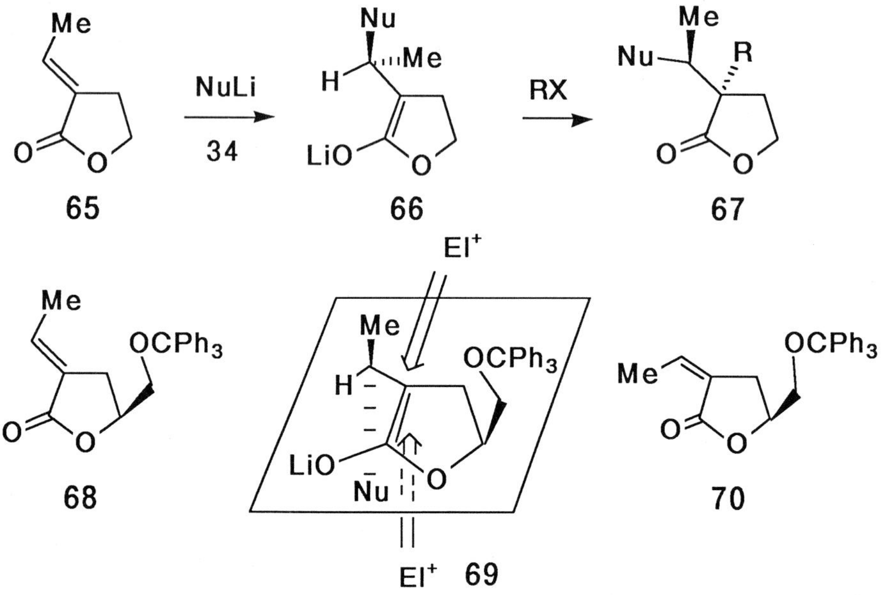

The endo-cyclic enolate (71) having a small R^2 group, and a relatively large R^1 group, produced 72 with very high diastereoselectivity (>99.5:0.5), while the enolate having a bulky R^2 and small R^1 group gave 73 in an extremely high selectivity (>99.5:0.5).[37]

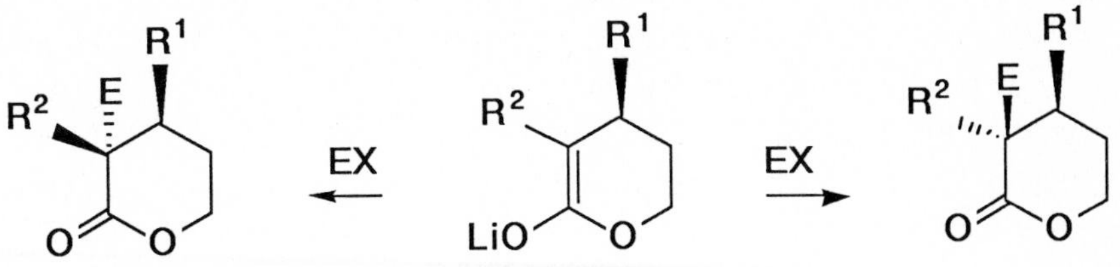

R^1=Ph, R^2=Me or
CH=CH$_2$

R^1=Me, R^2=CH$_2$C(SMe)$_2$SiMe$_3$
or CH$_2$CH(SMe)$_2$

It is now clear that the diastereofacial selection in 74 depends upon the balance of steric bulkiness between R^1 and R^3. When R^1 is bulkier than R^3, the electrophile attacks from the α-face of 74b. On the other hand, when R^3 is bulkier than R^1, the attack of the electrophile takes place from the β-face of 74a. It should be noted that the diastereofacial differentiation is controlled by the exo-allylic position for the specific conformation of the nonasymmetric carbon center of the enolate.

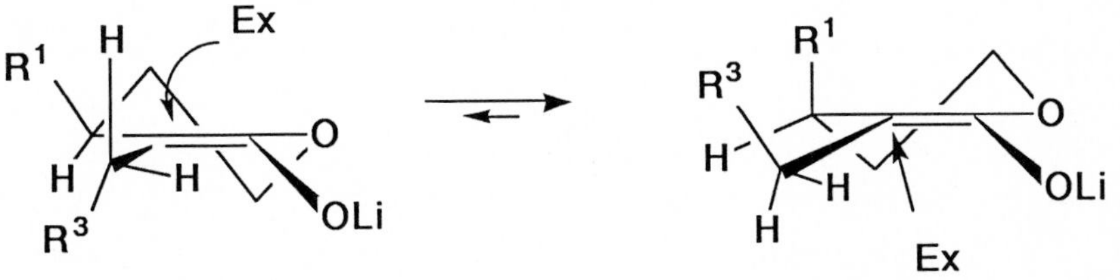

As mentioned in the acyclic system (48), the dianions of β-hydroxy esters are alkylated with better than 90% diastereoselectivity to produce the anti-isomer. The dianion of a cyclic analogue (75) underwent virtually complete anti-selective alkylation at C-2 to produce the trans-isomer.[38] This high stereoselectivity cannot be controlled by lithium coordination, as suggested for the acyclic case.[29] The developing bond, that is the electron density on C2, must form anti to the oxyanion due to electrostatic repulsion.[38a] Quite similarly, the dianion (75') underwent alkylation diastereoselectively to afford the trans-isomer in high yield; trans:cis = 20:1.[38c]

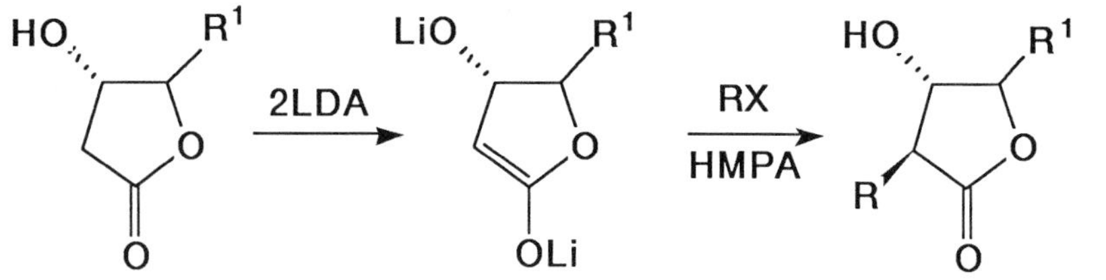

75

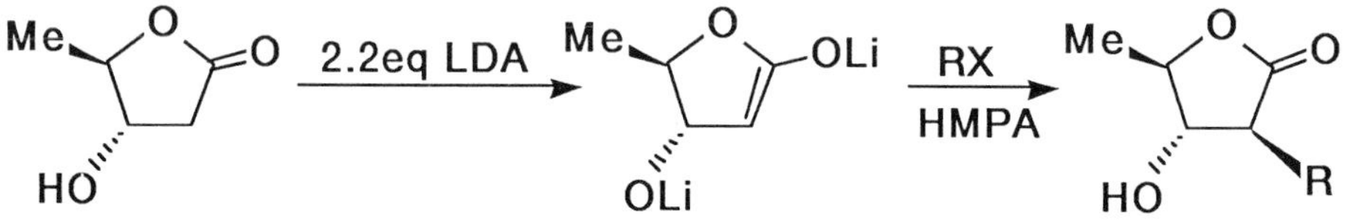

75'

The exocyclic enolates (76) derived from the trans-substituted dioxolanes (tartrate acetonides) gave the isomers (77) preferentially upon treatment with RX.[39] Methylation of the enolates (78), generated from the cis-substituted dioxolanes, led preferentially to the isomer(79).[40] Obviously, 76 and 78 underwent "contra-steric" alkylation syn to the neighboring substituents on the dioxolane ring (CO_2CH_3 in 76 and R in 78, respectively). The reason for the "contra-steric" course of the alkylation is not certain. The enolate (78) may have the conformation (80) in which the CH_3(a) and the R-group are maximally separated from each other, so that the methylation may take place anti to the CH_3 (b), namely syn to the R.[40]

The apparent "contra-steric" selectivity seems to be owing to the steric bias of the dioxolane ring. The methylation of 81 bearing the oxazoline ring gave 82 exclusively.[41] This is a normal stereoselectivity; the electrophile attacks from the α face of 81 which is not shielded by the substituent at the β-position. The endocyclic enolate (83) reacted with RCH_2X such as CH_3I, $CH_2=CHCH_2Br$, and

PHCH$_2$Br to give **84** in high yields (87~91%).[42] The diastereoselective formation of **84** can be explained in terms of the sum of the steric effects of the substituents. The attack of electrophiles occurs from the overall less-shielded plane. Ph and t-Bu groups are equatorial and, thus, their effect to the stereodifferentiation will be very small. The selectivity is dictated by the CO$_2$CH$_3$ group which is clearly anti-directing.

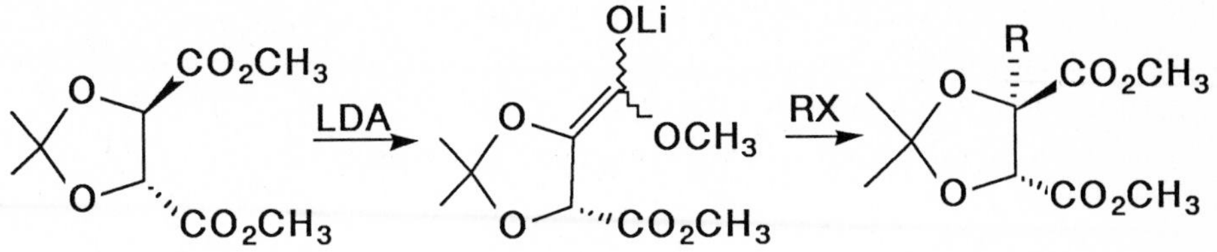

76   **77**

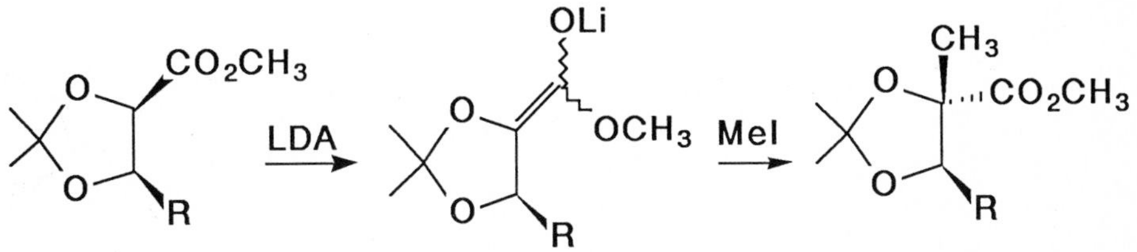

R=Me,Et, (epoxide)   **78**   **79**

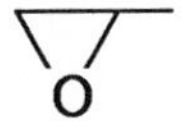

88-95% selectivity

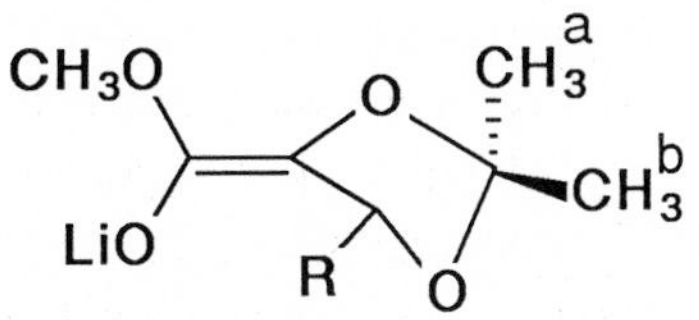

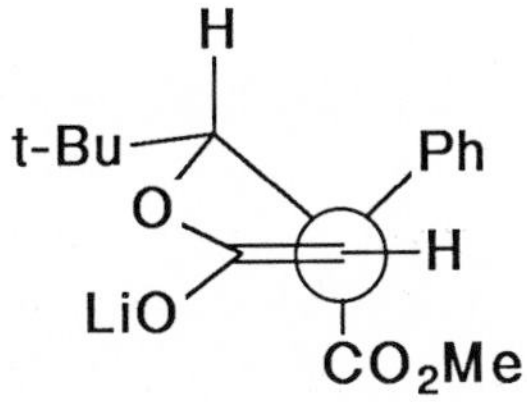

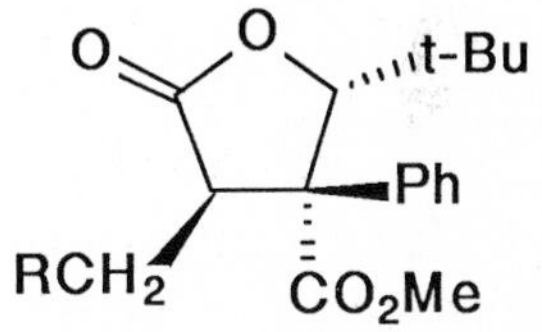

80   **83**   **84**

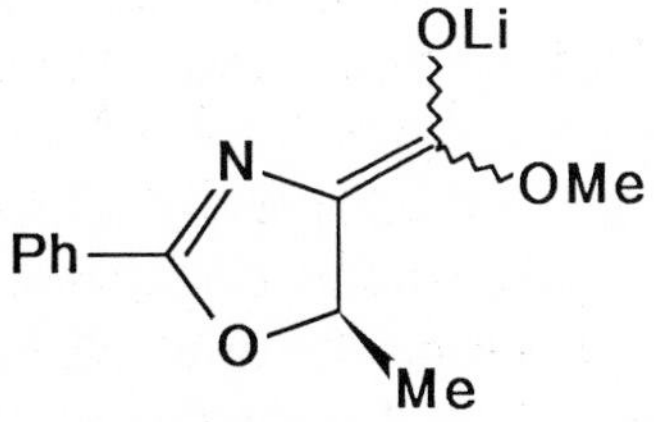

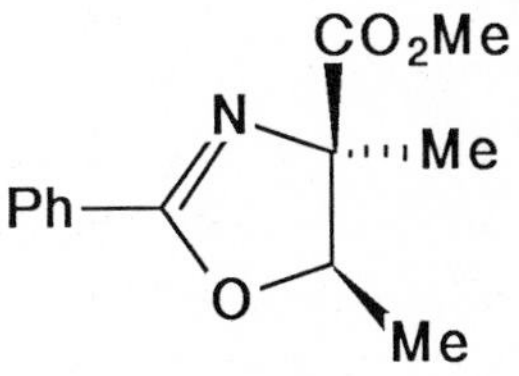

81   **82**

As mentioned in section 2.1, the alkylation stereochemistry of cyclic enolates is normally dictated by the steric approach control (less hindered side attack; for the alkylation of Li-enolate of 2-norbornenone, see ref 43). However, contra-steric alkylation was observed in the exocyclic enolates **76** and **78**. Quite recently, methylation under product development control has been reported. Treatment of **85** with dibutylcuprate, followed by the addition of MeI and subsequent cyclization, gave **86** and **87** in the ratio of 78:22 in 70% overall yield.[44] The major product (**86**) was produced by the cis-addition of the Me to Bu group. A similar, unusual, methylation has also been observed both, in the related five membered enolates, and in 2,3-disubstituted δ-lactone system.[44] The transition state developed from the less hindered side attack (path b in **88**) suffers from the eclipsing interaction between two groups (R^1 and R^2), while the transition state from the hindered side attack (path a) does not experience this unfavorable interaction. Consequently, methylation takes place syn to the Bu group in order to make the two adjacent larger groups align trans to each other (product development control).

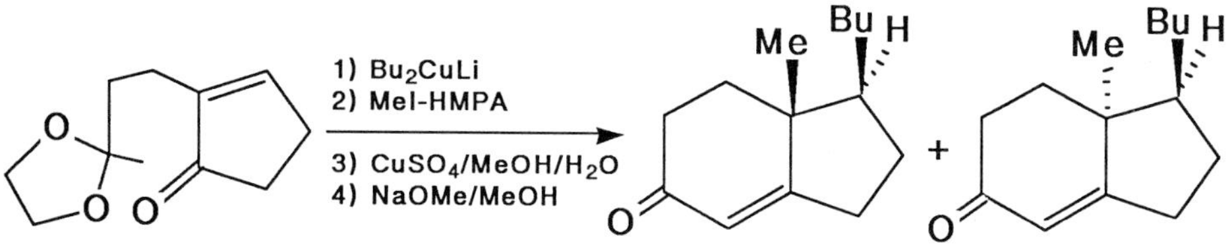

The regio-and stereoselective generation, and subsequent alkylation, of exo-lithiated Δ^2-isoxazolines are reported.[45] Treatment of 3-ethylisoxazoline (**89**) with lithium diethylamide, followed by benzyl bromide quench, provided **90** with very high diastereoselectivity. A variety of isoxazolines derived from both cis and trans olefins undergo similar diastereoselective alkylation. As mentioned in cyclic enolates, the observed stereoselectivity can be explained by diastereoselective alkylation of a stereochemically discrete azaenolate. Deprotonation of **89** will occur by approach of the base anti to the R group. Assuming that the proton being removed must be nearly coplanar with the C=N π-orbitals, two conformations are conceivable (**92** and **93**). Since there is a severe steric congestion between CH₃ and R groups in **93**, conformation **92** should be favored. Therefore, the azaenolate(**94**) would be produced selectively under kinetic control. Further, **94** is again more favorable under thermodynamic control than **95**. The alkylation will take place from the β-face (sterically less bulky face), giving **90** predominantly. It is important that the 4-substituent (R) controls both the enolate stereochemistry and the face selectivity in the alkylation. Although the 5-substituents (Rt and Rc) are not important to control the stereoselectivity, they play an important role in the selective formation of the exo-azaenolate. Presumably, Rt hinders the approach of a base to the endo hydrogen.

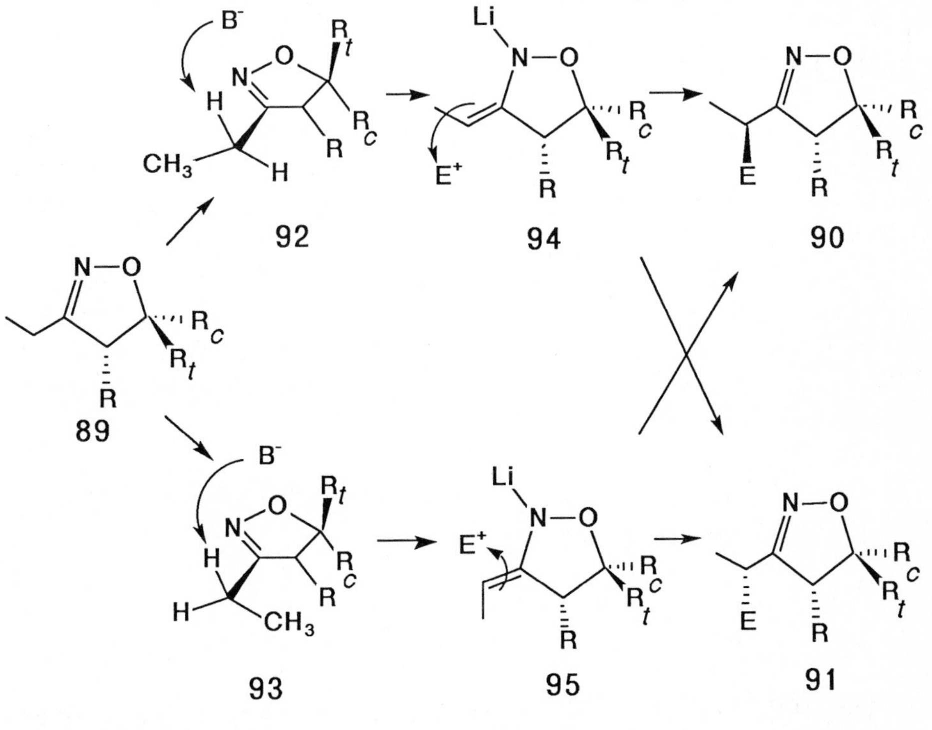

The almost complete diastereoselectivity in the capture of electrophiles by enolate anions from siloxy-substituted cyclopropanecarboxylates is interpreted as evidence that the true structure of the enolate is pyramidal, rather than an ordinary sp²-type plane.[45a] The electrophile adds cis to the Me₃Si group, whether this is larger or smaller than R¹.

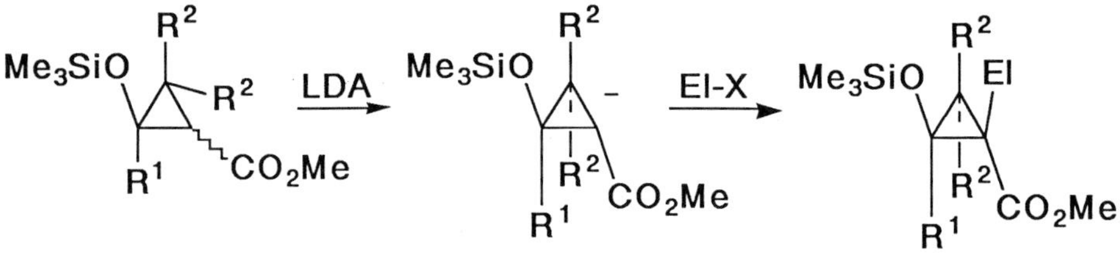

3. FORMATION OF ENOLATES, MECHANISM OF ALKYLATION, AND REGIOSELECTIVITY

3. 1 FORMATION OF ENOLATES. REGIO- AND STEREO-CHEMISTRY

In the above sections, the stereochemistry of alkylation reactions was discussed for lithium and alkali metal enolates. Actually, most synthetically useful metal enolates are lithium derivatives. In this chapter, we again treat lithium and alkali metal enolates along with silyl enol ethers, and in chapter 4 we will mention a variety of metal enolates including transition metals.

The most widely used base for deprotonation of ketones and esters is LDA, most frequently applied at -78°C in THF. Treatment of 2-heptanone with LDA at -78°C, followed by the addition of TMSCl and triethylamine, gave a mixture consisting of 84% of kinetic controlled enol ether 96, 7% of the Z-enol ether 97, and 9% of the E-enol ether 98.[46] The use of lithium t-octyl-t-butylamide (LOBA) under the same conditions did not improve the selectivity of enolate formation.[47] Since LOBA is a highly hindered base, deprotonation of the ketone becomes comparatively slower and, thus, proton transfer between enolates and ketone may take place, resulting in relatively low selectivity of enolate formation.

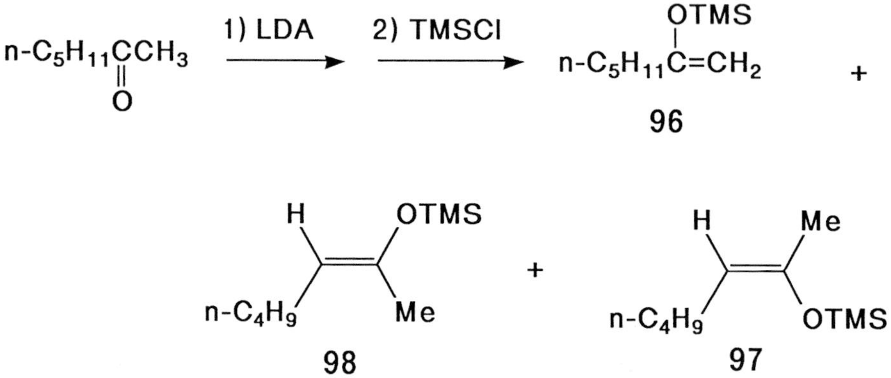

It was discovered that both LOBA and LDA are compatible with TMSCl at -78°C.[47,48] Two procedures could be used; 1) the ketone is added to a mixture of the base and excess LDA(5-10 equiv) at -78°C, or 2) a solution of the ketone and excess TMSCl at -78°C in THF is treated with the base. The efficiency of this internal quench method, as compared to the ordinary two step procedure, is shown in Scheme 1.[47] Highly regioselective, kinetically controlled enolate formation is realized by using $LiNR_2$ in the presence of TMSCl (internal quench method). Further, the yield of enol ethers in all internal quench reactions was quantitative and no condensation products between the enolates and starting ketone were observed

Scheme 1. Highly regioselective, kinetically controlled enolate formation using $LiNR_2$-Me_3SiCl

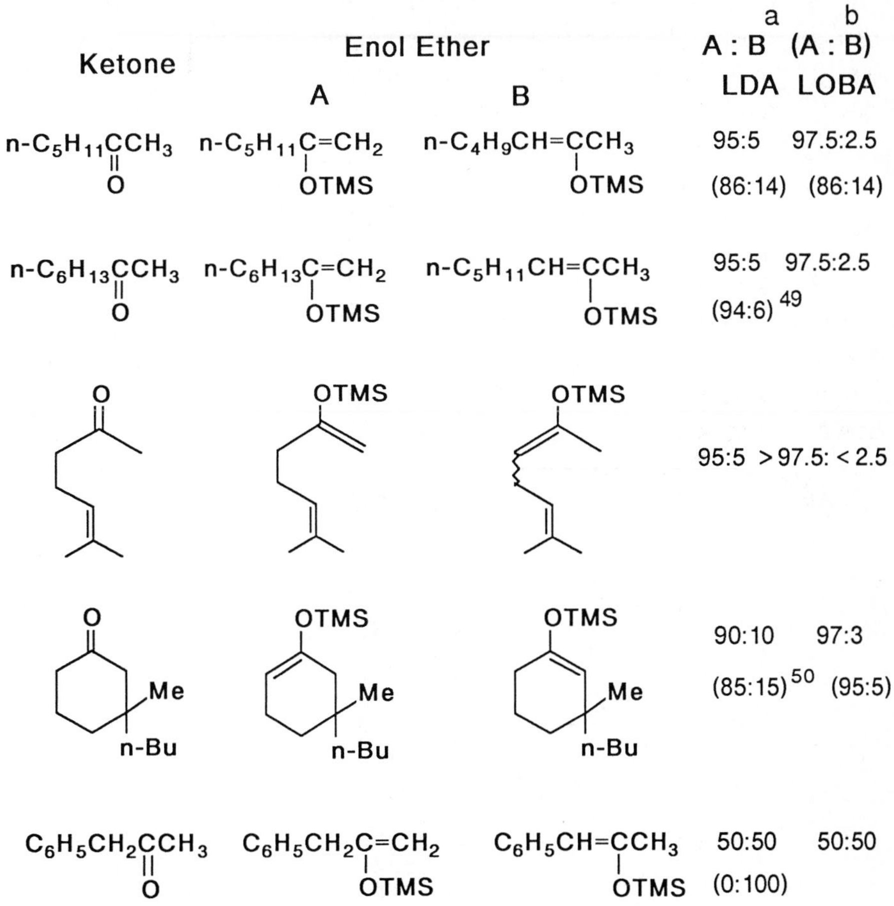

Ketone	Enol Ether		a $A:B$ LDA	b $(A:B)$ LOBA
	A	B		
$n\text{-}C_5H_{11}CCH_3$ $\overset{\|}{O}$	$n\text{-}C_5H_{11}C\text{=}CH_2$ $\overset{\|}{OTMS}$	$n\text{-}C_4H_9CH\text{=}CCH_3$ $\overset{\|}{OTMS}$	95:5 (86:14)	97.5:2.5 (86:14)
$n\text{-}C_6H_{13}CCH_3$ $\overset{\|}{O}$	$n\text{-}C_6H_{13}C\text{=}CH_2$ $\overset{\|}{OTMS}$	$n\text{-}C_5H_{11}CH\text{=}CCH_3$ $\overset{\|}{OTMS}$	95:5 (94:6)[49]	97.5:2.5
			95:5	>97.5: <2.5
			90:10 (85:15)[50]	97:3 (95:5)
$C_6H_5CH_2CCH_3$ $\overset{\|}{O}$	$C_6H_5CH_2C\text{=}CH_2$ $\overset{\|}{OTMS}$	$C_6H_5CH\text{=}CCH_3$ $\overset{\|}{OTMS}$	50:50 (0:100)	50:50

a) Internal quench   b) Two step procedure

Now it is clear that we can produce the "kinetic" enolates with very high regioselectivity regardless of cyclic and acyclic systems. Procedures for direct regioselective preparation of the more substituted "thermodynamic" enolates (or TMS enol ethers) are summarized in Scheme 2. As is evident from this scheme, the preparation of highly regioselective thermodynamic cyclic enolates under "kinetic" (non-equilibrating) conditions in aprotic media is realized with BMDA and BMHMDS.

Scheme 2. Regioselective, thermodynamically controlled enolate formation

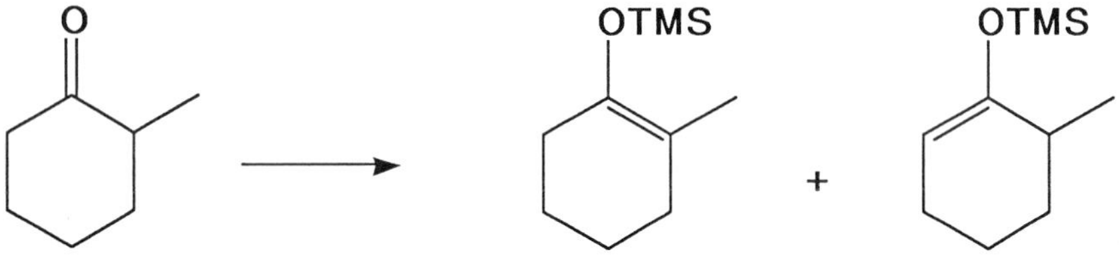

Reagent	t (thermodynamic)		k (kinetic)
	t	:	k
LDA/DME, then TMSCl[46]	1	:	99
NaH/DME, then TMSCl[51]	73	:	27
Et$_3$N/TMSCl/DMF, 130°[46]	78	:	22
KH/THF, then TMSCl[52]	67	:	33
(n-Bu)$_4$NF/TMSCH$_2$CO$_2$Et[53]	82	:	18
HMDS/TMSI[54]	90	:	10
BMDA[55]	97	:	3
BMHMDS[55]	96	:	4

The profound effects of conformation are found in medium rings on the enolization and, ultimately, the stereochemical outcome of alkylation.[56] Kinetic enolate alkylations were examined on a variety of mono substituted 8- to 12-membered macrocyclic ketones and lactones (Scheme 3). As indicated in Scheme 3, kinetic methylation (LDA/THF, then MeI at -60°C) of 8-membered rings proceeded with high stereoselectivity.

Scheme 3. Conformational effect of medium rings upon stereochemistry

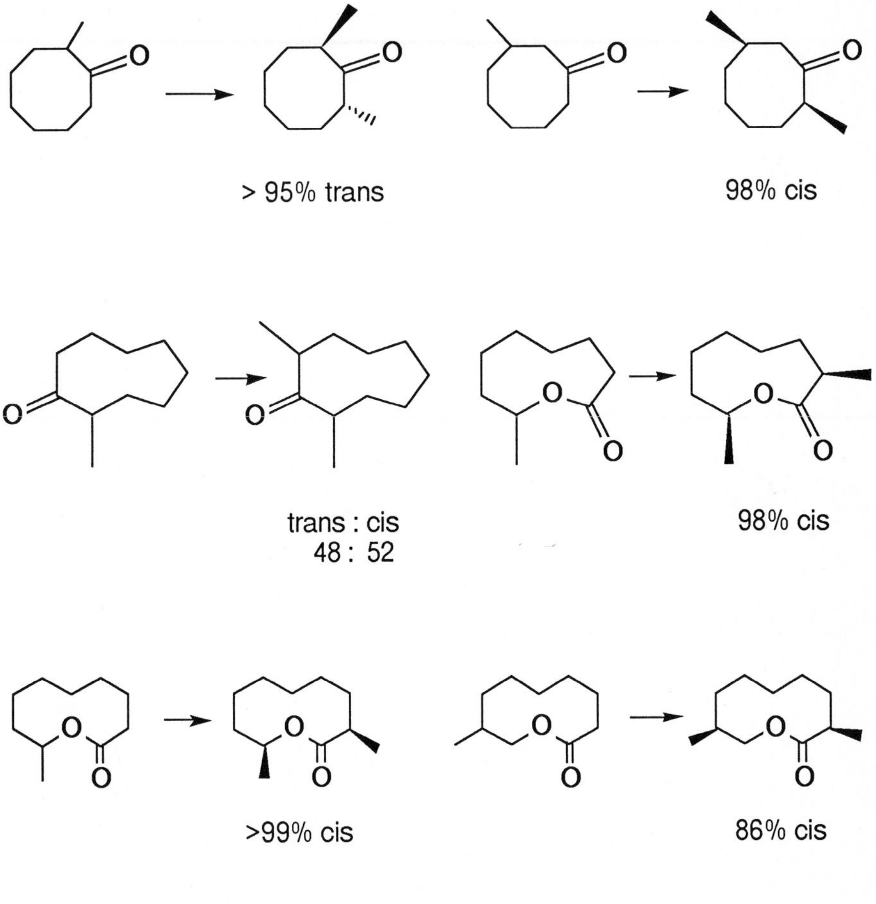

> 95% trans   98% cis

trans : cis   98% cis
48 : 52

>99% cis   86% cis

Although methylation of the 9-membered ketone resulted in low selectivity, that of the 9-membered lactone gave very high selectivity. Quite similarly, 10-, 11-, and 12-membered lactones produced high diastereoselectivity. This high stereoselectivity appears to be intimately associated with the conformational properties of the macrocyclic substrates, since simple molecular mechanics calculations allowed semiquantitative prediction of the product distributions in every case.[56] Another example for the importance of the ring conformation is shown in Scheme 4.[57] The N-benzyllactams were deprotonated with t-BuLi and then alkylated with RX, giving the ring alkylated product (r) and/or side chain alkylation product (s). For the 5- and 6-ring lactams, r was produced exclusively, whereas the 7- and 8-ring lactams gave only s. The 9-ring lactam afforded a 3~4:1 mixture of s : r. On the other hand, the 10-ring lactam led to an 87:13 ratio of alkylated material favoring r. The 11-ring lactam led to exclusive enolate alkylation, while the 13-ring derivative gave a 1:1 ratio of both products r and s. Here again, molecular mechanics calculations appear to be generally consistent with the experimental results.

Scheme 4. Conformational effect of rings upon regioselectivity of metallation

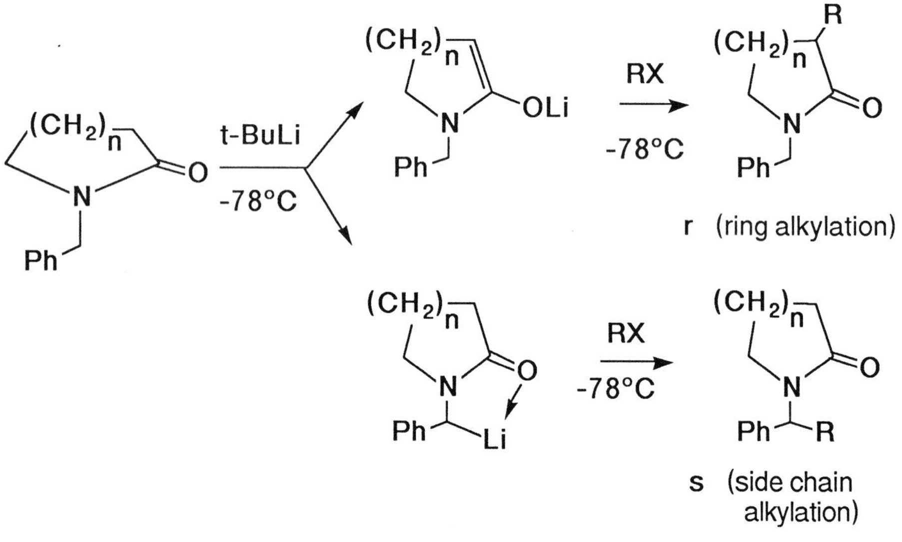

Even if the regioselective formation of the metal enolate is achieved, the alkylation step frequently causes some difficulties. Equilibration of enolates accompanying the alkylation process has resulted in loss of regioselectivity and in

the formation of polyalkylated products. If the kinetic enolate mixture could be substantially activated toward alkylation, proton transfer and the resulting equilibration and polyalkylation might not be an effective competing process in the presence of excess alkylating agent. The benzylation and n-butylation of the kinetic enolate mixture derived from 2-heptanone was studied in DME in the absence, and in the presence, of lithium ion complexing agents (triglyme, benzo-14-crown-4, DMF and HMPA).[57a] All ligands produced more terminal than internal alkylation; the effectiveness was in the order HMPA(4.9eq) > DMF(neat) > benzo-14-crown-4 (1.61eq) > DMF(6.18eq) > triglyme(1.94eq).

Under kinetic conditions, most of α-aminoketones, all of which contained fully substituted α-nitrogen atoms, gave regioisomeric mixtures of enolates ranging from 5:1 to 1:1.[57b] On the other hand, benzamidoacetone which upon deprotonation contains six potential sites (*) for alkylation, gave only the carbon-methylated product, 3-benzamidobutanone, when converted to its monoanion with one equivalent of LDA, and then CH3I in THF.[57c] Presumably, the amide NH is kinetically removed to generate the amide anion, and subsequent proton transfer generates the thermodynamically favored, dipole stabilized, ketone enolate anion, which undergoes α-C-methylation.

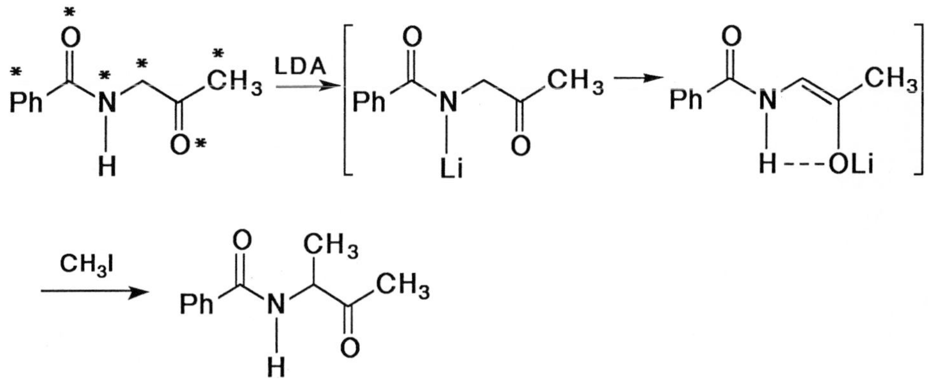

Since the regioselective formation of the Si-substituted enolates is achieved, the next problem is to control the stereochemistry of the enolate. The use of sterically bulky lithium dialkylamides would be expected to exert an influence on the stereochemistry. The results on the reactions of 3-pentanone are shown in Scheme 5. Reaction of internally trapped, sterically bulky LOBA led to a 98:2 ratio of the E and Z enol ethers[47](entry 5, cf. entries 1, 3, and 4).

Scheme 5. Highly stereoselective, kinetically controlled
enolate formation using **LOBA-TMSCl**

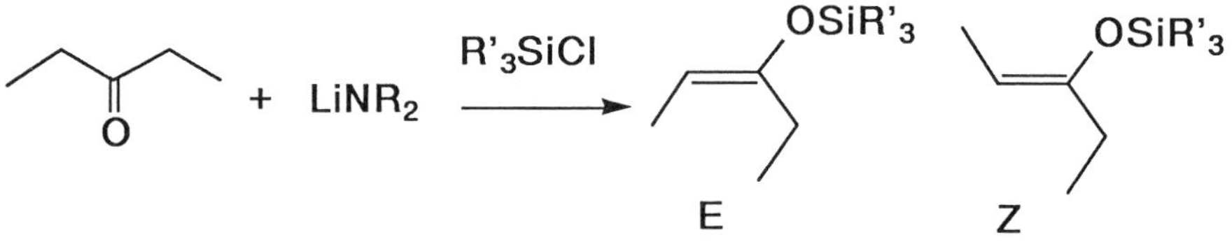

entry				E	:	Z
1	**LDA,** then **TBDMSCl,** in THF[58]			77	:	23
2	**LDA,** then **TBDMSCl,** in THF-HMPA[58]			5	:	95
3	**LiTMP,** then **TMSCl,** in THF[59]			86	:	14
4	**LDA-TMSCl** (internal quench), in THF[47]			77	:	23
5	**LOBA-TMSCl** (internal quench), in THF[47]			98	:	2
6	**LOBA-** 8 eq **TMSCl,** in THF-HMPA			37	:	63
7	**LOBA-** 17 eq **TMSCl,** in THF-HMPA			46	:	54
8	**LOBA,** then **TMSCl,** in THF-HMPA			18	:	82

The enhancement of stereoselectivity can be rationalized by considering the steric repulsion in the transition states **99** and **100**. A lithium dialkyl amide having a very bulky substituent, L, favors **99** over **100** since the large repulsion between L and R, in **100**, causes this transition state to be disfavored. Therefore, LDA-TMSCl gives a 77:23 ratio, and LOBA-TMSCl produces a 98:2 ratio. The reaction of 3-pentanone with LDA in THF-HMPA gave a 5:95 mixture of E and Z-enol ethers (entry 2). It was proposed that this selectivity was kinetically determined and that in the absense of HMPA lithium coordination to the carbonyl oxygen is strong enough to make the oxygen effectively bulkier than R, leading to the preference of **101** and, therefore, the E-enolate.[58] In the presence of HMPA this coordination is absent and **102**, in which R^1 becomes eclipsed with the now sterically smaller carbonyl oxygen during enolization, should be favored.[51] This argument, however, ignores the repulsion of L and R^1 in **102**. The presence of the pericyclic transition state in the presence of HMPA is also in question.

It was suggested that enolate formation in the presence of HMPA was thermodynamically controlled on the basis that high base to ketone ratios gave increased amounts of the E-enolate.[60] As compared to the ratio with a two step procedure (entry 8), greater amounts of E isomer were obtained with the in situ trapping procedure (entries 6 and 7). Consequently, it is concluded that stereoselectivity, in the presence of HMPA, is the result of equilibration to the more thermodynamically stable Z isomer.[47]

Molecular mechanics were used to calculate the energies of model transition states for the formation of cis and trans enolates from a number of ketones and esters.[61] Regression analysis was used to correlate these energies with the corresponding experimentally determined cis:trans enolate ratios reported in the literature for kinetic deprotonation with LDA in THF or in THF-HMPA. It was suggested that the transition state leading to the cis-isomer is destabilized by a steric interaction with the base, thus supporting a cyclic transition state. That the steric interaction is greatly diminished in the presence of HMPA suggests that the transition state is much more loosely organized in HMPA-THF. In THF, the transition states are reactant-like, cyclic and contain a steric interaction between R^1 and the base and/or countercation which disfavors formation of the cis-isomer (see **101**). According to the calculations, the transition states in THF-HMPA are still cyclic, but are expanded by comparison, and are less reactant-like. The interaction which destabilizes the transition state leading to the cis-isomer is diminished, and the inherent stability of the eclipsed orientation between R^1 and O results in the preferred formation of the Z-isomer.

Stereo- and regio-chemistry of deconjugative protonations and alkylations of α, β-enolates were studied in detail.[61a] It has been clearly established that the deconjugative alkylation of the dienolate derived from 2- or 3-alkenoate esters occurred at the 2- rather than 4-position. Concerning the stereochemistry, the following conclusions were reached: (1) electrophilic discharge of the Li dienolate from the ester of a (Z)-2-alkenoate (disubstituted double bond) leads stereoselectively to the (E)-3-alkenoate ester. (2) electrophilic discharge of the Li-enolate from the ester of an (E)-2-alkenoate (disubstituted double bond) leads stereospecifically to the (Z)-3-alkenoate ester, unless the C-4 carbon bears a substituent larger than CH_3, beyond which the reaction becomes increasingly stereorandom. (3) alkylation (or protonation) of a Li dienolate from either a (Z)- or (E)-3-alkenoate ester is stereospecific, with retention of precursor double bond position and geometry.

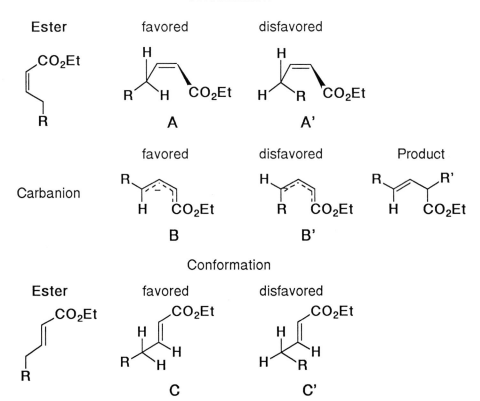

favored	disfavored	Product

The observed inversion of stereochemistry in the conversion of the 2-Z to the 3-E series is consistent with stereoelectronic control (orbital overlap) in formation of the enolate intermediate. Deprotonation of the 2-Z ester takes place from the favored conformation **A**, rather than from **A'**, to give the minimum energy carbanion **B**, which undergoes alkylations to give the 3-E enoate. Quite similarly, deprotonation of the 2-E ester produces the enolate **D** through **C**, leading to the 3-Z enoate.

The stereochemistry of the lithium dienolate (**103'**) was determined by using NMR data.[62] Treatment of **104**, in which the E-configuration is known to be thermodynamically favored, with LDA in THF, followed by the addition of Me_3SiCl, gave **103**. The Z-geometry for the ketene acetal portion of **103** was determined both by a 20% NOE between the vinyl proton at C-2 and the O-Me group and by the observation that the O-silyl residue of **103** undergoes a first-order migration to the C-4 atom. The relatively more planar structure **105** (also Z-isomer) was excluded in this enolate formation. Although most of carbonyl compounds give the E-enolate under kinetic deprotonation with LDA, this particular dienolate provides the Z-enolate. The cis-selectivity of the kinetic deprotonation was also observed in the reaction of dithiopropanoates.

Dithiopropanoates were deprotonated under kinetic conditions: LDA, THF, -78°C. The resulting enolate was trapped with methyl iodide. The Z-isomer was produced as a major isomer.[63] The cis-selectivity increased with the size of the alkylthio group and with the ability of the alkylthio group to chelate the lithium cation. In summary, the stereochemistry of the kinetic deprotonation of various carbonyl compounds with $LiNR_2$ is summarized as follows. (1) E-lithium enolates are formed from aldehydes,[64] esters,[58,65] ketones,[66] imines,[67] and thioesters.[68] (2) Z-lithium enolates are produced from amides,[68] and Z-lithium thioenolates from thioamides[69] and dithiopropanoates.[63] It was also found possible in particular cases to form E and Z boron enolates selectively.[70]

R = Et, iPr, t-Bu, Ph, CH_2OMe, $C(Me)_2Ph$

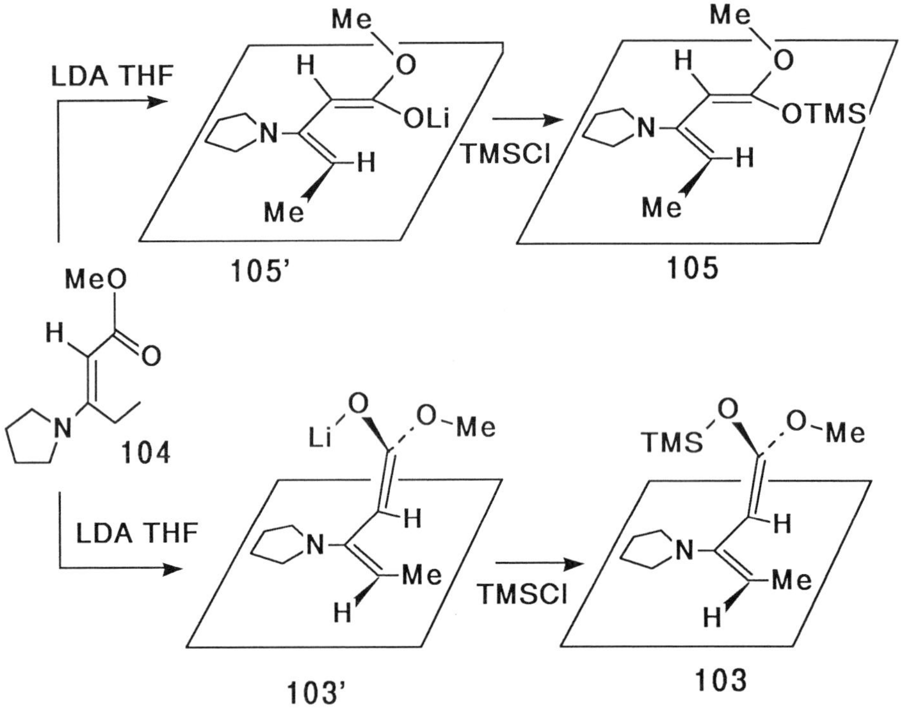

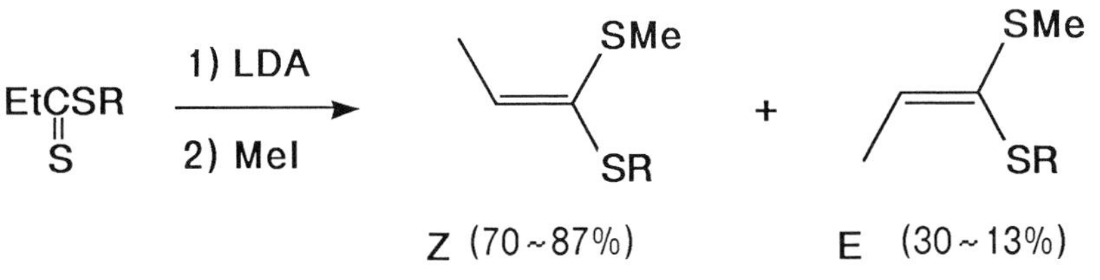

General procedures for the stereoselective generation of tetrasubstituted enolates, or silyl enol ethers, by proton abstraction from secondary carbons are not available (eq 1). As mentioned above, however, the stereocontrol of trisubstituted enolates or silyl enol ethers is realized with various methods (eq 2). A different approach to the synthesis of more highly substituted enolates involves addition of Grignard reagents to N, N-dimethyl-α- methacrylothioamide (eq 3).[71]

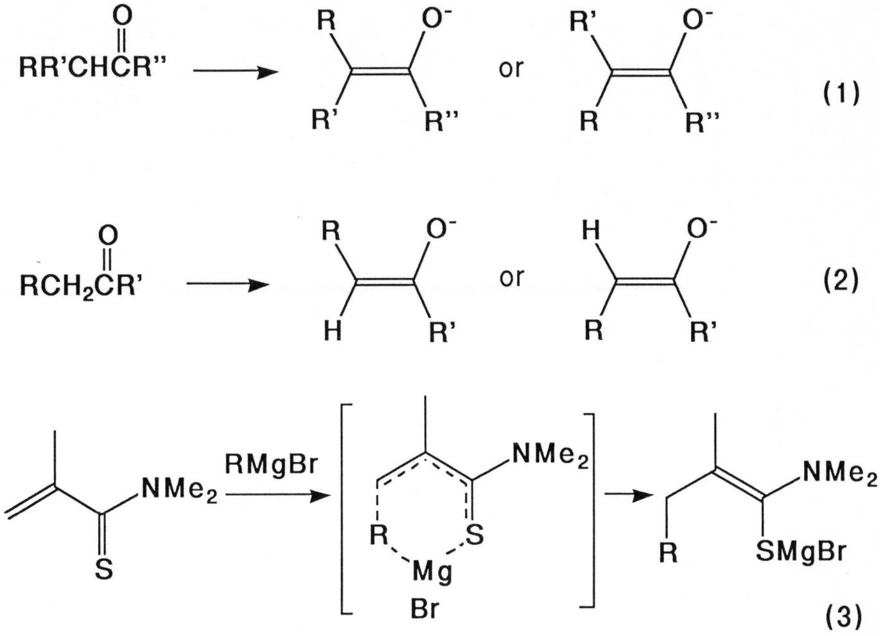

$$\text{RR'CHCR''} \longrightarrow \quad \text{or} \quad \tag{1}$$

$$\text{RCH}_2\text{CR'} \longrightarrow \quad \text{or} \quad \tag{2}$$

$$\tag{3}$$

It was found that the addition of alkyllithium reagents to ketenes provides a regioselective synthesis of enolates capturable as silyl enol ethers.[72] These Si-enol ethers are not the favored isomers formed from ordinary ketone enolization under either kinetic or thermodynamic conditions (106)[72a,b] All attempts to produce 107 by proton abstraction from the corresponding ketone have so far been unsuccessful.[72c] However, the new method can generate 107 without any problems. Reaction of unsymmetrical ketenes with organolithiums, followed by silylation with TMSCl, gave stereospecific formation of a single silyl enol ether(108, 109).[72d] This result is interpreted in terms of preferential attack by the organolithium reagent in the plane of the ketene from the side opposite the more sterically demanding group.

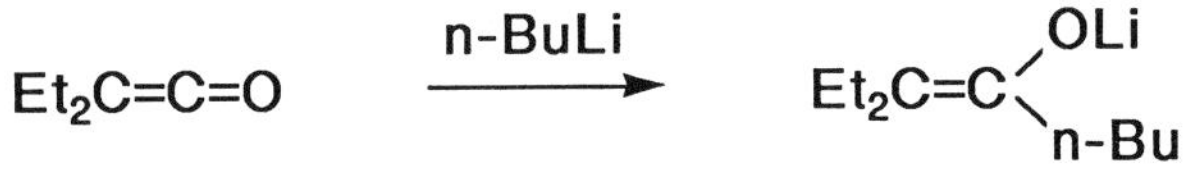

$$Et_2C=C=O \xrightarrow{\text{n-BuLi}} Et_2C=C\begin{smallmatrix} OLi \\ n\text{-Bu} \end{smallmatrix}$$

106

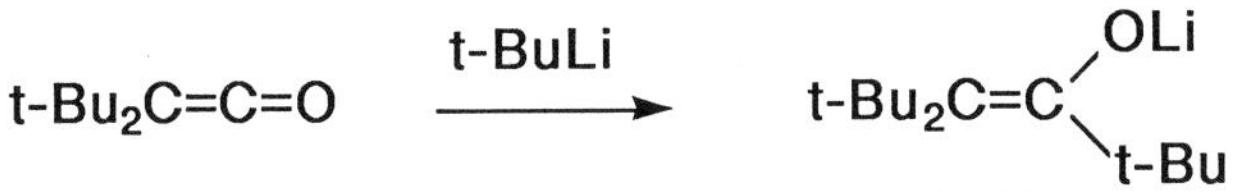

$$t\text{-Bu}_2C=C=O \xrightarrow{\text{t-BuLi}} t\text{-Bu}_2C=C\begin{smallmatrix} OLi \\ t\text{-Bu} \end{smallmatrix}$$

107

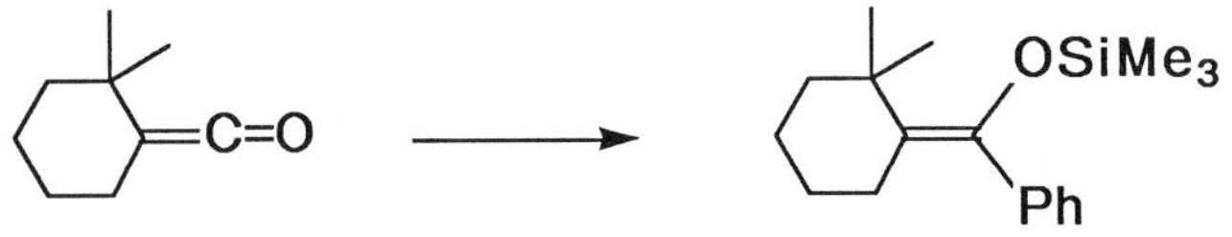

108

$$PhMeC=C=O \longrightarrow \begin{smallmatrix} Ph \\ Me \end{smallmatrix}C=C\begin{smallmatrix} OSiMe_3 \\ Ph \end{smallmatrix}$$

109

Haloenol silyl ethers are prepared in high yield and excellent stereochemical purity via the hydrogen migration in halocarbenes which are generated by treatment of the trihaloderivative with n-BuLi (eq 4).[73] The initial attempts to add silylstannane to cyclohexanone in the presence of Lewis acids were unsuccessful. However, the use of "naked" cyanide ion catalysis (KCN/18-crown-6, Bu4NCN, TASCN) efficiently catalyzed an exothermic Michael addition to produce the γ-stannyl silyl enol ether in high yields(eq 5).[74]

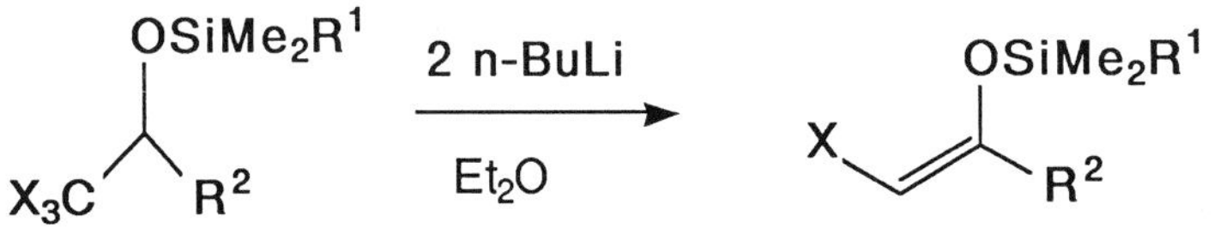

$$(4)$$

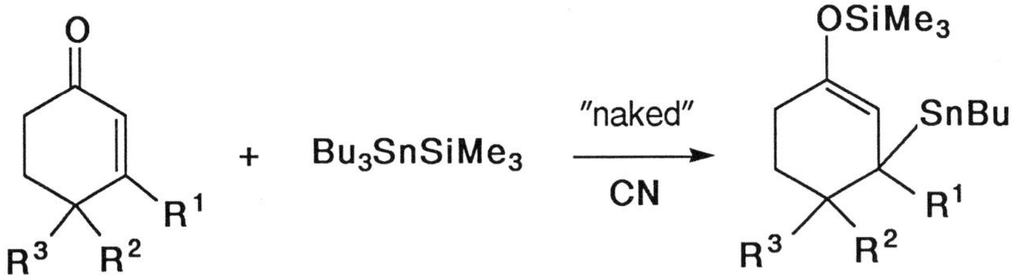

$$(5)$$

Crystal structure analysis was carried out for lithium enolates of the following three esters; t-butyl propionate (**A**), t-butyl 2-methylpropionate (**B**) and methyl 3, 3-dimethylbutanoate (**C**). The enolates were generated by treatment with LDA in THF. Enolates (**A** and **B**) are dimeric (with one TMEDA per Li atom), whereas **C** is tetrameric (with one THF per Li atom).[74a] The Z-configuration of **A** and **C** established by X-ray analysis is in agreement with the results shown in Scheme 5.

$$A \qquad\qquad B \qquad\qquad C$$

3.2 MECHANISM OF ALKYLATION

It is widely accepted that the reaction of an enolate anion with an alkyl substrate (halide or tosylate) proceeds by an S_N2 process. However, in certain cases, a different mechanism may be involved. For reactions involving p-nitrobenzyl chloride, a $S_{RN}1$ type radical-radical anion chain mechanism is involved.[75] The $S_{RN}1$ mechanism is also involved in the reaction of $XCMe_2NO_2$ (X=Cl, NO_2, or p-$MeC_6H_4SO_2$) with lithium enolates.[76] Under irradiation, ketone enolate ions in ammonia solution are phenylated by reaction with halobenzenes.[77] These reactions occur by the radical-chain $S_{RN}1$ mechanism. Similarly, dihalopyridines react with pinacolone potassium enolate to give disubstituted products in excellent yields.[78] These reactions and $S_{RN}1$ mechanism are summarized in Scheme 6. The proposed initiation step (1) involves a photoassisted electron transfer from enolates (YK) to aryl halides (XQX). The propagation cycle begins with step 2, where radical anion XQX˙ produces radical XQ˙ and continues (step 3) by combination of XQ˙ with enolate to give radical anion XQY˙. The next propagation step 4 involves fragmentation of XQY˙ to form radical QY˙, which then combines with enolate to afford radical anion YQY˙ (step 5). Completion of the propagation cycle (step 6) involves electron transfer from YQY˙ to substrate XQX to give the disubstituted product and radical anion XQX˙, the latter of which can reenter the radical chain cycle at step 2.

A long time ago, it was suggested that typical aliphatic halides could be reacting with enolate anions by an electron transfer process.[79] Evidence for a radical process in the reaction of the lithium enolate of propiophenone with a primary alkyl iodide was obtained by the observation of cyclization of an appropriate radical probe, by trapping of the radical intermediate and by the comparison of the relative rates of reactions of the alkyl iodide probe with those of the corresponding bromide and tosylate.[80] The results of these studies indicate that single electron transfer is the major reaction pathway involved in the reaction of the enolate with the alkyl iodide, in HMPA, and that the corresponding bromide and tosylate react by a S_N2 process. Evidence consistent with the involvement of a single electron transfer mechanism in aldol reactions involving enolate nucleophiles with aromatic ketones is also reported.[81] In summary, there is a possibility that an electron transfer process may be involved in certain alkylation reactions of enolates. The effects of leaving group, solvent, structures of enolates and substrates are important in gauging the extent of such a radical processes.

Scheme 6. $S_{RN}1$ mechanism of enolate ions.

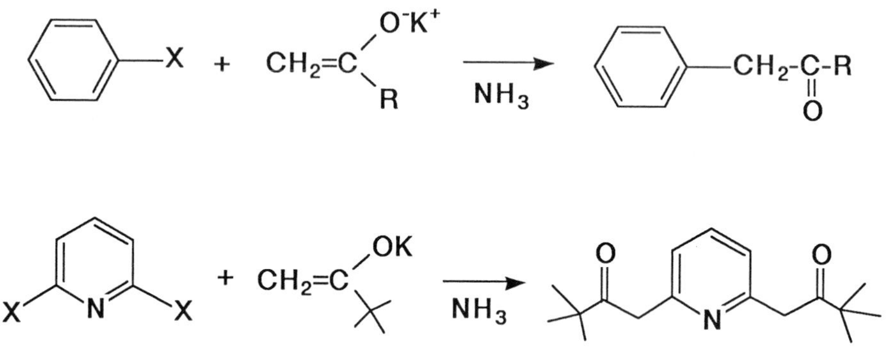

$$XQX + YK \longrightarrow XQX^{\cdot -} + Y^{\cdot} + K^+ \qquad \text{(initiation)} \quad (1)$$

$$XQX^{\cdot -} \longrightarrow XQ^{\cdot} + X^- \qquad \text{(propagation)} \quad (2)$$

$$XQ^{\cdot} + YK \longrightarrow XQY^{\cdot -} + K^+ \qquad (3)$$

$$XQY^{\cdot -} \longrightarrow QY^{\cdot} + X^- \qquad (4)$$

$$QY^{\cdot} + YK \longrightarrow YQY^{\cdot -} + K^+ \qquad (5)$$

$$YQY^{\cdot -} + XQX \longrightarrow YQY + XQX^{\cdot -} \qquad (6)$$

3.3 REGIOSELECTIVITY. C- AND O- ALKYLATION

The alkylation of enolate anions (conjugate bases of aldehydes, ketones, esters, amides) may in principle occur on either carbon or oxygen. The actual regiospecificity depends strongly on solvent, temperature, and counterion.[10] Alkylation of preformed enolates of aldehydes and ketones takes place exclusively on the carbon in less polar solvents.[82a,c] The use of more polar solvent such as Me_2SO and HMPT increases the extent of O-alkylation.[82b] The alkylation of crowded Li-enolates of aldehydes and ketones, such as 110 and 107, gave C alkylation only in the former and mixtures of O and C in the latter, while 111 afforded only O-alkylation product (112).[72c] These results indicate that steric hindrance with bulky alkyl groups does not produce regioselectivity. The alkylation of enolates of carboxylic acid derivatives proceeds, virtually always, exclusively with C-alkylation.[83]

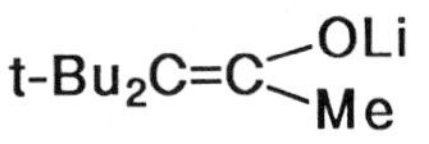

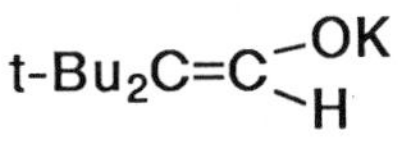

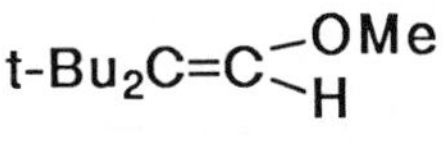

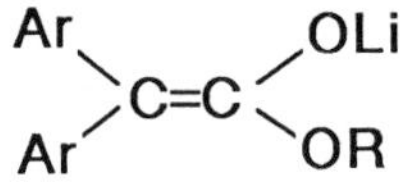

On the other hand, esters of diarylcarboxylic acid (113) underwent exclusively O-alkylation to give the diarylketene acetals.[84] Introduction of pentasubstituted aryl groups into the α-carbon of an ester leads to exclusive O-alkylation of the ester enolate, and the resulting ketene acetals are highly acid-resistant.

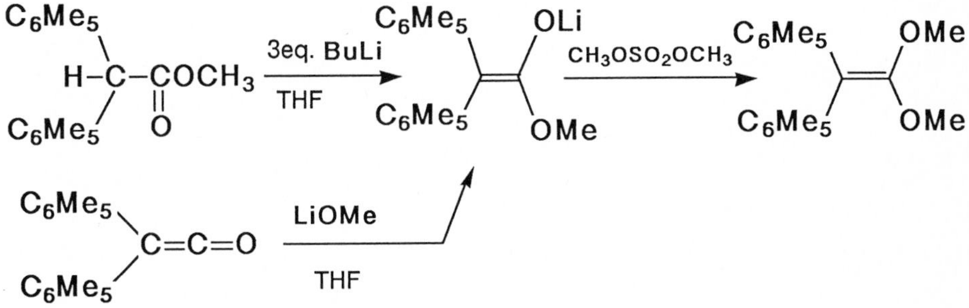

The regiochemistry of alkylation of enolate ions, generated by ion-molecule reaction in the gas phase, was studied.[85] The reaction of cyclohexanone enolate (naked ion) with methyl bromide produced only the product resulting from O-alkylation. In solution, polar aprotic solvents tend to favor the alkylation of the more electronegative O atom since they can complex the positive counterion and leave the oxygen relatively unhindered. Therefore, it is expected that O-alkylation would predominate in the gas phase since the negative charge resides primarily on the more electronegative oxygen atom and it is unhindered by solvent or counterion. The experimental result is in agreement with this expectation. It was shown that the formation of 2-methylcyclohexanone is exothermic by 43.8 kcal/mole, while the formation of 1-methoxycyclohexene is less energetic (-26.9 kcal/mole). Therefore, the C-alkylation product would be expected to be favored on the basis of thermodynamics.

The transition structures for the gas phase C- and O-alkylation of acetaldehyde enolate with methyl fluoride have been located with ab initio RHF calculations and the 3-21G basis set.[86] The activation energy for O-alkylation is lower than that for C-alkylation, even though the latter is favored thermodynamically. These results parallel those found for the reactions of Me⁻ and OH⁻ with MeF and arise from the lower intrinsic barrier for reactions of oxygen-centered nucleophiles (or leaving groups) than carbon-centered nucleophiles in S_N2 reactions. The geometry of the transition structures indicates that stereoelectronic factors favor product-like conformations, even for relatively early transition states.

As mentioned above, it is generally believed that the alkylation of enolates by most primary and secondary alkyl halides and tosylates, in weakly polar solvents, occurs by the S_N2 mechanism. Lithioisobutyrophenone reacted with (+)-2(S)-iodooctane to give the corresponding enol ether (114) and 1-phenyl-2,2,3-trimethylnonanone (115). In dimethoxyethane and dioxane, the C- and O-alkylation products were formed in the ratios 0.56 and 0.85, respectively, with net inversion of configuration.[87] The qualitative result for both solvents is that both C- and O-alkylation proceed with net inversion of configuration.

Silyl enol ethers almost always undergo C-alkylation.[88] The electron rich C-C double bond of Si-enol ethers is easily attacked by electrophiles to give a trimethylsilyloxy stabilized carbocation. The nucleophilic anion associated with the electrophile attacks the silyl center to give a C-substituted ketone. Tertiary alkyl halides which ionize to carbocations on treatment with Lewis acids react with Si-enolates to give α-alkylated ketones. Other electrophiles such as CH_3OCH_2Cl, $PhSCH_2Cl$, and $ClCH_2N(CH_3)CO_2CH_3$ also undergo α-C-alkylation in the presence of Lewis acids.[88a-e] Interestingly, cobalt-complexed propargylic ethers react with silyl enol ethers in the presence of Lewis acids.[88f]

In connection with the regiochemistry of alkylation, we briefly mention the regiochemistry of acylation of enolates. Detailed discussions are given in the previous reviews.[13] When acid derivatives such as acid chlorides and acid anhydrides are used as acylating agents, O-acylation and/or C-acylation take place depending on the reaction conditions. When the acylation occurs under kinetic conditions (i. e., when the metal enolate is added to a large excess of the acylating agent), the O- and C-acylation ratio depends upon (i) the nature of the metal cation, (ii) the solvent, (iii) the structure of enolates, and (iv) the structure of the acylating agent. In general, O-acylation is preferred with metal enolates having solvent separated ion pairs, such as Li^+, Na^+, K^+, and Zn^{2+} enolates, in DME or more polar solvents. On the other hand, Mg^{2+} enolates in less polar solvents such as diethyl ether, in which the metal enolate takes the form of tight ion pairs or covalently bonded species, give C-acylation predominantly. This tendency is quite similar to that found in the C- and/or O-alkylation of enolates.

Although the dianion of 2,4-diketones can be acetylated with ethyl acetate to give 2,4,6-triketones, acetylations of the trianions of triketones and a number of other multiple anions have failed because the strongly basic trianions abstract a proton from the ester rather than attacking the carbonyl group. The use of N-methoxy-N-methylacetamide(116) solved this problem and efficient acetylation of the multiple anions of poly-β-carbonyl compounds was achieved.[89] The C-alkylation of alkyl, alkenyl, alkynyl, and aryl lithium reagents by ketenes generates lithium enolates regio -and stereo-selectively (**106~109**). Ketenes react with lithium enolates mainly by O-acylation, giving vinyl ester enolates (**117**).[90]

$$n = 1\text{-}3$$

116

117

Asymmetric acylation is realized with chiral enolates. Deprotonation of the N-acyloxazolidinone[91](**118**) with LDA in THF gave the corresponding lithium enolate which was reacted with t-Boc-D-alanine anhydride (inverse addition, see above) afforded ketone **119** in 78% yield.[92] Formation of the diastereomer **119'** was less than 1%. This result is consistent with the alignment of the chelated (Z)-lithium enolate of **118** with the acylating agent such that the latter is situated trans to the isopropyl group. By using a similar procedure, both enantiomers of β-dicarbonyl adducts can be prepared (**120, 121**).[93] The direct acylation of these chiral imide enolates provides a direct entry into chiral β-dicarbonyl synthons.

Reaction of the β-ketosulphone with NaH in THF followed by addition of pivaloyl chloride gave the O-acylated product (**122**).[94] The enolate derived from methyl 4,6-O-benzylidene-2-deoxy-α-D-erythro-hexopyranosid-3-ulose underwent O-acylation exclusively with a variety of reagents, whereas C-acylation was successful only with the corresponding C-glycosyl derivatives.[95]

The trialkylsilyl derivatives of esters (silyl ketene acetals) are readily C-acylated with acid chlorides (eq 6). The less reactive trialkylsilyl derivatives of ketones (silyl enol ethers) are C-acylated only by reactive polyhalogenated acid chlorides (eq 7). Simple acid chlorides either do not react, or react in the presence of catalysts to give O-acylated products (eq 7). A variety of trialkylsilyl enol ethers are readily C-acylated with acetyl tetrafluoroborate to produce the corresponding 1,3-diketones (eq 8).[96] Reaction with acid chlorides in the presence of zinc chloride, or antimony trichloride, results in good to excellent yields of 1,3-diketones (eq 9).[97] Even by using this method, some to trace amounts of O-acylation products are observed in most cases.

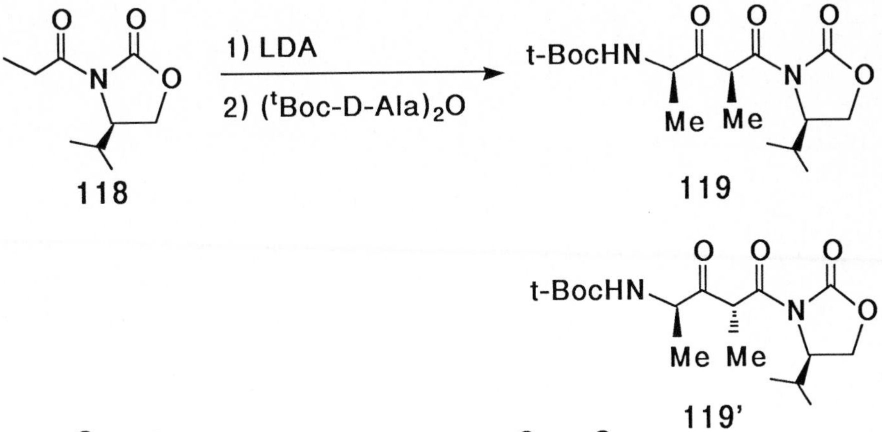

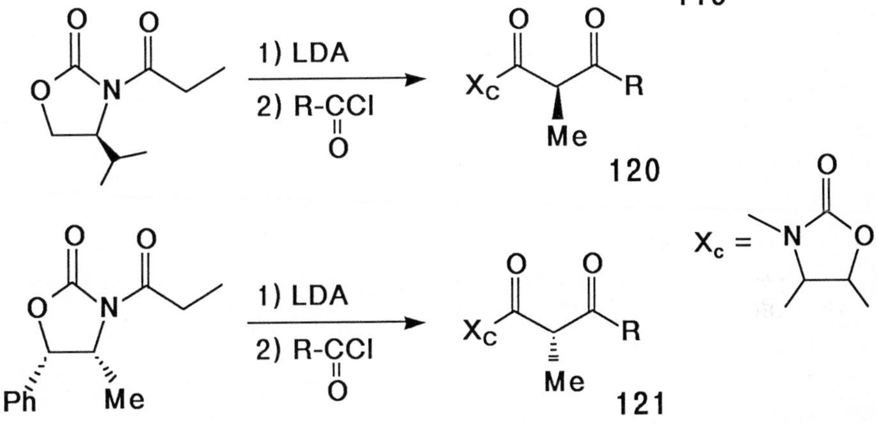

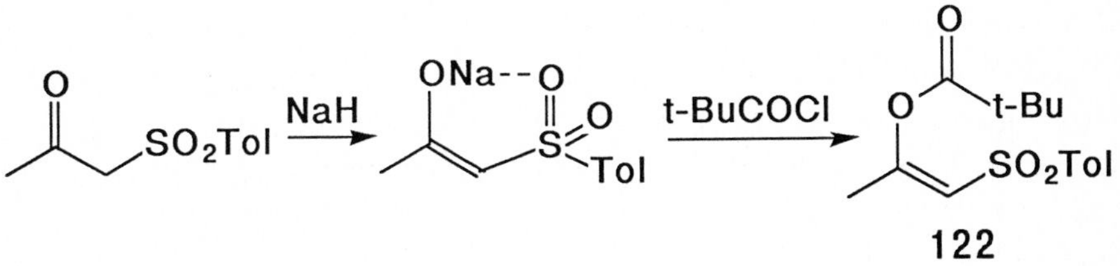

$$\underset{\text{(6)}}{\begin{array}{c} \diagdown \\ \diagup \end{array} C=C \begin{array}{c} \diagup OSiR_3 \\ \diagdown OR \end{array} \quad + \quad R'COCl \quad \xrightarrow{\quad} \quad \xrightarrow{H_3O^+} \quad R'\underset{\underset{O}{\parallel}}{C}-\overset{|}{\underset{|}{C}}-CO_2R}$$

$$\begin{array}{c} \diagdown \\ \diagup \end{array} C=C \begin{array}{c} \diagup OSiR_3 \\ \diagdown R \end{array} \quad \begin{array}{c} \xrightarrow{Cl_3CCOCl} \quad \xrightarrow{H_3O^+} \quad Cl_3C-\underset{\underset{O}{\parallel}}{C}-\overset{|}{\underset{|}{C}}-\underset{\underset{O}{\parallel}}{C}-R \\ \\ \xrightarrow[CH_3COCl]{HgCl_2} \quad \xrightarrow{H_3O^+} \quad CH_3\underset{\underset{O}{\parallel}}{C}OC=C\overset{R}{\diagup}\diagdown \end{array} \quad \text{(7)}$$

$$\begin{array}{c} \diagdown \\ \diagup \end{array} C=C \begin{array}{c} \diagup OSiR_3 \\ \diagdown R \end{array} \quad + \quad CH_3COBF_4 \quad \xrightarrow{CH_3NO_2} \quad \xrightarrow{NaOAc} \quad R-\underset{\underset{O}{\parallel}}{C}-\overset{|}{\underset{|}{C}}-\underset{\underset{O}{\parallel}}{C}-CH_3 \quad \text{(8)}$$

$$\begin{array}{c} \diagdown \\ \diagup \end{array} C=C \begin{array}{c} \diagup OSiR_3 \\ \diagdown R \end{array} \quad + \quad R'COCl \quad \xrightarrow[SbCl_3]{ZnCl_2 \text{ or } H_2O} \quad \xrightarrow{\quad} \quad R-\underset{\underset{O}{\parallel}}{C}-\overset{|}{\underset{|}{C}}-\underset{\underset{O}{\parallel}}{C}-R' \quad \text{(9)}$$

4. OTHER METAL ENOLATES INCLUDING TRANSITION METALS
4.1   B, Zn, Ti, Zr, Bi, Sn, AND Pd ENOLATES

We have discussed the formation of alkali metal enolates mainly derived from lithium; also, we discussed Si-enolates, and the regio- and stereo-chemistry of their alkylation and related reactions. In this chapter, we describe the reaction of other metal enolates such as B, Zn, Ti, Zr, Bi, Sn, and transition metals.

The reaction of potassium enoxyborates, (readily obtainable by treating potassium enolates with a trialkylborane such as triethylborane; with allylic electrophiles such as allylic chlorides and acetates in the presence of a catalytic amount of a palladium-phosphine complex, e.g., Pd(PPh3)4) proceeded readily at room temperature to produce the corresponding α-allylated ketones in high yields, with essentially complete retention of both the enolate regiochemistry, and the allyl geometry (eq 10).[98]

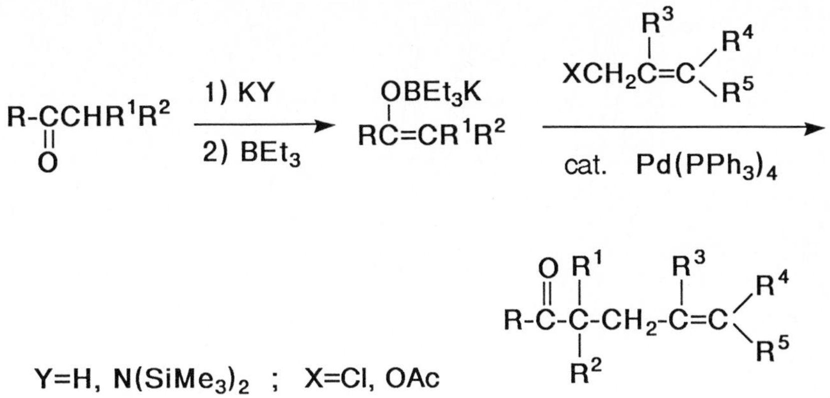

$$R\text{-}\underset{\underset{O}{\|}}{C}\text{-}CHR^1R^2 \quad \xrightarrow[\text{2) BEt}_3]{\text{1) KY}} \quad \underset{RC=CR^1R^2}{OBEt_3K}$$

$$\xrightarrow[\text{cat. } Pd(PPh_3)_4]{XCH_2\underset{\underset{R^3}{|}}{C}=C\underset{R^5}{\overset{R^4}{\diagup}}}$$

(10)

$$R\text{-}\underset{\underset{O}{\|}}{C}\text{-}\underset{\underset{R^2}{|}}{\overset{\overset{R^1}{|}}{C}}\text{-}CH_2\text{-}\underset{\underset{R^3}{|}}{C}=C\underset{R^5}{\overset{R^4}{\diagup}}$$

Y=H, N(SiMe_3)_2 ; X=Cl, OAc

Lithium enolates derived from ketones did not react with trialkylboranes to form the corresponding enoxyborates.[99] However, the lithium enolate-R_3B mixture exhibited higher threo selectivity in aldol reactions than the lithium enolate itself.[100] The potassium enolates produced the corresponding enoxyborates[101], which selectively reacted with various alkylating agents to give α-monoalkylated ketones.[102] Unfortunately, this procedure did not work with allylic electrophiles except allyl bromide. Allylation of alkali metal enolates with allylic acetates catalyzed by Pd(PPh_3)_4 was generally unsatisfactory.[103] The Pd-catalyzed method using potassium enoxy borates (eq 10) solves this problem. Further studies revealed that the palladium catalyzed allylation of enolates works well with other counterions such as zinc enolates and lithium enolates -BEt_3.[104] All other counterions suffered from low product yields, e.g., MgCl, AlMe_2, LiAlMe_3, KAlMe_3, SiMe_3, and TiCp_2Cl, and/or low regio- or stereo-specificity, e.g., Li and SnBu_3. Overall retention with respect to the allylic carbon center was observed in the reaction with either KBEt_3 or ZnCl counterions (eq 11).

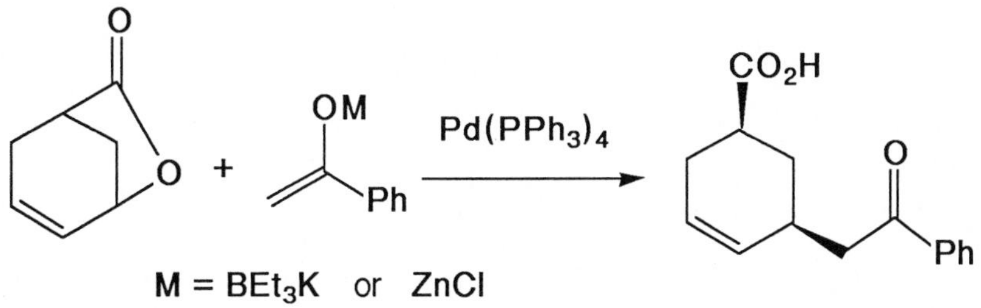

M = BEt_3K or ZnCl

(11)

A reactive palladium (II) enolate intermediate is involved in the decarboxylation of palladium(II) β-ketocarboxylate. A catalytic amount of Pd(PPh₃)₄ induced the decarboxylation of allylic β-ketocarboxylates to produce α-allylic ketones via π-allylpalladium (II)-enolate complexes (eq 12).[105,106]

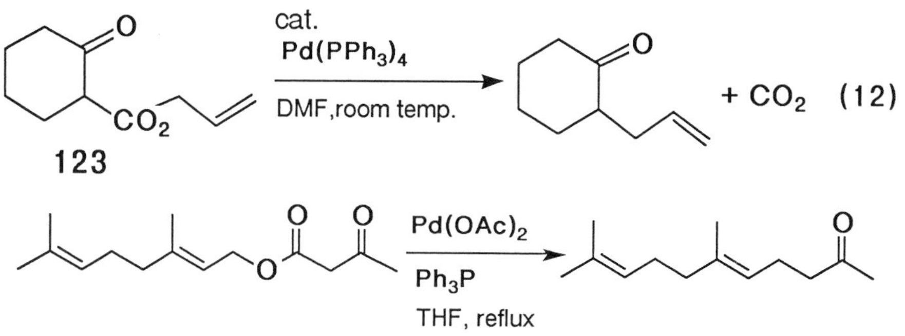

$$+ CO_2 \quad (12)$$

Scheme 7. Catalytic allylation of enolates.

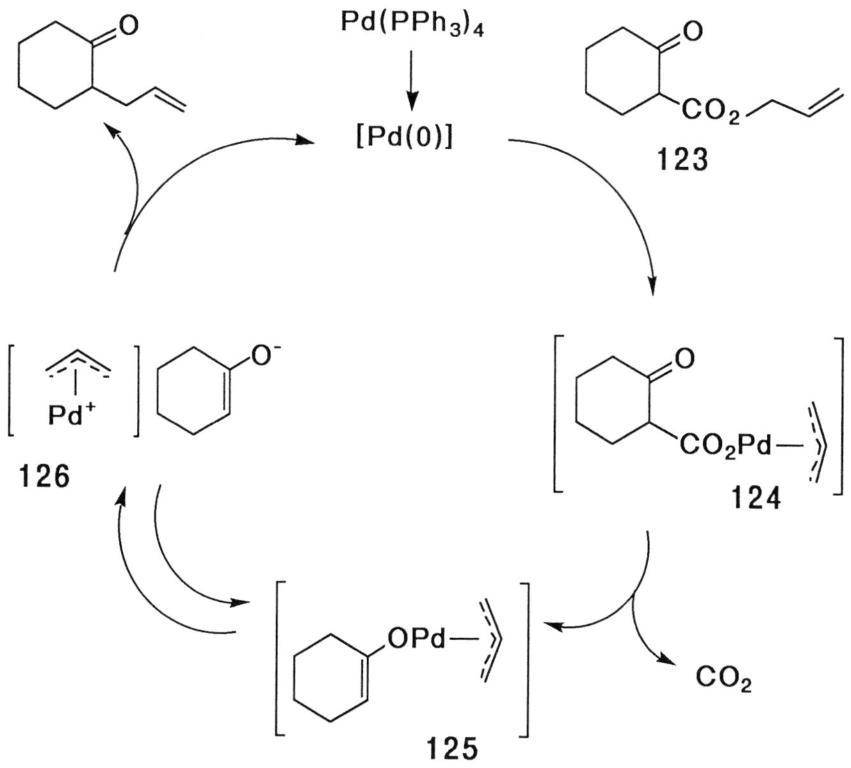

The proposed reaction path is shown in Scheme 7. Oxidative addition of **123** to Pd(O) produces a π-allylpalladium (II)-cyclohexanone-2-carboxylate complex (**124**) which undergoes decarboxylation to generate a π-allylpalladium (II) enolate intermediate (**125**). Dissociation of the enolate ligand in **125** takes place forming a cationic π-allylpalladium (II) complex (**126**) followed by nucleophilic attack of the enolate anion to produce the allylated cyclohexanone.

Early-transition metal alkylidene complexes react with acid chloride to yield the corresponding enolate complex (eq 13).[107] The easily generated titanocene methylidene complex gives Cp_2TiCl-enolates upon treatment with acid chlorides (eq 14).[108] This stable enolate can be isolated and characterized by spectroscopic and chemical methods. Titanium enolates are also prepared by the reaction of Cp_2TiCH_2 with hindered ketones,[109] by the reaction of lithium enolates with $Cp_2TiMeCl$, Cp_2TiCl_2,[110] or $TiCl_4$,[111] and by the reaction of trimethylsilyl enol ethers with $TiCl_4$.[112] Recently, it was reported that the reaction of $Cp_2*Ti=CH_2$, generated in situ by thermolysis of the dimethyl compound, with epoxides affords titanium enolates(eq 15).[113] Some of these titanium enolates react rapidly with aldehydes, but normally the alkylation reaction is very sluggish. Thus, we cannot find, so far, the synthetic usefulness for these enolates in alkylation reactions.

$$LnM=CHR \;+\; R^1\text{-COCl} \longrightarrow R^1\text{-C(OMLnCl)}=CHR \qquad (13)$$

$$(14)$$

$$Cp_2^*TiMe_2 \xrightarrow{\Delta} Cp_2^*Ti=CH_2 \qquad (15)$$

Zirconium enolates are readily prepared by the reaction of lithium enolates with Cp_2ZrCl_2 (eq 16).[114] Although these zirconium enolates exhibit an interesting erythro selectivity in aldol reactions, regardless of the geometry of the starting enolate, the alkylation reaction is sluggish and thus syntheticaly not useful.

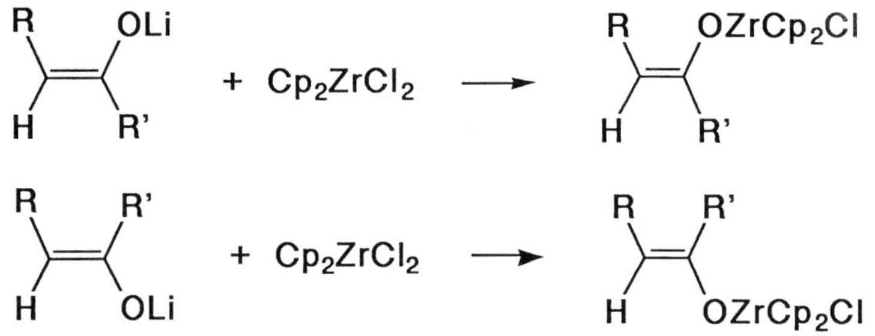

$$\text{(structure: } R,\ OLi\ /\ H,\ R'\ ) + Cp_2ZrCl_2 \longrightarrow (\text{structure: } R,\ OZrCp_2Cl\ /\ H,\ R')$$

$$\text{(structure: } R,\ R'\ /\ H,\ OLi\ ) + Cp_2ZrCl_2 \longrightarrow (\text{structure: } R,\ R'\ /\ H,\ OZrCp_2Cl)$$

(16)

Arylation reactions of enols with pentavalent organobismuth reagents involve initial formation of an intermediate (127) possessing a covalent Bi-O bond. The breakdown of this intermediate may be controlled by the nature of the ancillary groups around the bismuth atom, with electron withdrawing groups favoring reductive α-elimination. The reaction of β-naphthol with $Ph_4BiOCOCF_3$ under neutral conditions gives the enol ether. On the other hand, the reaction of β-naphthol anion (generated by treatment with N-t-butyl-N',N"-tetramethylguanidine as base) with $Ph_4BiOCOCF_3$ completely changes the regiochemical course; a new carbon-phenyl bond is formed exclusively.[115]

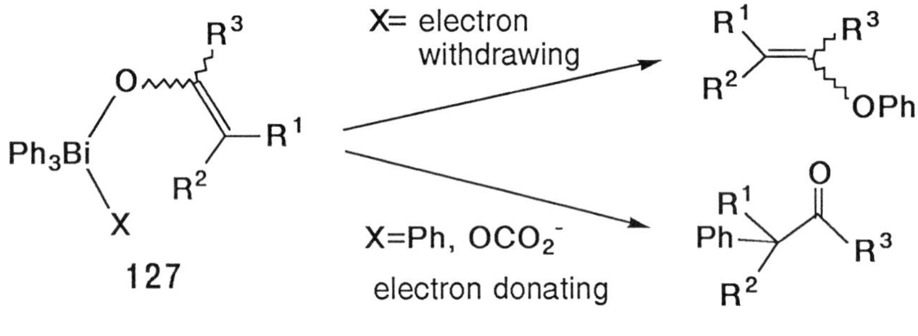

Tin enolates undergo clean C-monoalkylation (eq 17).[116] Tin enolates can be obtained by the transesterification of enol acetates (accessible from the corresponding ketones). Tin enolates can also be generated in situ from the corresponding lithium enolates and R_3SnCl. By using this latter method, the polyalkylation of lithium enolates is avoided.[117,118] The stereochemical aspects of the alkylation of triphenyltin enolates formed in situ were studied.[118] As shown in Scheme 8, enhancement of diastereoselectivity is observed via Ph_3Sn enolates in comparison with the direct alkylation. Consequently, alkylation with Ph_3Sn enolates gives the monoalkylation product in good yield without formation of polyalkylation by-products, and in some cases produces high stereoselectivity. Tin enolates generated in situ from lithium enolates exhibited erythro aldol stereoselectivity,[119] though the isolated tin enolates prepared from transesterification showed threo diastereoselectivity.[120] This discrepancy is due to the co-presence of lithium halides in the former case.[121] Trimethylgermanium enolates, prepared in situ from lithium enolates and Me$_3$GeX, undergo a rapid aldol condensation with benzaldehyde to give the erythro-aldol, while the same enolates produce the threo-aldol if the coexisting lithium halides were removed before condensation.[121]

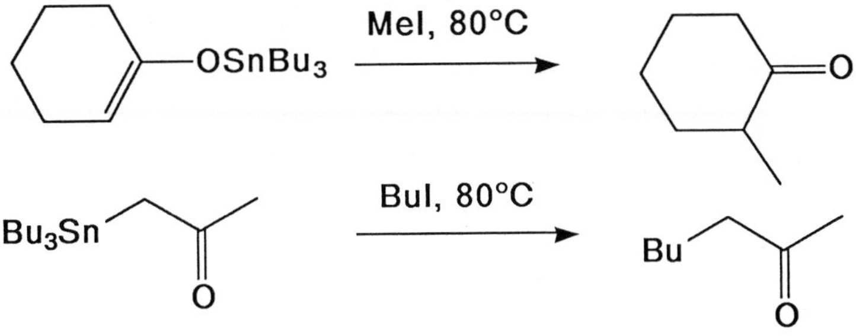

(17)

Scheme 8. Stereoselectivity of Ph$_3$Sn-enolates formed in situ.

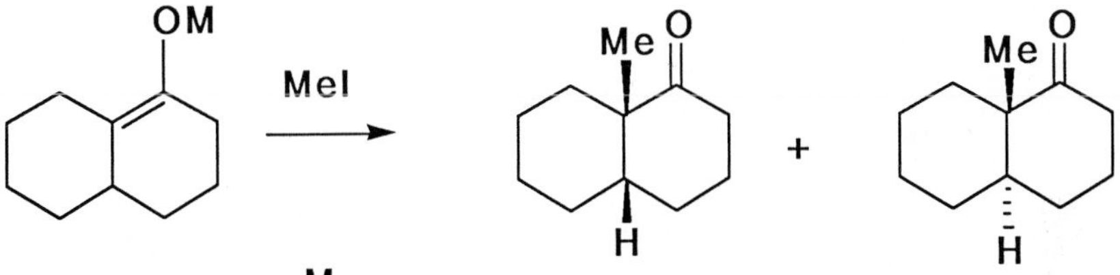

M			
Li	70	:	30
Ph$_3$Sn	100	:	-

M			
Li	80	:	20
Ph$_3$Sn	85	:	15

An important reaction of tin enolates is the palladium catalyzed alkylation.[116] As shown below, C-alkylation takes place with aryl halides, vinyl halides, α-haloketones, allylic acetates, and allylic carbonates. The reaction with allylic carbonates proceeds with a catalytic cycle similar to that shown in Scheme 7.

Organotin enolates react with acyl halides and give, exclusively, O-acylated product.[116]

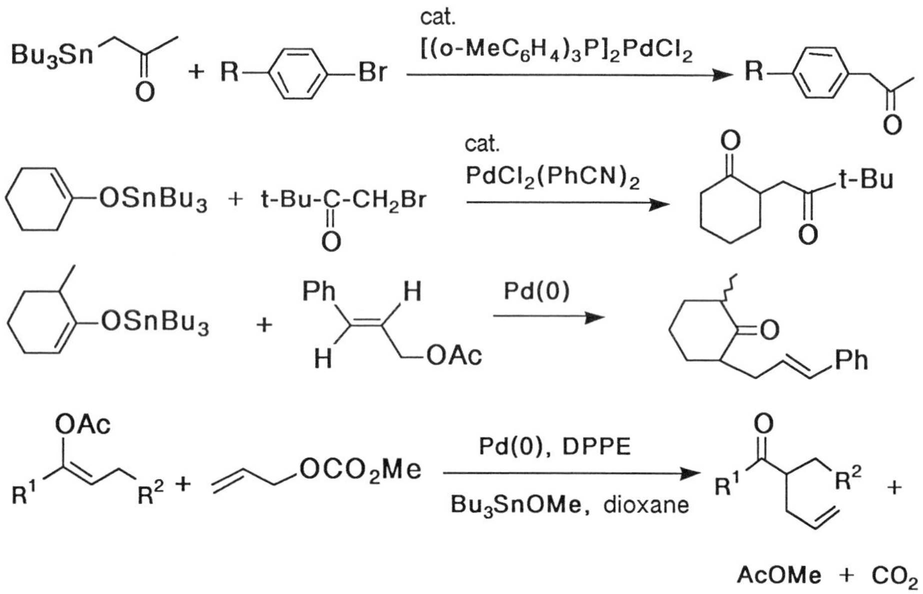

4.2 TRANSITION METAL ENOLATES

Enolates derived from acyl ligands attached to the chiral auxiliary Fe complex (X) undergo highly stereoselective alkylation reactions.[122,125] This type of enolate is not a transition metal enolate where the metal is bonded to the oxygen atom, but a lithium enolate (e.g., 133) having the Fe complex at the α-position. For the synthetic importance, we discuss these pseudo transition metal enolates.

The methoxycarbene complex (128) is readily prepared by treatment of the bromide (129) with the lithium acetylide, followed by protonation, and subsequent addition of methanol. Deprotonation of 128 at -78°C with t-BuOK occurs stereoselectively (>95%) to yield the Z-methoxyvinyl complex (130). Alkylation of 130 proceeds stereospecifically to give 131; the benzylation takes place from the unshielded face of the methoxyvinyl ligand of 130'.[122] Deprotonation of 132 with n-BuLi produces the lithium enolate (133), which undergoes benzylation again from the unshielded face to give 134.[122,125] It is to be noted that opposite diastereoisomers (**131** and **134**) are produced via the two methods because of the opposite geometries of the enolate and methoxyvinyl ligand.

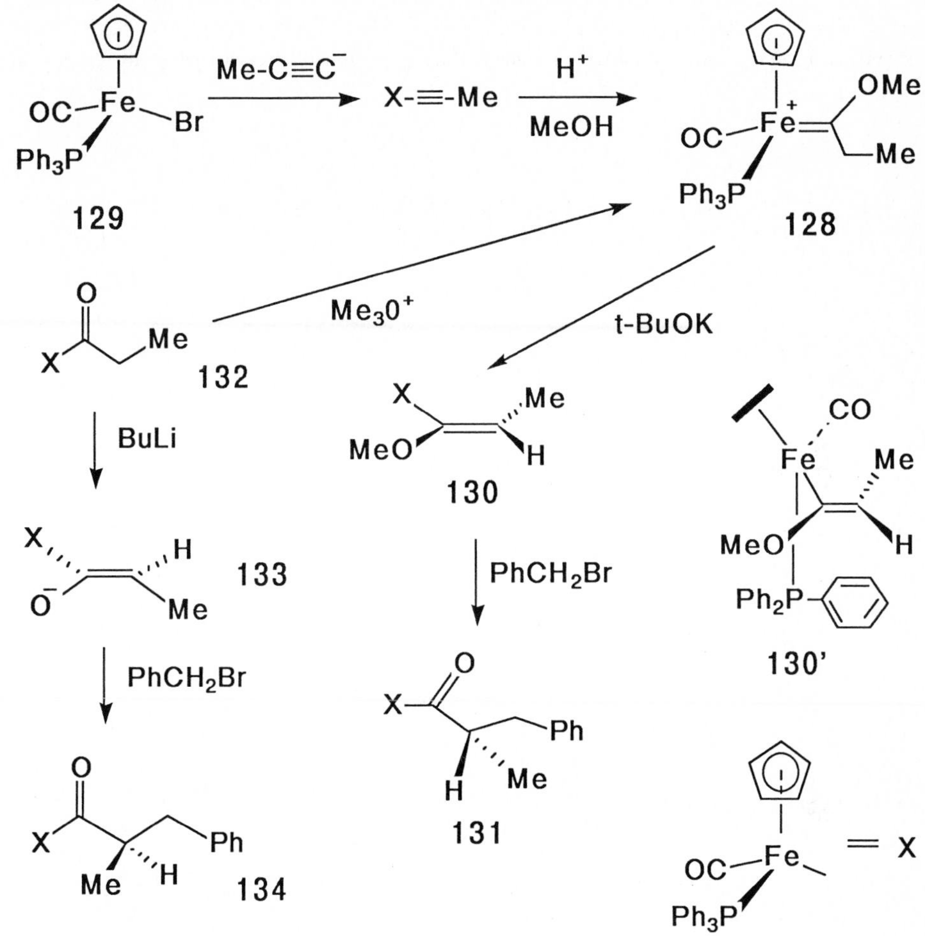

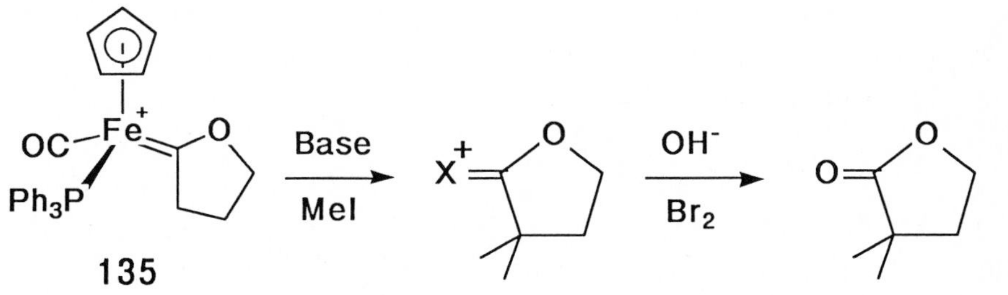

Repeated deprotonation and alkylation of **135** allows the stereoselective synthesis of quarternary carbon centers to be achieved with decomplexation, leading to 2,2-dialkylbutyrolactones.[123a] The aluminium enolate derived from X-COMe undergoes stereoselective aldol condensations with aldehydes to generate β-hydroxy acyl complexes which yield β-hydroxy acids on decomplexation.[123b,124]

Reaction of **136** with two equivalents of n-BuLi, followed by addition of excess MeI, gives **138** and **140** in the ratio of 3:2.[126] The major diastereomer **138** is that expected from methylation of the Z-enolate **137** from the unhindered face with the minor diastereomer **140** coming from the E-enolate **139**. Presumably, lactone formation via the CO ligand, prior to enolate formation, leads to the otherwise disfavored (Z)-enolate. α,β-Unsaturated acyl complexes like **141** undergo tandem stereoselective Michael addition, and subsequent methylations, which result in the stereocontrolled synthesis of α- and β-substituted iron acyl complexes.[127] The aluminium and copper enolates derived from **132** react with aldehydes and ketones diastereoselectively.[125,128]

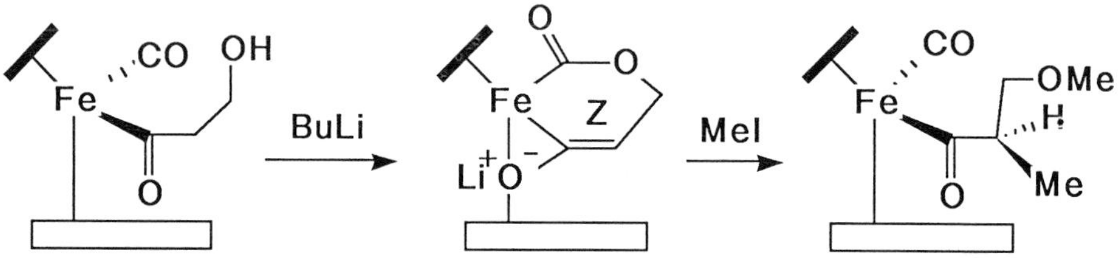

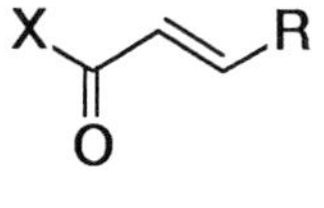

Although **136** assumes a predominantly Z-geometry due to lactone ring formation, the acyl ligand bound to X undergoes deprotonation by n-BuLi to give exclusively (>200:1) the corresponding E-enolate (**142**). This E-enolate undergoes methylation exclusively (>200:1) in the anti acyl-oxygen to CO conformation, from the unhindered face opposite the PPh3.[129a] Crotyl iron complexes (**142'**) were also prepared, and their deprotonation-alkylation reactions exhibit a

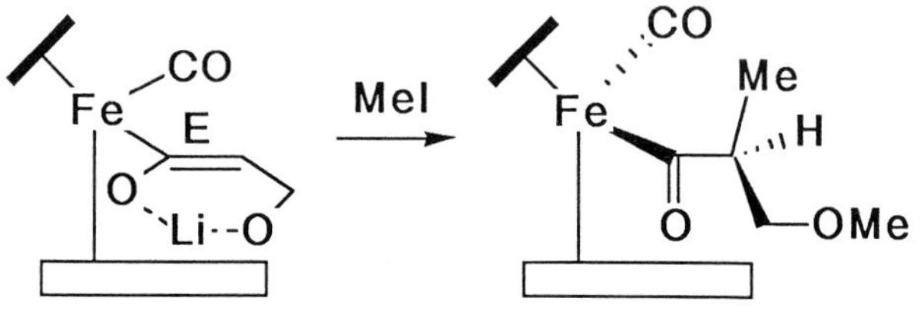

stereoselectivity related to that of 142.[129b] The conjugate additions to 142'
followed by alkylations were also studied.[129b,130] The lithium enolate derived
from $XCOCH_3$ can be trapped in either the enol or keto form by using either
oxophylic, i.e., Cp_2ZrCl_2, or carbophylic, i.e., $Au(PPh_3)Cl$, metal fragments.[131]

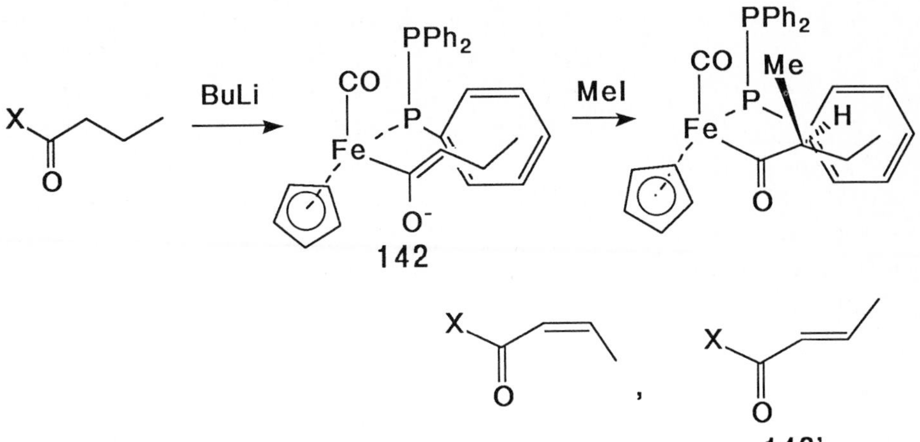

The cobaltacyclopentanone (143) can be converted to the corresponding lithium
enolate, which reacts rapidly with electrophilic reagents to give products typical of
alkylation with organic halides and aldol condensations with aldehydes.[132] The
chirality of the metal center results in a significant amount of asymmetric
induction in these reactions.

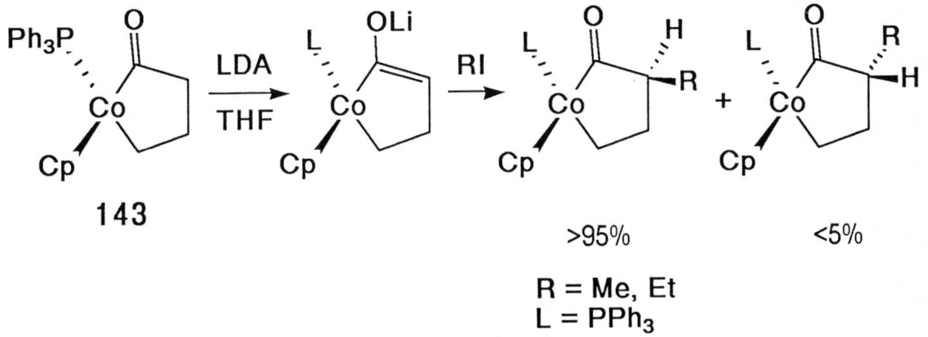

The molybdenum acyl complex (144) can be deprotonated with n-BuLi or KH to
form the corresponding enolate.[133] The enolate has been characterized by IR, and
^{1}H, ^{13}C, ^{31}P NMR spectroscopies. The results suggest that II is the major

contributing resonance form of the enolate in solution. When the enolate is treated with alkyl halides, alkylation at the α-carbon takes place in good yield.

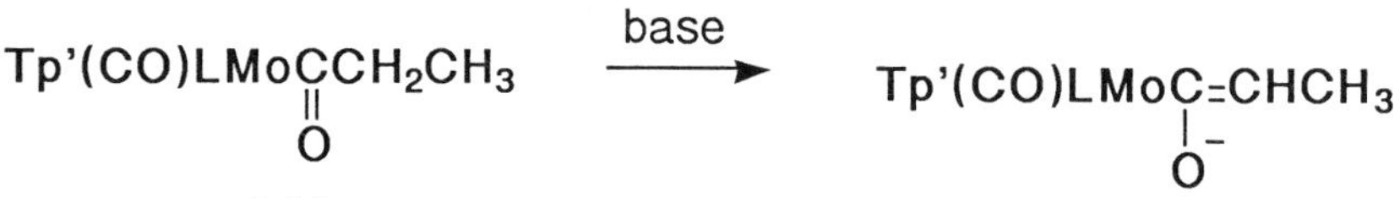
Tp'(CO)LMoCCH$_2$CH$_3$
O
base
Tp'(CO)LMoC=CHCH$_3$
O$^-$

144

Tp' = hydridotris(3,5-dimethylpyrazolyl)borate

L = $P(OPh)_3$

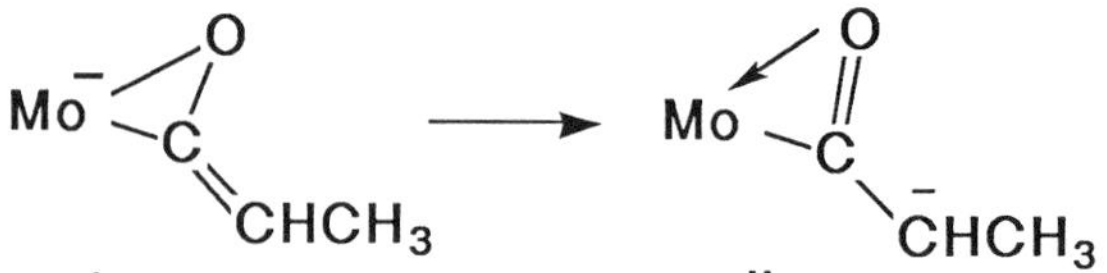
Mo
O
C
CHCH$_3$
Mo
O
C
CHCH$_3$

I II

Treatment of the gem-dibromocyclopropane with the silylamine in the presence of $Ni(CO)_4$ in DMF, produces the cyclopropanecarboxamide in which hydrogenolysis of the C-Br bond, and introduction of the carbamoyl group are performed.[134] In this reaction, formation of the nickel enolate as an intermediate is proposed. In the first step, $Ni(CO)_4$ may react with the silylamine to generate a carbamoyl nickel intermediate which reacts with the dihalide, as shown below. Finally, the enolate or α-Ni-carbonyl compound undergoes protonolysis to give the product.

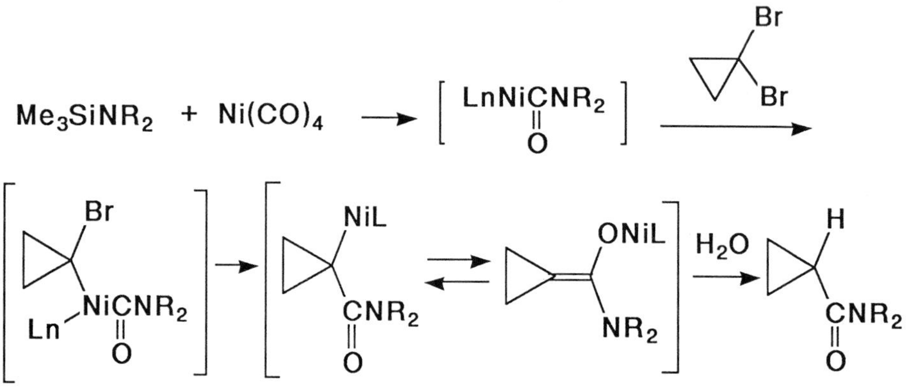
Me$_3$SiNR$_2$ + Ni(CO)$_4$
LnNiCNR$_2$
O
Br
Br
Br
NiCNR$_2$
Ln
O
NiL
CNR$_2$
O
ONiL
NR$_2$
H$_2$O
H
CNR$_2$
O

Several different types of organotransition-metal enolates are summarized in Scheme 9.[135] Type A enolates are generated in situ by deprotonation of metal-acyl complexes of iron, cobalt and molybdenum as mentioned above. Type B enolates are generated by reactions involving complexes of Zr,[114] Ti,[107,113] Pt,[136] Pd,[137] Mo,[138] Th[139] and U.[139] Most main group metal enolates belong to type B. Type C enolates are synthesized or invoked as reaction intermediates for Mo,[135,140] Fe,[140] Mn,[141] Co,[142] W,[135,140,143] Re,[144] Ni,[134] Pd,[137,145] Pt,[146] Rh,[147] Cu,[148] and Hg.[149] Type D enolates (oxa-π-allyl complexes) are also proposed as intermediates in some of the above metal enolates. Unfortunately, alkylation reactions from type C transition metal enolates are very sluggish and synthetically not so important.

Scheme 9.   Types of Metal Enolates

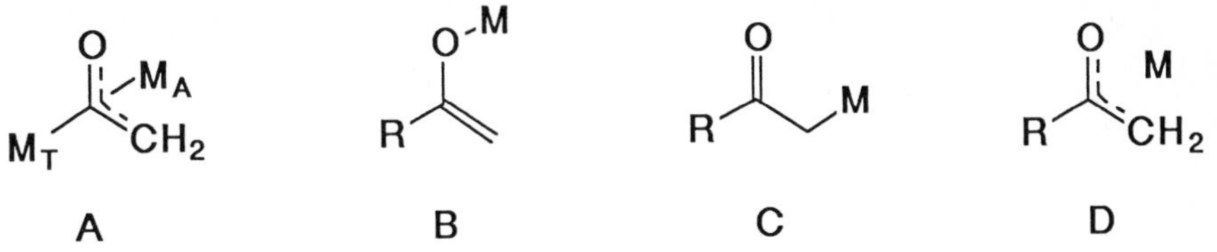

M_A= alkali metal, M_T=transition metal, $M=M_T$ or M_A

It is not clear that the copper enolate, generated from the conjugate addition of organocopper reagents to α, β-unsaturated esters, takes either a type B or C enolate structure. Further, there is also a possibility of its acquiring a type D structure. To clarify this problem, a stereochemical probe was used, as shown in Scheme 10.[150] Conjugate addition to **145** produces E, while the 1,4-addition to **146** gives F. The intermediate copper-oxygen bonded enolates (E and F) are protonated through the eclipsed conformation (E' and F', respectively) as mentioned previously (**10**). The electrophilic attack to E' and F' from the less hindered side results in predominant formation of **147**, along with minor amounts of **148**. Consequently, 1,4-addition to **145** and **146** should give the same isomer ratio. On the other hand, 1,2-addition to **145** produces G, while 1,2-addition to **146** gives H (Scheme 11). The α-cuprio ester intermediates G and H are diastereomeric in comparison with the enantiomeric intermediates E and F. Therefore, protonolysis of the C-Cu bond of G and H must give a different isomer ratio regardless of the mode of the electrophilic cleavage. The experimental results showed that the ratio of **147**/**148** was identical, irrespective of the starting materials. Consequently, the ultimate destination of copper in the conjugate addition to enoates is the oxygen atom rather than the α-carbon atom.

Scheme 10.   Structure of Copper Enolates (Type B)

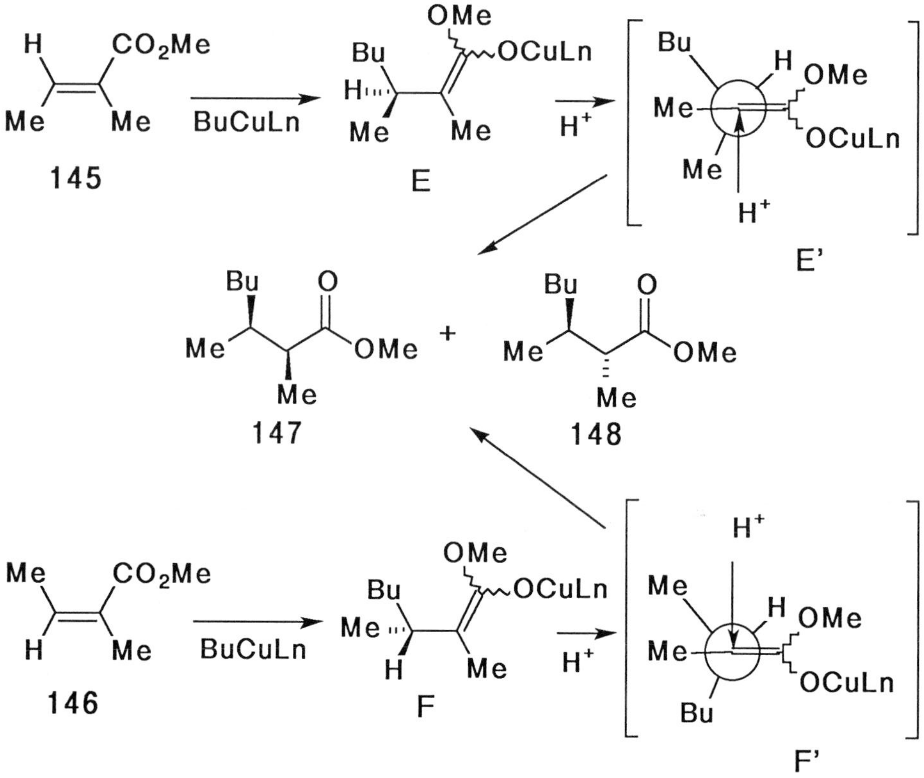

Scheme 11.   Structure of Copper Enolate (Type C)

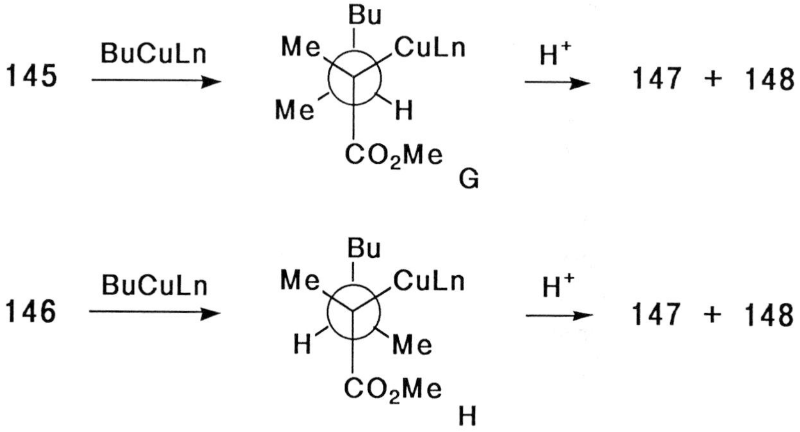

5. ASYMMETRIC ALKYLATION
5.1 ESTER ENOLATES

Molecular design of a chiral auxiliary is important in order to achieve high asymmetric induction in the alkylation. Generally speaking, chiral auxiliaries (R*) are introduced to either the ester portion (RCO$_2$R*) or the alkyl chain portion (R*CH$_2$CO$_2$R). In some cases, chiral bases such as LiNR$_2$* are used without introduction of a chiral auxiliary to the substrate (RCO$_2$R).

Very high diastereoselective alkylations of propionates of chiral alcohols derived from (+)-camphor are reported (Scheme 12).[151] Deprotonation of **A** with LICA (lithium cyclohexylisopropylamide) in THF at -80ºC produced **B** stereoselectively (see Scheme 5), which was trapped with R'X from the unshielded side to afford **C**. On the other hand, deprotonation of **A** with LICA in THF-HMPA gave **D** stereoselectively which underwent alkylation from the less hindered side to produce **E**. A number of camphor derivatives including exo and endo isomers were developed as the chiral auxiliary. By using this procedure, the side chain of α-tocopherol (**F**) is prepared in 58% overall yield via 9 steps.[152] This high chiral induction through **B** and **D** can be applied to O-protected glycolates such as **G** and **H**; but, compared to the propionates, a completely different mechanism is involved.[153] Regardless of the deprotonation procedures (LDA in THF or LDA in THF-HMPA), O-benzylglycolates **G** and **H** give almost identical product ratios. Under both conditions, **G** gives **I** and **H** affords **J** stereoselectively. Presumably, the enolates take an E-configuration, like **D**, under both conditions owing to chelation of O-Li...OBn. Camphor derivatives are also used as chiral auxiliaries in the conjugate addition-enolate trapping reactions.[154] Alkylations of the anion (**K**) derived from a camphor-based oxazoline proceed in good yield, and hydrolysis affords the corresponding α-hydroxy acids in high enantiomer excess (77~92%ee).[155]

Dioxolanes (**149**) and oxazolidinones (**150**) derived from pivalaldehyde and lactic acid, mandelic acid and proline, respectively, furnish chiral enolates of type (**151**) by deprotonation of LDA. Reactions of these enolates with alkyl halides are highly diastereoselective.[156] Thus, the overall enantioselective α-alkylation of chiral, non-racemic α-heterosubstituted carboxylic acids is realized. The oxazolidinones (**152**) are deprotonated with LDEA (lithium diethylamide) in THF and alkylated (MeI, benzyl bromide) to 4,4-disubstituted oxazolidinones (**153**) with high diastereoselectivity (9:1 to 50:1); **152(S)** gives **153(S)** and **152(R)** produces **153(R)**.[157] Hydrolysis of these oxazolidinones to amino acids of known configuration and optical purity indicates that little if any recemization occurs in the process. The heterocyclic acids (**154**) are doubly deprotonated by LDA or

LDEA to give the corresponding enolates (155), which are reacted with electrophiles (MeOD, MeI, C₆H₅CH₂Br) to give the crystalline products (156).[158]

Scheme 12.   Asymmetric Alkylation

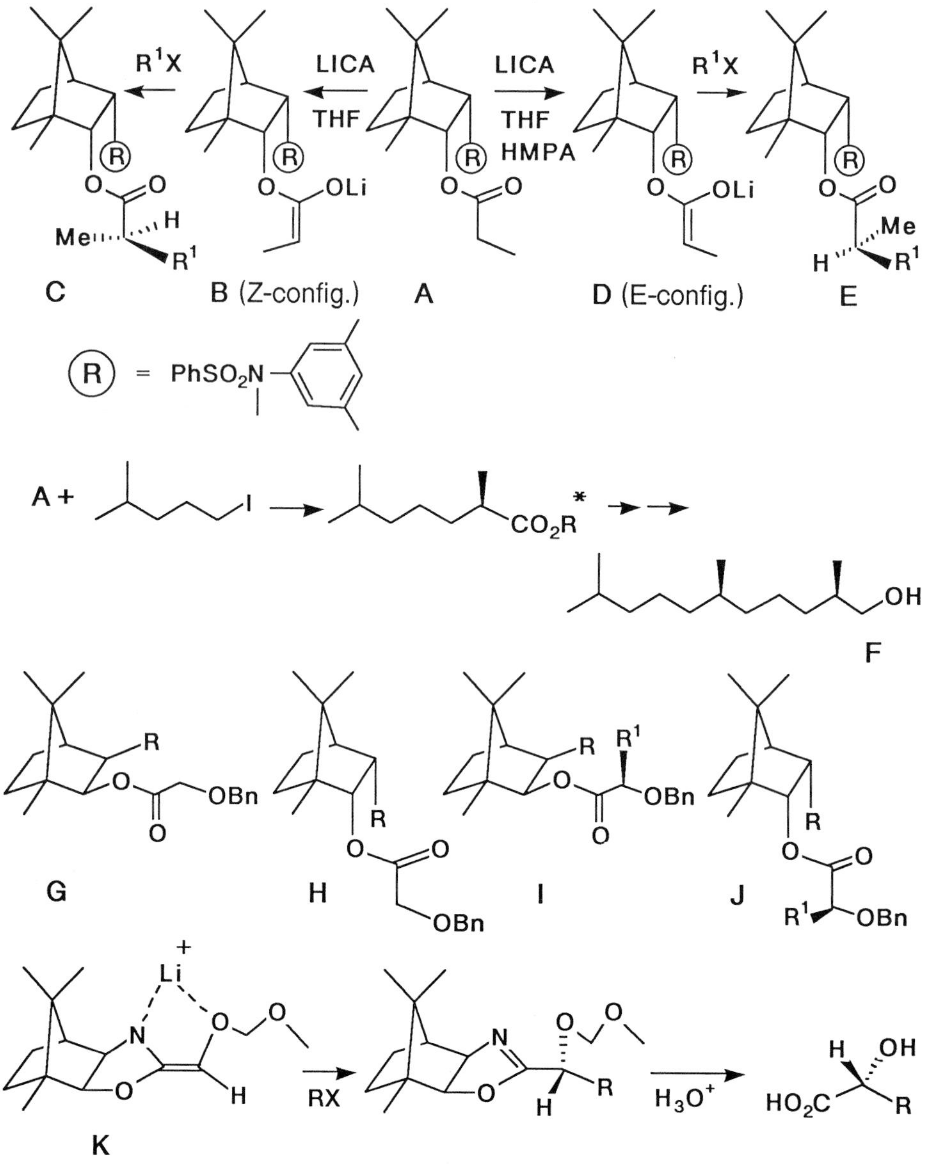

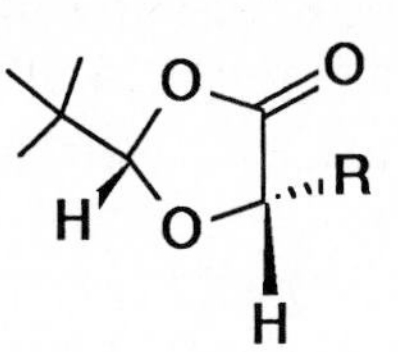

149 (R=CH₃,C₆H₅) **150** **151**

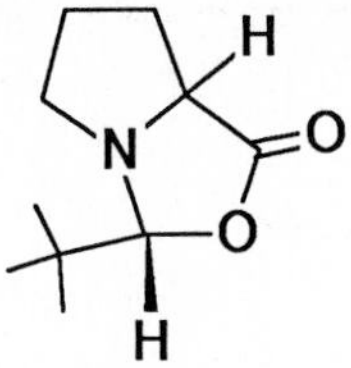

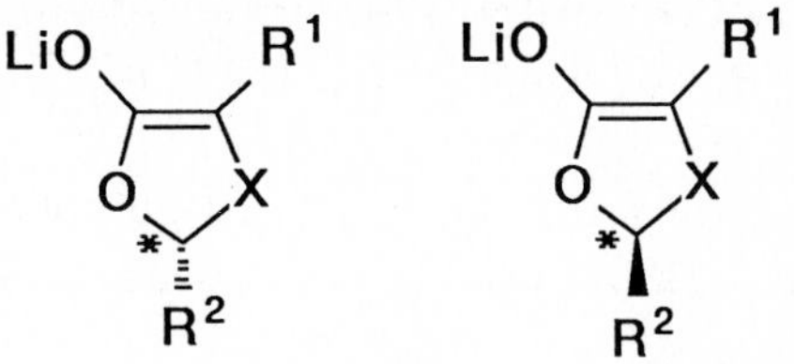

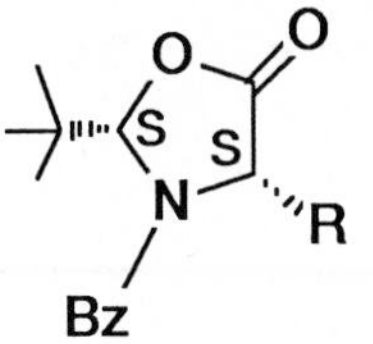

$Bz = COC_6H_5$

152 (S) 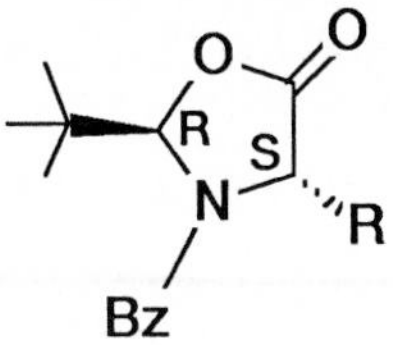 **152 (R)** 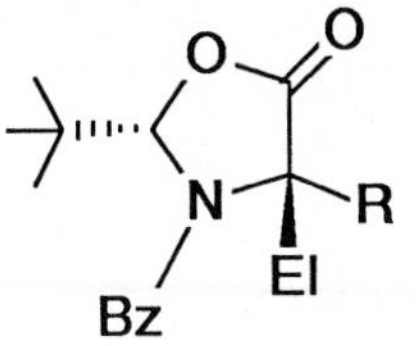 **153 (S)** 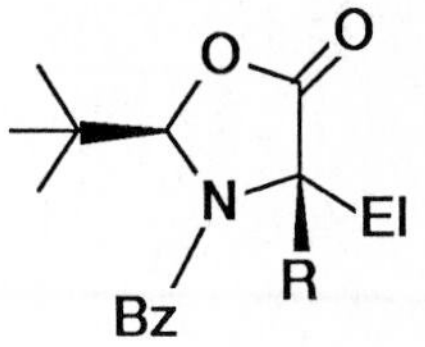 **153 (R)**

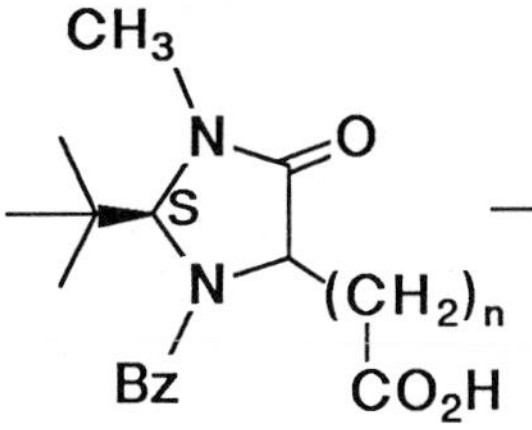

154 (n=1,2) 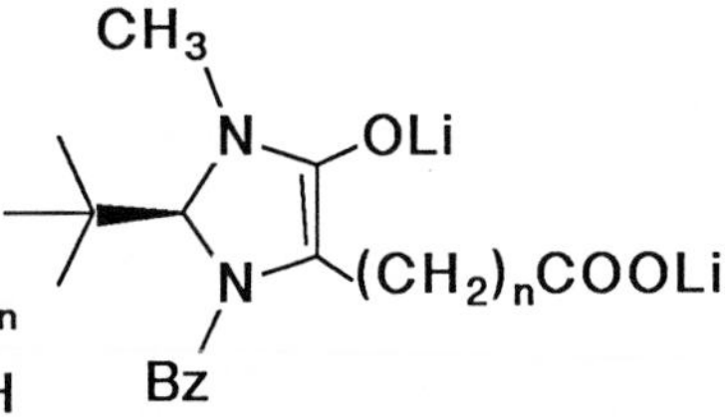 **155** 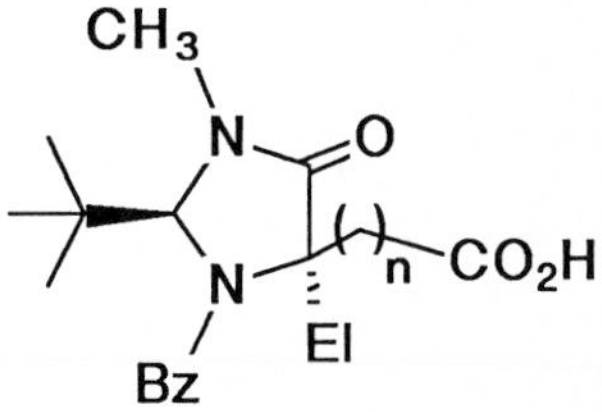 **156**

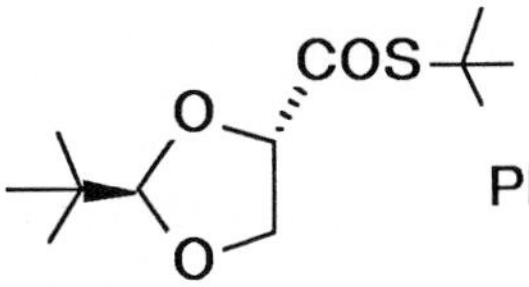

157a 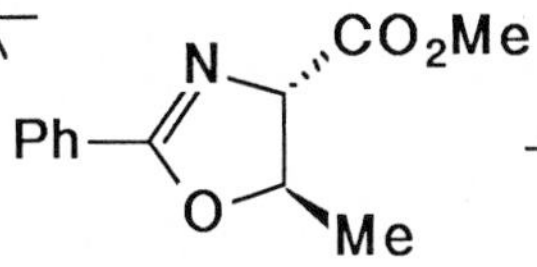 **157b** 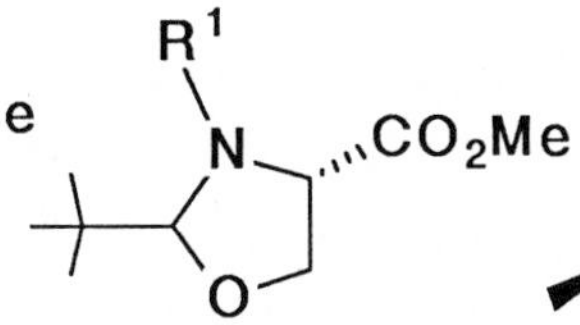 **157c** 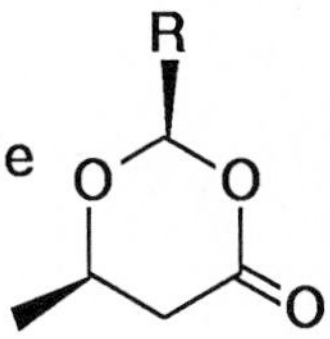 **158**

Hydrolysis of the imidazolidinone ring of these products gave the corresponding aspartic and glutamic acids. The chiral, non-racemic, serine, glyceric acid, threonine, and tartaric acid were converted to methyl dioxolane-(cf **157a**), oxazoline-(**157b**), and oxazolidine-(cf **157c**) carboxylates. Similarly, deprotonation by LDA gave solutions of the lithium enolates with exocyclic enolate double bonds. Alkylations proceeded with diastereoselectivities above 90%.[159] The products of alkylation could be hydrolyzed to give α-branched serins and allothreonines; (R)-3-hydroxy-butyric acid and pivalaldehyde gave the crystalline (R,R)-2-(t-butyl)-6-methyl-1,3-dioxan-4-one (158), the enolate of which was alkylated exclusively in trans manner.[160] The enolate (159), derived from 150 which was produced from proline and pivalaldehyde, was reacted with a number of electrophiles including (benzene)(tricarbonyl) chromium and diphenyl disulfide (phenylating reagents).[161] Cleavage of the products furnished α-alkylated proline derivatives (160), and the overall process is an electrophilic substitution of the α-proton of proline with retention of configuration at the asymmetric carbon atom. Since no external chiral auxiliary is necessary to achieve this transformation without loss of enantiomeric purity, it is called a self-reproduction of chirality.[161]

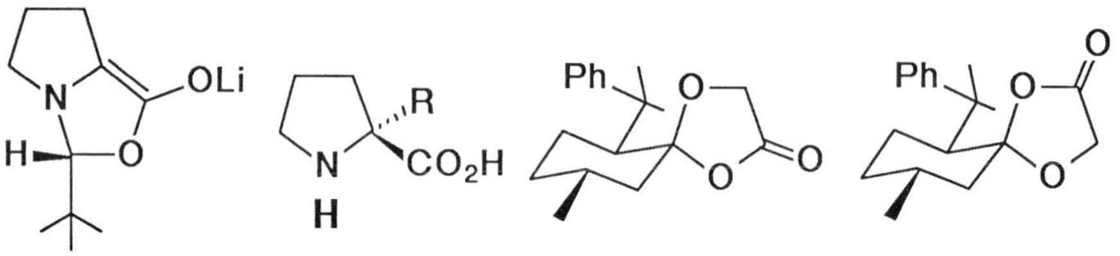

159   160   161a   161b

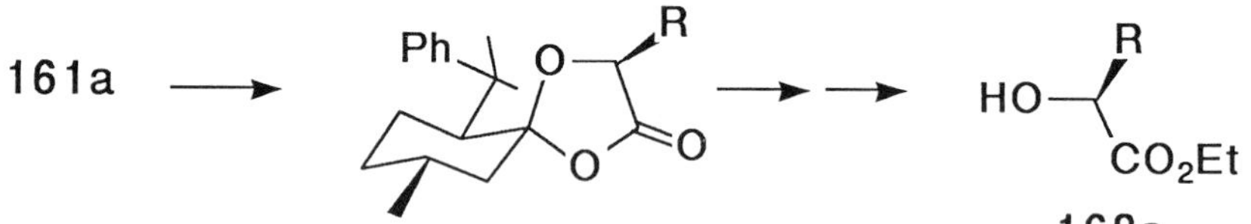

161a ⟶ ... 162a

161b ⟶ ... 162b

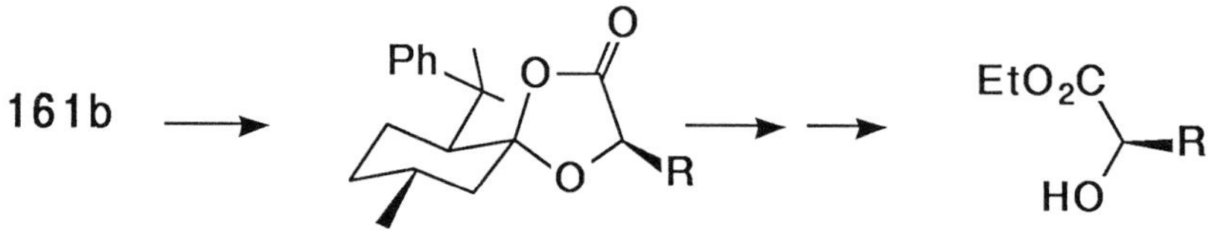

The enolates of spiro-fused dioxolanones (161a, 161b) served as chiral glycolate enolate equivalents (HOCH$_2$CO$_2$H→HOCH=C(OLi)OH), providing either enantiomer of the α-substituted α-hydroxy esters (162) upon alkylation and hydrolysis.[162] The alkylation of acyclic amino acids (alanine and phenylalanine) proceeded stereospecifically with retention of configuration.[163] The method involves (i) conversion of the amino acid to the predominantly cis 2-aryl-3-

carbobenzyloxy oxazolidinones (like **152**), (ii) alkylation of the potassium enolate with CH_3I or $PhCH_2Br$, (iii) base hydrolysis and hydrogenolysis to afford the alkylated amino acid.[163] Optically active alanine was obtained by an asymmetric, one carbon transfer reaction.[164] The imine was deprotonated with LDA, and the resulting enolate was methylated. Hydrolysis gave alanine in good yields. Variation of the nature of the imine moiety of Schiff bases derived from glycine caused the enantiomeric excess to shift from 0 to 70%.

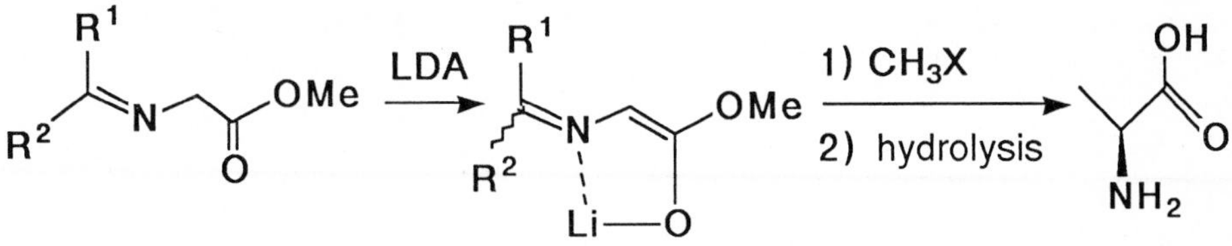

Asymmetric synthesis of amino acids with high optical purities was achieved through stereoselective alkylations of chiral β-lactam ester enolates followed by the reductive cleavage of the β-lactam rings.[165] The β-lactam enolate (**163**) was generated by treating the β-lactam (**164**) with LDA in THF at 0 5°C, and then the solution was cooled to -78 -90°C. The addition of RBr gave **165** with excellent diastereoselectivity. The reductive cleavage of the lactam ester (**165**) or the corresponding carboxylic acid, through either hydrogenolysis on Pd-C or reduction with dissolving metal, gave the corresponding dipeptide derivatives. The enolate presumably takes the conformation depicted in **163**, and the electrophile attacks from the less shielded side of the enolate plane.

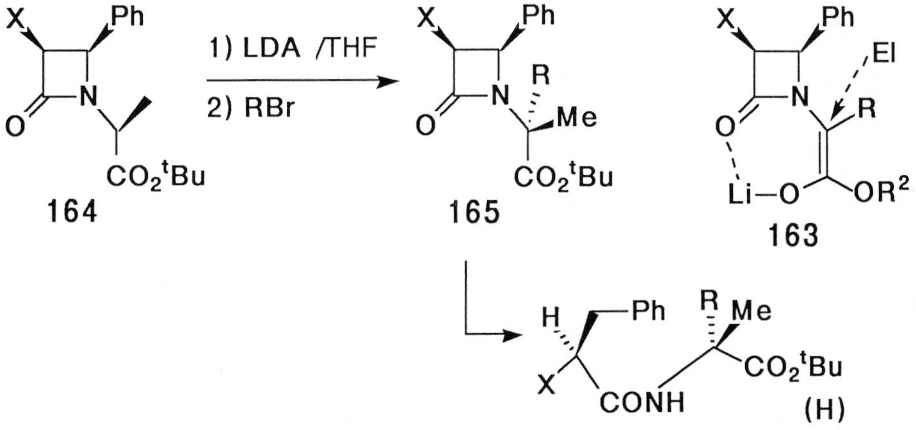

Treatment of (-)-dimenthyl succinate with LiTMP in THF, followed by the addition of 1,ω-dihalides, gave the carbocyclic derivative having diester groups with very high ee.[166] Alkylation of the enolate anion of 2-methoxycarbonyl-1-indanone with an optically active dialkylphenyl selenonium salt resulted in the formation of optically active C-alkylated products.[167]

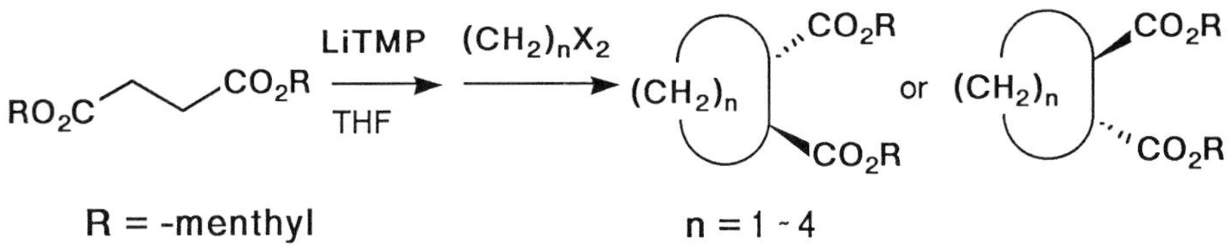

R = -menthyl   n = 1 - 4

5.2 AMIDE AND IMIDE ENOLATES

Deprotonation of amides or imides produces (Z)-enolates owing to the allylic strain conformational control elements (Scheme 13). This same conformational control element in the enolization process assumes also an important role in preventing product racemization (via enolization) in the bond-construction process.[91]

Scheme 13. Amide deprotonation

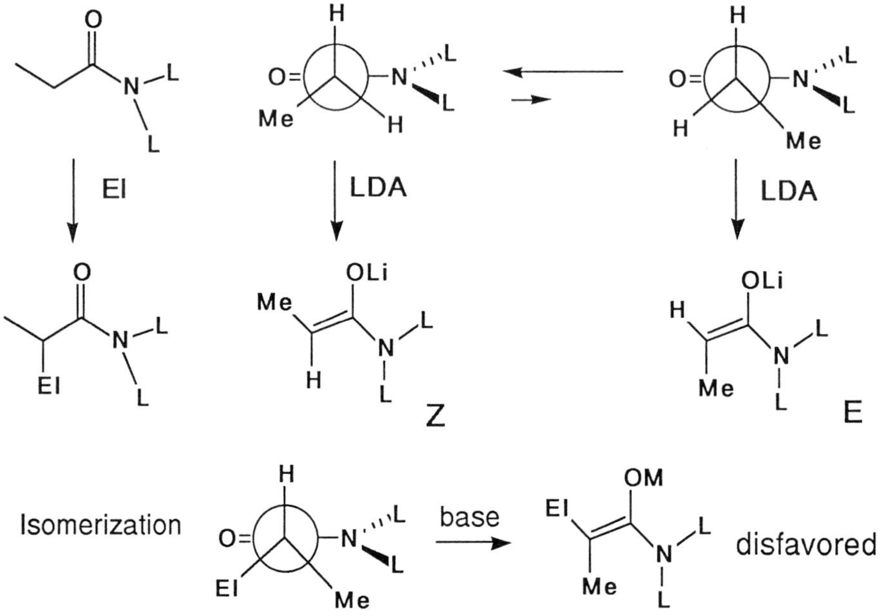

Scheme 14. Conformational lock of amide enolate

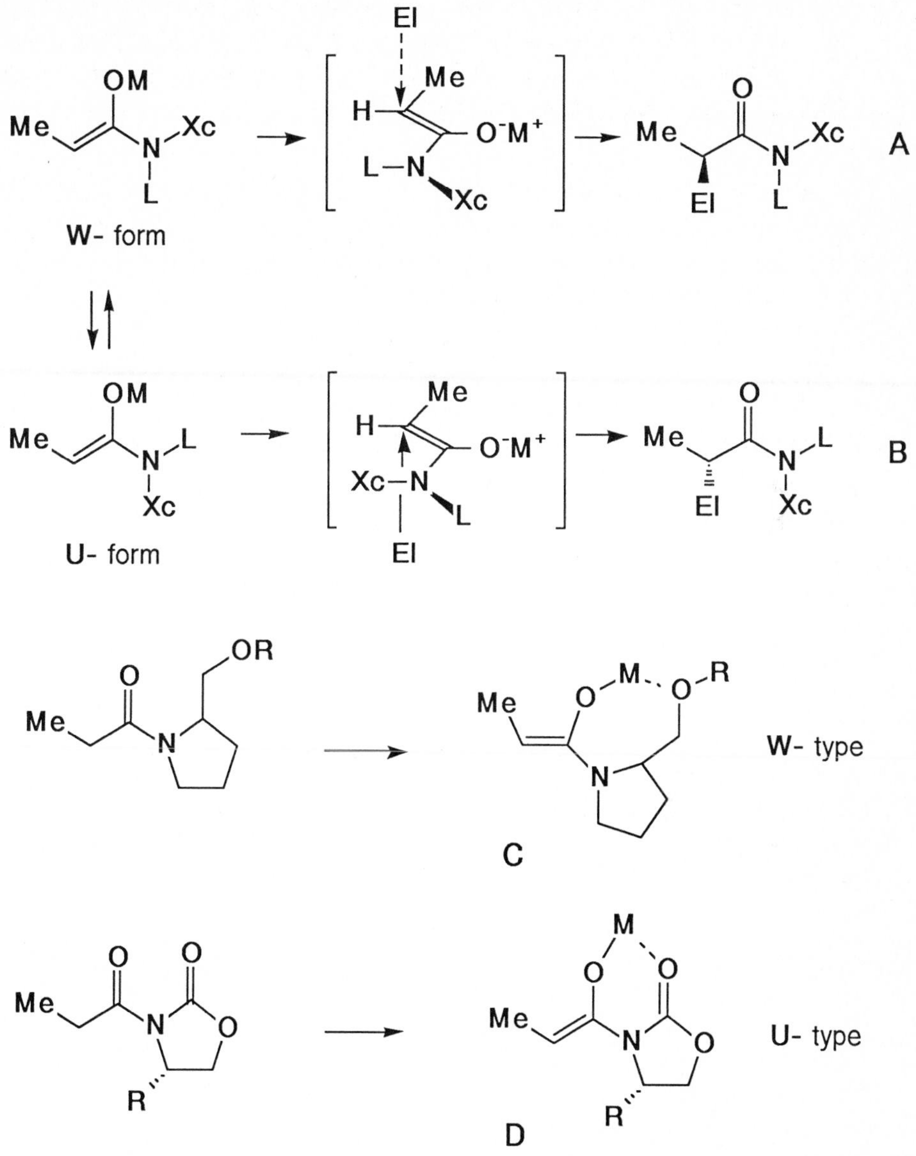

Since the geometry of amide enolates is fixed as Z owing to the conformational bias, the next problem is how to lock the chiral auxiliary **Xc** (Scheme 14).[91] If the enolate takes a W-form, El comes from the less hindered β-face to give **A**, while the enolate having U-form produces **B** by the attack of electrophiles from the α-face. Therefore, to produce high asymmetric induction, the following three criteria are important; (i) to immobilize a W or Z conformation, (ii) to immobilize **Xc**, and (iii) to construct maximal facial bias. Chelated chiral enolates (**C** and **D**) are typical examples of W- and U-type geometry. Acylation reactions with these amide enolates are shown in **118-121**.

Asymmetric alkylation reactions are summarized in Scheme 15.[168] The chiral O-silylated enolates (M=Si in **D**, R=iPr) could be alkylated or sulphenylated with high diastereoselectivity (~20:1) by 1,3-dithienium fluoroborate or PhSCl.[169] However, phenylthioalkylation resulted in low diastereoselectivity. α-Ketoamides were deprotonated twice with LDA/HMPA to give the dianions, which reacted with alkyl halides to yield α-amino tertiary alcohols.[170] This reaction was also employed in asymmetric induction by using the amide enolate having a chiral auxiliary like **C**.

Scheme 15. Asymmetric alkylation

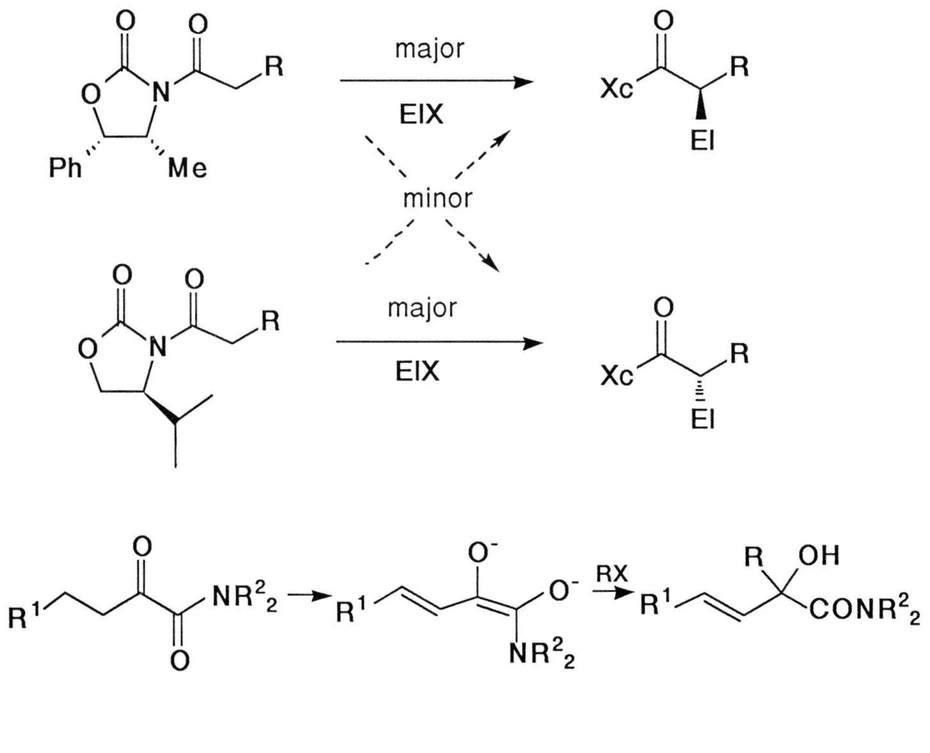

Asymmetric acylation of carboxamides (**166**) having trans-2.5-bis (methoxymethoxymethyl) pyrrolidine moiety as a chiral auxiliary proceeded with very high diastereoselectivity (>98%).[171] Asymmetric alkylation via the chiral enolate derived from **167** gave the quarternary asymmetric center with high d.e.[172] Stereocontrolled (trans) introduction of ethyl, isopropyl and acetyl groups at the 3-position of 2-azetidinone (**168**) was achieved through the enolate anion generated by treatment with LDA-HMPA.[173]

166   **167**   **168**

5.3 OTHERS

As mentioned above (section 4.2), some transition metal enolates having a chiral auxiliary are effective for asymmetric alkylations and acylations. Chiral transition metal ligands such as CpFe(CO)(Ph₃P)[125,129] and CpCo(PPh₃)[132] are utilized for such purpose. Palladium catalyzed allylations of metal enolates are also described in section 4.1.

New chiral phosphine ligands containing a chiral functional group remote from the phosphino-groups were found to be effective in the palladium catalyzed reaction of allylic acetate with the enolate anion of 2-acetylcyclohexanone and gave the allylated product in significant enantiomeric enrichment.[174] Nucleophiles attack the face of the π-allyl group opposite to the palladium; thus, a low diastereoselectivity may result from type **A** ligands since the distance between the inducing moiety on the ligand and the developing asymmetric center is large (**A**). On the other hand, the type **B** reaction may cause high asymmetric induction due to the intimate interaction between R* and M⁺. In fact, a ligand like L* caused higher asymmetric induction.[174]

The chiral imine, prepared from (-)-or(+)-erythro-2-methoxy-1,2-diphenylethylamine and cyclohexanone was metallated and alkylated to give the corresponding 2-alkylcyclohexane in 79-92% optical purity.[175] This is not the chemistry of metal enolates but rather that of metal azaenolates.

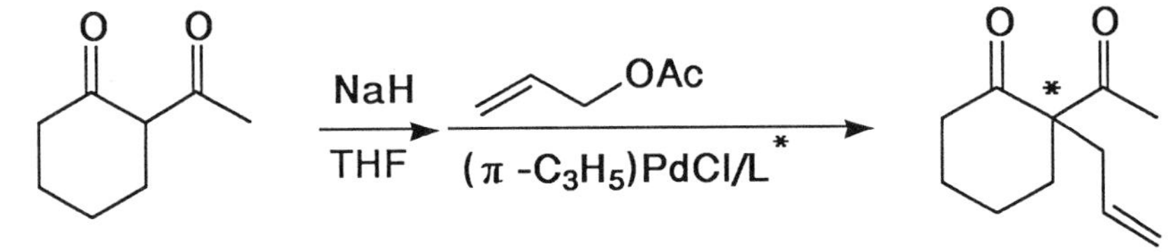

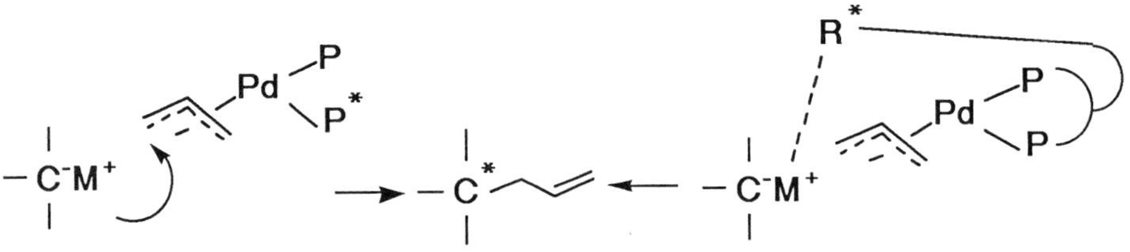

NaH

OAc

THF $(\pi\text{-}C_3H_5)PdCl/L^*$

Pd

P

P*

$-\overset{|}{\underset{|}{C}}^-M^+$

$-\overset{|}{\underset{|}{C}}^* \diagdown \diagup$

$-\overset{|}{\underset{|}{C}}^-M^+$

R*

Pd

P

P

A

B

L* : Ph$_2$P

Ph$_2$P

$\overset{O}{\overset{\|}{N\text{-}C}}(CH_2)_n \overset{O}{\overset{\|}{C}}NH \overset{iPr \quad H}{\diagup} CO_2Me$

Ph   Ph

N   OMe

1) LDA

2) ZnX$_2$

Ph   Ph

N   OMe

ZnX

1) RX

2) H$_3$O$^+$

O

R

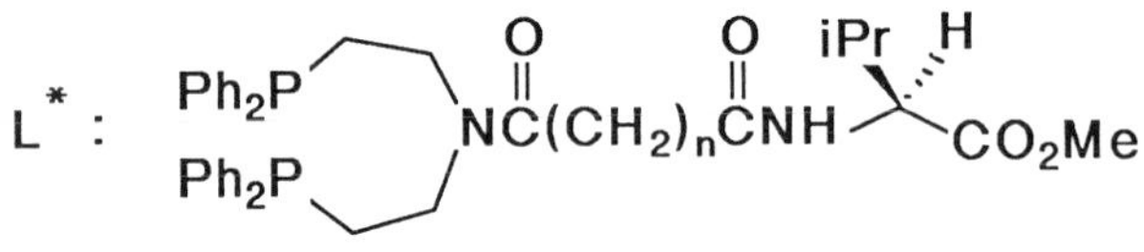

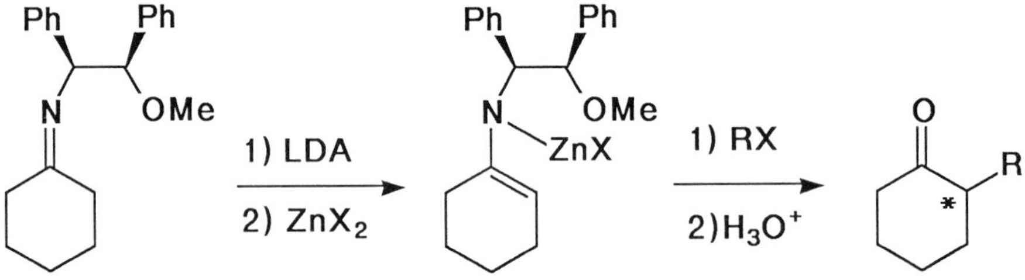

Chiral imidate ester enolates were generated by deprotonation of **169** using n-BuLi, t-BuLi or LDEA, and were then alkylated with various alkyl halides.[176] The electrophiles attacked from the less hindered side of the π-face as shown in **170**, producing high stereoselectivity.

Chiral lithium amides are useful for generating lithium enolates having chiral properties and for asymmetric alkylation of these enolates.[177] The prochiral lithium enolate (**171**) was prepared from the corresponding ketone by deprotonation with lithium (S,S)-α,α'-dimethyldibenzylamide.[178] Reaction of **171** with carbon dioxide, followed by methylation with MeI, yielded the corresponding ester with an e.e. of 67%.

Enantioselective alkylation of carboxylic acids (172) at the α-position was performed using the chiral lithium isopropylamides, which function as both a strong base and a chiral auxiliary.[179]

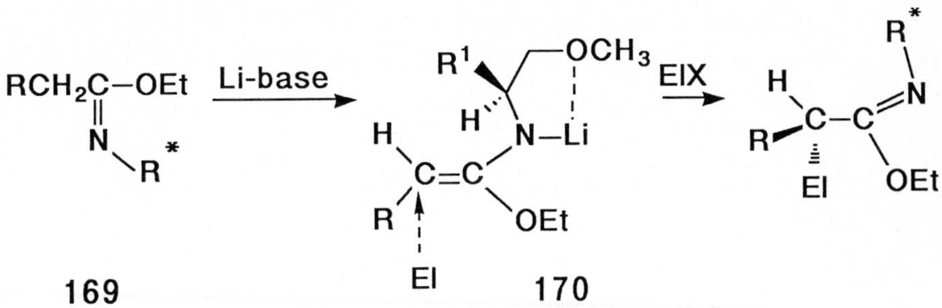

169   El   170

R = CH₃, C₂H₅

$R = CH_3, C_2H_5$

$R* = CH(CH_2Ph)CH_2OCH_3, CH(Ph)CH_2OCH_3$

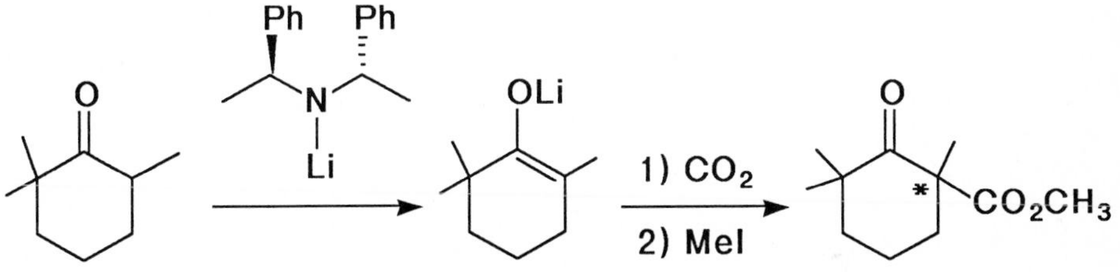

171

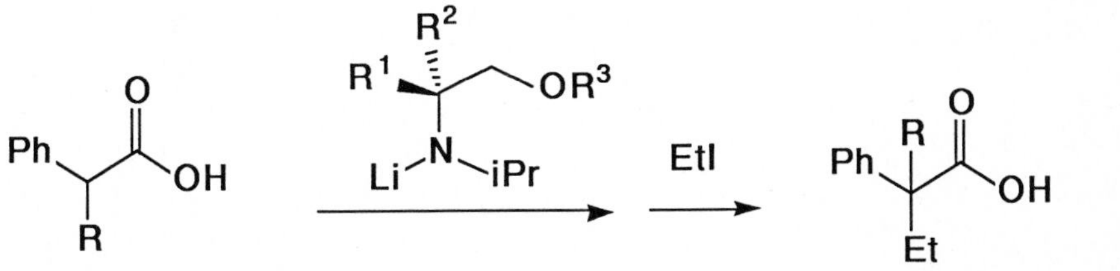

172   0~24% e.e.

The chiral lithium amides (173) were prepared from the corresponding amine and n-BuLi in THF. HMPA was added to 173 at -78°C and then TMSCl was added. t-Butylcyclohexanone was then added (internal quench method). **173a** gave (**S**)-**174** with 26% e.e. **173b** produced (**R**)-**174** with 84% e.e. **173c** afforded (**R**)-**174** with 97% e.e. Therefore, by using this procedure, highly enantioselective deprotonation of prochiral 4-alkylcyclohexanones is achieved with chiral lithium amides.[180] The interaction between lithium enolates and secondary amines formed after deprotonation are discussed both in solution and in crystal.[181]

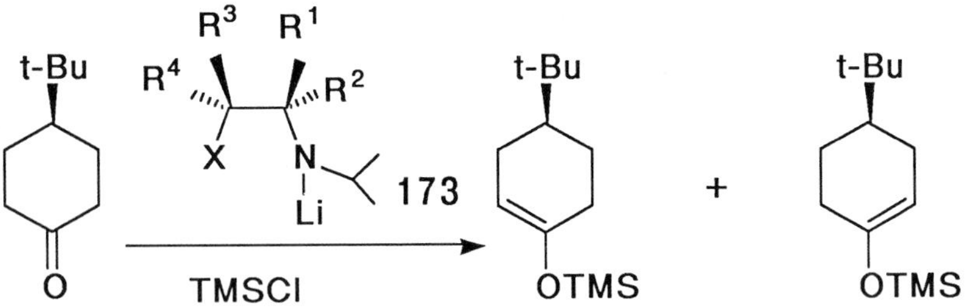

	R¹	R²	R³	R⁴	X
173a	H	iPr	H	H	OMe
173b	Ph	H	H	H	 N⟩
173c	Ph	H	H	H	 N⌒NMe

6. INTRODUCTION OF HETEROATOMS AND PROTON AT THE α-POSITION

6. 1 PROTONATION

Like the alkylation of metal enolates, protonation preceeds in such a way that steric hindrance to approach of the proton donor is a major factor in controlling from which face a proton is delivered to the α-carbon.[182] The effect is enhanced with large proton donors. The enolate of 4-phenylcyclohexanone undergoes preferential kinetic protonation from the equatorial side to afford the less stable cis product. The selectivity range from 61% to 79%, depending on the steric demands of the proton donor (cf. **59-61**).[182]

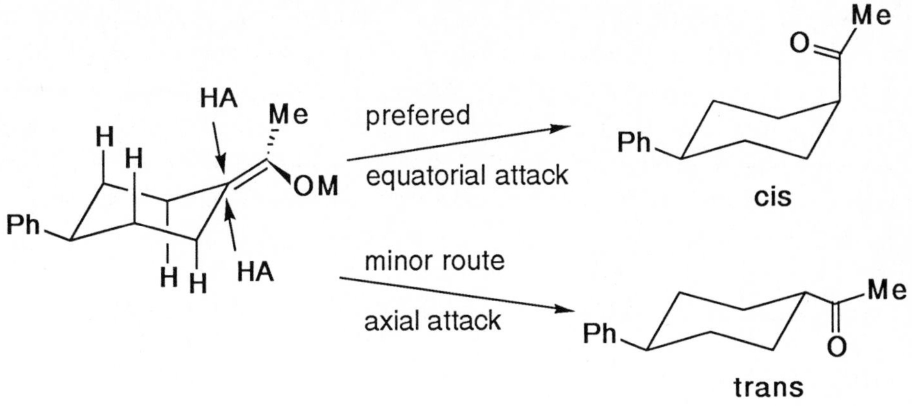

Unlike the alkylation, there is a possibility that O-protonation takes place first and then prototropy occurs from the oxygen of the resulting enol to the α-carbon. The lithium enolates (175) generated from the corresponding γ-lactones underwent protonation to give 176(cis) predominantly, though the ratio of 176(cis)/176(trans) varied with substrates and proton sources.[183]

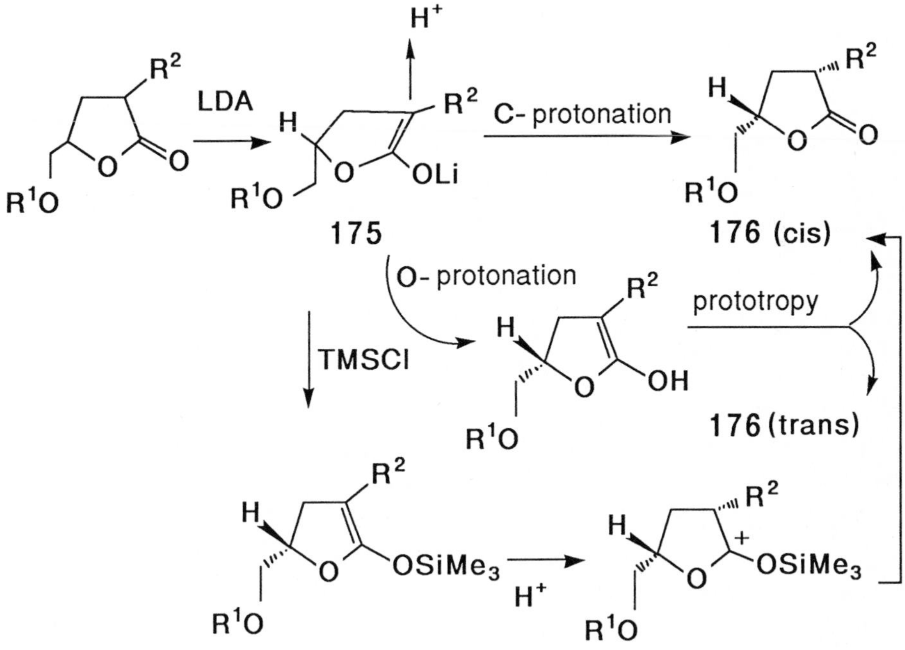

The experimental results suggested that (i) the protonation occurred competitively at α-carbon and at the enolate oxygen and (ii) a proton source with appropriate chirality would enhance asymmetric induction in chiral γ-lactone substrates. When 175 was once trapped with TMSCl and then the resulting silyl ketene acetal was treated with H+, much higher stereoselectivity (cis/trans) was obtained. The reason for this enhancement is presumably that protonation takes place only at the α-carbon. By using this procedure (stereocontrolled protonation), a new synthesis of a steroid side chain was achieved; (-)-desmosterol was prepared in high yield.[184]

The enantioselective protonation of enolates of α-amino-acid derivatives was realized with the chiral proton source, (2R,3R)O,O-dipivaloyltartaric acid (DPTA).[185] The deracemization of racemic 177 was carried out by using LHMDS(path a) or LDA(path c).

After metallation with LHMDS, diisopropylamine was added before the asymmetric protonation (path a→path c). The optical enrichment of the material recovered was identical to that of the experiment with LDA (path c), indicating that ligand exchange takes place before the asymmetric protonation, if amines were added to the enolates. In some cases, a higher ee was observed with this ligand exchange procedure.

Protonation of the bicyclobutane bridged enolate anion **178**, formed from the corresponding racemic ketone and lithium (S,S)-α,α'-dimethyldibenzylamide, produced the optically active ketone with 48% ee.[186] Treatment of the ketone with LDA, followed by protonation with chiral proton donors such as R(-)-mandelic acid, (-)-ephedrine, and (-)-menthol did not produce the optically active ketone. It is likely that the intermolecular transfer of chirality is due to a multi-coordination of the lithium atom involving diastereomeric complexes.

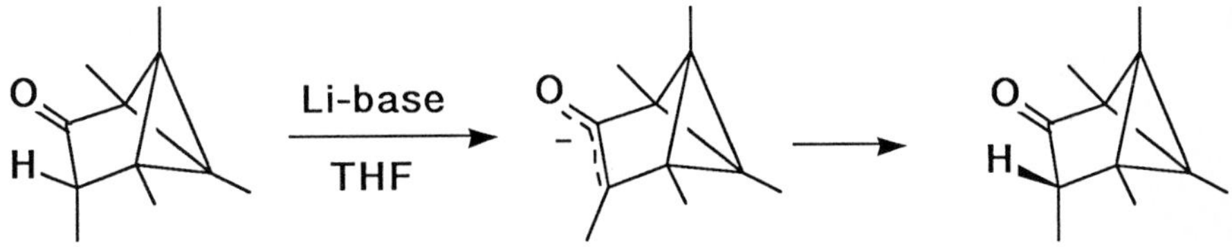

racemic 178 optically active

6.2 AMINATION

Asymmetric electrophilic amination of the E silyl ketene acetal (**179**) was achieved by using di-tert-butyl azodicarboxylate (DTBAD or DBAD) in the presence of $TiCl_4$.[187] This process fulfills the following requirements: (a) enantiomeric excess in the range 78-91%; (b) good chemical yields; (c) both enantiomers of the chiral auxiliary are commercially available, and can be recycled; (d) the absolute configuration of the products is predictable. By using this procedure, rare, and unnatural α-amino acids can be easily prepared. Quite a similar approach was made with **179-a** and DTBAD in TiLn.[187-a]

The lithium enolates derived from the N-acyloxazolidones readily reacted with DTBAD to provide the hydrazide adducts in excellent yield and with very high diastereoselectivity(97:3~ >99:1). The oxazolidone auxiliary of the adducts was then removed to give the amino acid derivatives with high ee (normally >99% ee).[188]

Quite similarly, **180** was treated with LDA and then with various dialkyl azodicarboxylates.[189] The resulting aminated carboximides were converted to the amino acids. The substitution of both the dialkyl azodiformate and the acyl side chain influenced the selectivity. As the size of the R' group on the aminating reagent increased, the diastereomer ratio improved (Me<Et<CH$_2$Ph<t-Bu). Similarly, greater bulk of the acyl side chain also increased the diastereoselectivity (Me<CH$_2$Ph<iPr).

180

6. 3 OXYGENATION

Direct enolate oxygenation is achieved with molecular oxygen (O_2),[190] or molybdenum peroxide-pyridine-HMPA (MoOPH).[191] With O_2 oxidation, α-carbon cleavage may occur as well as α-dicarbonyl formation. While enolate oxidation using MoOPH is more general, oxidation of 1.3-dicarbonyl enolates fails and overoxidation to α-dicarbonyl compounds does occur. Further, the stereoselectivity via these reagents is sometimes, but not always, poor.

Direct oxidation of ketone and ester enolates, using 2-sulfonyloxaziridine (**181**), afforded α-hydroxy carbonyl compounds (acyloins) in high yield with excellent stereoselectivity.[192] The direct oxidation of chiral amide enolates to optically active mandelic acid was achieved by using this method.[194]

Asymmetric oxygenation of chiral imide enolates were realized by treatment with **181**.[193] The carboximides were transformed into the Z-sodium enolates upon treatment with NaN(Me3Si)2 in THF and treated with **181** at -78°C, resulting in formation of the α-hydroxycarboximides with high diastereoselectivity. The sense of asymmetric induction parallels the observation made in the alkylation (Scheme 15). Lithium, sodium and potassium enolates all reacted rapidly with **181**, but the hydroxylation of lithium enolates accompanied aldol addition with the product sulfonyl imine(**183**). It was concluded that the tetrahedral intermediate, hemiaminal **182**, may be involved in the reaction and that the subsequent equilibration (**182** **183**) would be counterion dependent (Keq>1 for M=Li,, Keq<1 for M=Na).[193,194]

The sulfonamide shielded O-silylated ketene acetal (179-a) underwent π-face selective α-acetoxylation on successive treatment with Pb(OAc)₄ and NEt₃•HF to give after recrystallization the corresponding α-acetoxy ester in 55-67% yields and in 95-100% de.[195]

Asymmetric oxidation of the sodium enolates of ketones using oxaziridines (+)(2R,8aS)-184 and (-)-(2S,8aR)-185 afforded α-hydroxy ketones in high optical purity (69-95% ee).[196a] Asymmetric oxidation of the chiral acyclic tetrasubstituted enolate (186) with 184/185, in the presence and absence of HMPA, afforded optically active tertiary α-hydroxy amide in high optical purity (88-91% de).[196b]

(+)(2R,8ₐS) 184   (-)(2S,8ₐR) 185   186

The α-hydroxylation of chiral esters of 3-phenylpropionic acid by MoO₅•Py•HMPA was optimized to 98% de and 73% yield by systematic variation of the reaction conditions.[197] The addition of at least 3 equiv. of K(sec-BuO) proved to be essential. By using this procedure, highly efficient synthesis of natural (-)-verrucarinolactone was achieved.

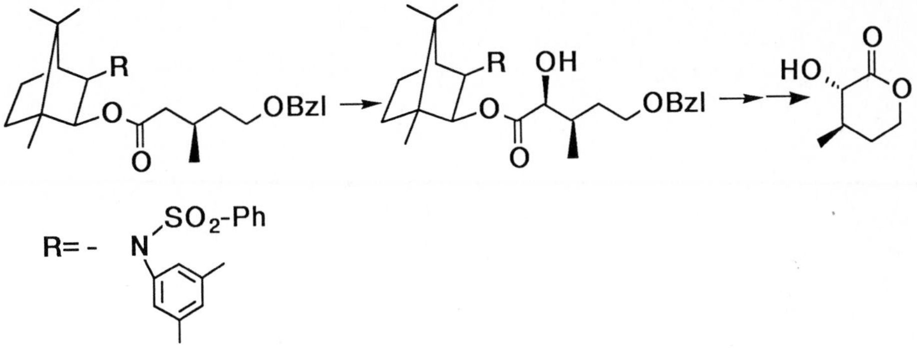

R= - (structure shown with SO₂-Ph group)

The reaction of the enolate, derived from 3-methyl-4,4,4-trifluorobutyrate and LDA, with MoO₅-Py-HMPA complex provided ethyl (2S*,3S*)-2-hydroxy-3-methyl-4,4,4-trifluorobutyrate with good to excellent diastereoselectivity.[198] In the course of asymmetric synthesis of an enantiomerically pure prostaglandin building block, the α-hydroxylation of the lithium enolate was carried out by using MoOPH[199] or O₂/P(OMe)₃.[199,200]

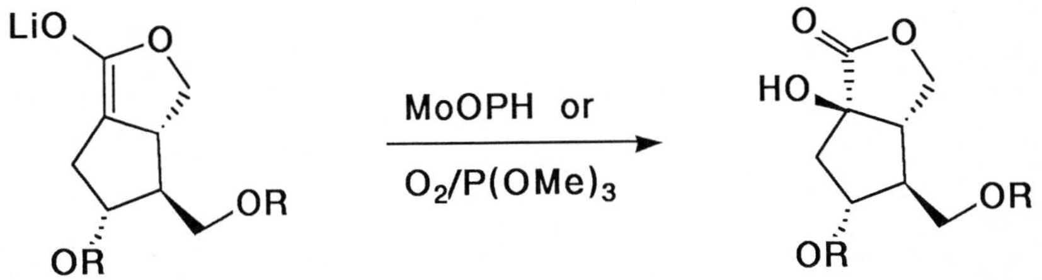

α-Hydroxylation of enolates are also achieved by MCPBA/n-hexane[201a] and Mn(OAc)₃ / C₆H₆.[201b] Heteroatom introduction such as halogen, sulfur, and selenium atoms are omitted in this review article.

7. APPLICATION TO NATURAL PRODUCTS SYNTHESIS.

In this section, we demonstrate how the alkylation and related reactions mentioned above were applied to the synthesis of natural products. We will present the key step (alkylation of enolates) in detail while other processes will be omitted. A number of enolate alkylation reactions are used in natural product synthesis, and these reactions are senerally used only as a step among numerous synthetic steps. So, it is difficult to summarize all examples of alkylation reactions in natural product synthesis. Typical examples are shown below. Some of the examples have been already mentioned in the above sections.

Cyclization of the bromoketone with LDA gave a 1:1 mixture of the C- and O-alkylation products.[202] A molecular model of the enolate indicates that there is some strain in achieving the proper trajectory required for 6-(enolendo)-exo-tet cyclization, and therefore, 6-exo-tet cyclization (O-alkylation) becomes a viable alternative. This O-alkylation product was converted to the optically active trans-hydrindenone derivative which has potential as an intermediate in steroid synthesis.

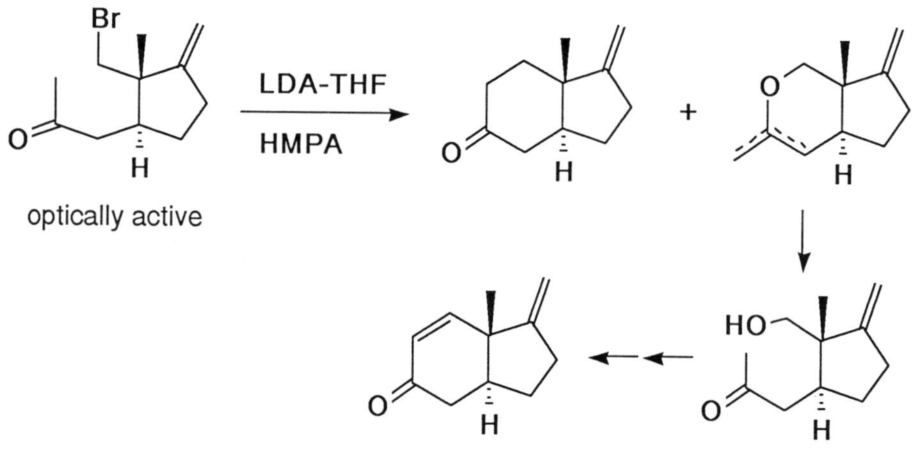

A short stereoselective synthesis of triquinane sesquiterpene hydrocarbon (±)-pentalenene was realized starting from commercially available 1,5-dimethylcycloocta-1,5-diene.[203] Kinetically controlled allylation of the lithium enolate of the 8-membered ring exclusively furnished the trans-allylation product (see Scheme 3). This was converted to the bicyclic enone, and then to the desired target molecule.

Alkylation of the enolate of **187** with the allylic chloride in the presence of KI yielded the allylation products in 97% yield as a pair of diastereomers.[204] The adduct was converted to the C(1)-C(17) half of boromycin.

187

Diastereoselective methylation of the dilithio derivative of ethyl (R)-(-)-3-hydroxybutanoate was followed by consecutive Wittig reactions to achieve chain elongation by four carbon atoms[205](cf **48**). The diolide (**188**),the macrocyclic component of Elaiophylin, was prepared via 11 steps with an overall yield of 9.5%. (S)-(+)Lactic acid and pivalaldehyde were condensed to yield **189**. This was deprotonated with LDA to give the (S)-enolate which was alkylated with the dimethyl acetal of 5-iodo-2-pentanone. Further transfomation gave (R)-(+)-

frontalin in an overall yield of 73% from lactic acid.[206] This is a nice application of the principle of self-reproduction of chirality in natural product synthesis (cf.149, 159, 160).

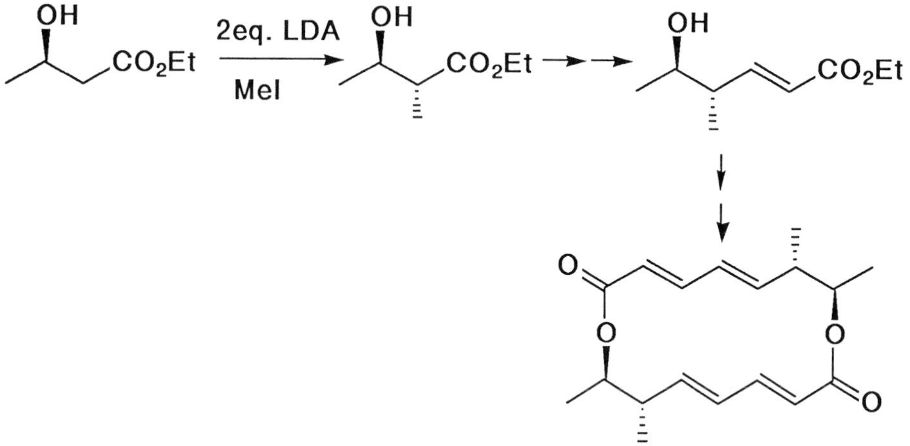

188

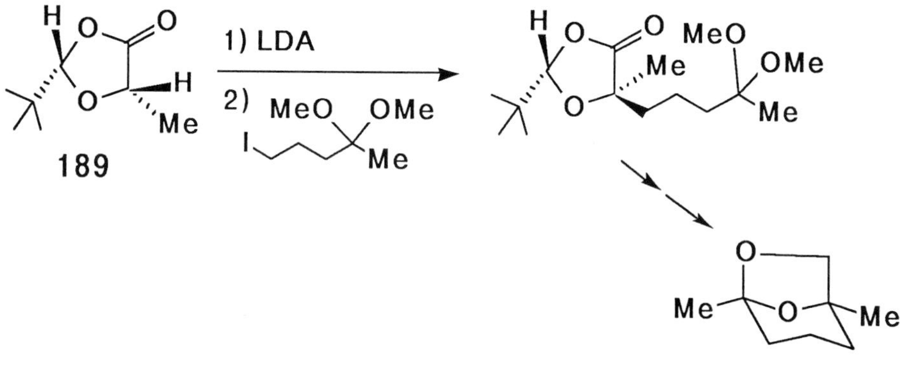

189

(R)-(+)- frontalin

Asymmetric alkylation of the chiral propionimide enolate (190) was applied to an efficient synthesis of (+)-Prelog-Djerassi lactone[207](cf. Scheme 15). Independent enantioselective synthesis of (R)- and (S)-thiophane, the enkephaline inhibitor, was achieved in which the pivotal step was the enantioselective alkylation of the chiral oxazolidyl carboximide-derived enolates with bromomethyl benzyl sulfide. The diastereoselection in these processes was in excess of 95%.[208a] Treatment of the sodium propionimide enolate (Li was replaced by Na in 190) with the allylic iodide gave the allylation product with 98:2 diastereoselection.[208b] This alkylation was used for the asymmetric synthesis of the C-1 - C-10 synthon of the ionophore antibiotic ionomycin. The reaction of **190** with 2-methylallyl iodide was also utilized in the asymmetric synthesis of the C-1 - C-9 ferensimycin synthon.[208c]

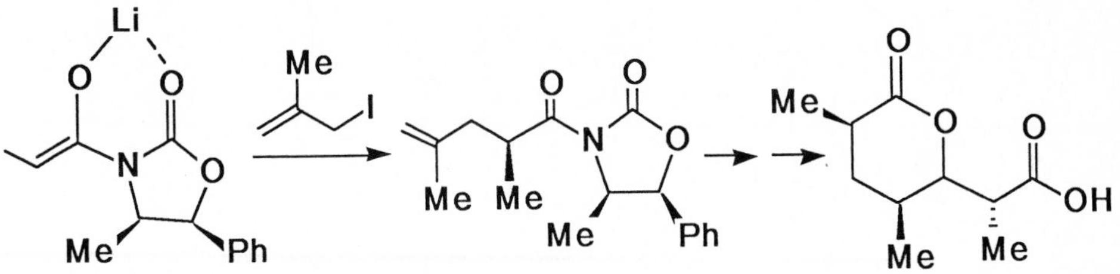

190

Eight different C-glycosidic derivatives of 3-deoxy-D-manno-2-octulosonic acid were prepared by reacting the enolates, derived from **191**, with electrophiles such as iodomethane, 3-bromopropyne, benzyl bromide, formaldehyde, carbon dioxide, acetyl chloride, acetic anhydride, phenyl acetate, t-butyl 2-bromoacetate, and methyl acrylate.[209] All C-glycosides were formed with the β-configuration predominating; the β to α ratio varied from 70:30 (formaldehyde, phenyl acetate) to >95:5 (alkyl halides).

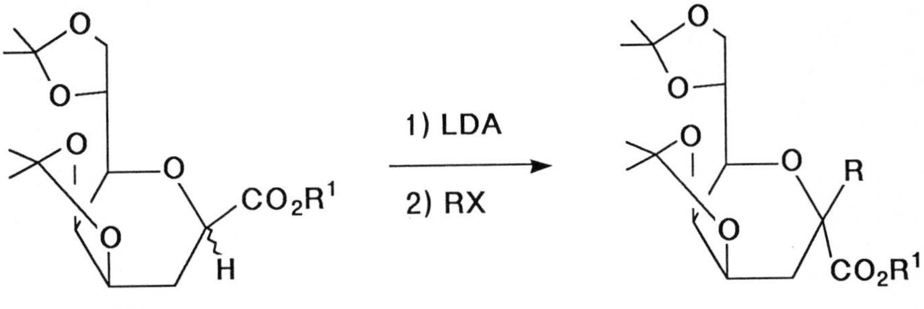

191

Alkylation of 3-ethoxy-2-cyclohexenone with LDA and 1-chloro-3-iodopropane gave **192**, which was converted to **193**. Treatment of **193** with LDA-THF produced the bicyclic enone (intramolecular enolate alkylation), which was converted to dl-erythro-juvabione.[210a] This efficient construction of bicyclo [3.3.1]non-3-en-2-ones was also applied to the synthesis of dl-clovene.[210b]

juvabione

The lithium enolate derived from 194 was trapped with methyl bromoacetate to produce 195 in 87% yield (trans:cis = 95:5).[211]   The trans ring-fusion stereochemistry of the methylhydrinden derivative is established at this stage. Further transformation of 195, via four steps, gave trans-7a-methylhydrind-4-en-1,6-dione. α,β-Unsaturated esters having a diquinane framework were produced through regiocontrolled alkylation of cyclopentanone enolates with methyl 4-bromo-3-methoxycrotonate, base-promoted cyclization, and ketalization.[212] The stereoselective alkylation by cinnamyl bromide of the enolate derived by conjugate addition of the anion of allyl phenyl sulphide to 2-methylcyclopent-2-enone was one of the key steps in a simple synthesis of an aromatic steroid.[213] Here again, the alkylation takes place trans to the substituent at the β-position.

The potassium enolate derived from 196 underwent the alkylation with 3-chloro-1-iodobut-2-ene to give the keto-olefin as a 4:1 mixture of β- and α-epimers.[214]

The capnellenediol was obtained from this keto-olefins. The dianion (197) was generated from the corresponding α-methyltetronic acid by treatment, first with sodium hydride in THF-HMPA, and then with n-BuLi. The alkylation gave 198 in good yield, in which the diastereoselectivity was not determined.[215] Futher transformation of 198 gave (9Z)-trisporol(A) and trisporol(B).

Alkylation and aldol reactions of the lithium enolate (199) of the chiral 4-allylazetidinone gave 3,4-trans-azetidinones as major products, in which 200 was converted in several steps to the optically active 6-epicarpetimycins.[216]

199   R=Me,Et,C(OH)Me₂   200
                (200)

As previously mentioned, the enolates (78) yielded 79 predominantly, under kinetic control, on reaction with methyl iodide. The reaction of the enolate (+)-78

(R=Me) with acetone, under thermodynamic control, gave almost exclusively compound **201**.[217] This was transformed into (+)-viridifloric acid in four steps.

201

A variety of 16α-substituted 17β-estradiol derivatives were prepared by a convenient two step procedure. The lithium enolate of estrone 3-O-benzyl or 3-O-tert-butyldimethylsilyl ether underwent clean, stereospecific alkylation with a variety of allylic, benzylic, or propargylic bromides (some bearing additional functionality) to furnish the 16α-substituted estrone ethers.[218] Even with relatively bulky 16α-substituents, reduction of the C-17 ketone with LiAlH$_4$ proceeded with very high stereoselectivity to give the 16α-substituted 17β-estradiol 3-O-ethers.

8. CONCLUDING REMARKS

Alkylations and aldol reactions of enolates are very old reactions in organic chemistry. During the past decade, rapid progress has been made in this field. Regio-, stereo-, and chemo-selective reactions have been developed. Especially, highly efficient asymmetric syntheses have been made via these reactions. It is not too much to say that these reactions are one of the most frequently encountered reactions in organic chemistry. Since there had been a number of excellent review articles in this field, we intended to summarize the alkylation reactions and heteroatom introduction to metal enolates, which were not so frequently reviewed in comparison with aldol reactions though they were synthetically important. We emphasize the stereo- and regio-chemistry both of metal enolates formation and of their reactions with RX. Although most of alkylation and heteroatom introduction reactions are carried out with alkali metal (mostly lithium) enolates or with silyl enol ethers, we summarize other metal enolates including transition metals. These metal enolates, especially transition metal enolates, may be promising for the future investigation; new type of reactions, or new selectivity might be created, and catalytic use of transition metals may become possible (actually, this proved to be possible; see eq 17 and Scheme 7).

We do not cover all the literatures on alkylation and heteroatom introduction since 1980. We rather concentrate on the synthetically or mechanistically interesting reactions. Enolate formation is limited primarily to deprotonation of carbonyl compounds. Other important procedures, such as deprotonation of imines and hydrazones (Schiff's base derivatives)[219] or conjugate addition to α,β-unsaturated carbonyl derivatives,[220] are not mentioned. Finally, a number of interesting and useful reactions of metal enolates other than alkylations are not cited; for example, cyanation,[221] the reaction of enolates with nitriles,[222] enolate reactions with halogenated olefins and chloroacetylenes[223] (a sort of Michael addition of enolates to Michael acceptors), and reactions with imines and related substrates.[224] These types of reactions with π-electron bearing substrates should be treated in aldol reactions.

REFERENCES

1) D. A. Evans in "Asymmetric Synthesis (J. D. Morrison, ed.)", Vol 3, p. 2. Academic Press, New York (1984).

2) D. A. Evans, J. V. Nelson, T. R. Taher, in "Topics in Stereochemistry (N. L. Allinger, E. L. Eliel, S. H. Wilen, ed.)", Vol 13, p. 1. John Wiley & Sons, New York (1982).

3) C. H. Heathcock, in "Comprehensive Carbanion Chemistry, Part B (E. Buncel and T. Durst, ed.)", p. 177. Elsevier, Amsterdam (1984).

4) C. H. Heathcock, in "Asymmetric Synthesis (J. D. Morrison, ed.)", Vol 3, p.111 Academic Press, New York (1984).

5) T. Mukaiyama, Org. React. (N. Y.) 28, 203 (1982).

6) S. Masamune, W. Choy, J. S. Petersen, L. R. Sita, Angew. Chem. Int. Ed. Engl. 24, 1 (1985).

82

7) (a) Y. Yamamoto, K. Maruyama, Heterocycles, **18**, 357 (1982). (b) R. W. Hoffmann, Angew. Chem. Int. Ed. Engl. **21**, 555 (1982). (c) M. T. Reetz, Angew. Chem. Int. Ed. Engl. **23**, 556 (1984). (d) D. Seebach, V. Prelog, Angew. Chem. Int. Ed. Engl. **21**, 654 (1982). (e) P. A. Bartlett, Tetrahedron **36**, 2 (1980). (f) G. J. McGarvey, M. Kimura, T. Oh, J. M. Williams, Carbohydr. Chem. 3, 125 (1984).

8) R. Gompper, Angew. Chem. Int. Ed. Engl. 3, 560 (1964).

9) S. A. Sheveiev, Russ. Chem. Rev. **39**, 844 (1970).

10) H. D. House, Modern Synthetic Reactions, 2nd Edition, Benjamin, Menlo Park, Park, Califormia (1972).

11) J. d'Angelo, Tetrahedron **32**, 2979 (1976).

12) Z. G. Hajos, Carbon-Carbon Bond Formation Vol 1, Ed. by R. L. Angustine, Marcel Dekker, 1979, p 1.

13) D. Caine, Carbon-Carbon Bond Formation Vol 1, Ed. by R. L. Angustine, Marcel Dekker, 1979, p 85.

14) L. M. Jackman, B. C. Lange, Tetrahedron, **33**, 2737 (1977).

15) M. N. Paddon-Row, N. G. Rondan, K. N. Houk, J. Am. Chem. Soc., **104**, 7162 (1982).

16) D. J. Cram, F. A. A. Elhafez, J. Am. Chem. Soc., **74**, 5828 (1952).

17) M. Chérest, H. Felkin, N. Prudent, Tetrahedron Lett., 2199 (1968); M. Chérest, H. Felkin, ibid., 2205 (1968). N. T. Anh, O. Eisenstein, Nouv. J. Chem. **1**, 61 (1977).

18) Y. Yamamoto, K. Maruyama, J. Chem. Soc. Chem. Commun., 904 (1984).

19) I. Fleming, J. J. Lewis, J. Chem. Soc. Chem. Commun., 149 (1985).

20) W. Bernhard. I. Fleming, D. Waterson, J. Chem. Soc. Chem. Chommun., 28 (1984).

21) I. Fleming, J. H. M. Hill, D. Parker, D. Waterson, J. Chem. Soc. Chem. Commun., 318 (1985).

22) T. Hayashi, Chemica Scripta, **25**, 61 (1985); J. Syn. Org. Chem. Jpn., **43**, 419 (1985), and references cited therein. H. Wetter, P. Scherer, Helv. Chim Acta, **66**, 118 (1983). G. Wickham, W. Kitching. J. Org. Chem., **48**, 612 (1983). I. Fleming, N. K. Terrett, Tetrahedron Lett., **24**, 4153 (1983).

23) E. L. Eliel, N. L. Allinger, S. J. Angyal, and G. A. Morrison, "Conformational Analysis", Wiley, New York 1967, p 44; W. Kitching, H. A. Olszowy, G. M. Drew, W. Adcock, J. Org. Chem., **47**, 5153 (1982).

24) G. J. McGarvey, J. M. Williams, J. Am. Chem. Soc., **107**, 1435 (1985).

25) W. Kitching, D. Doddrell, J. B. Grutzner, J. Organomet. Chem. **107**, C5 (1976). W. Kitching, H. A. Olszowy, K. Harvey, J. Org. Chem. **47**, 1893 (1982).

26) H. Kawasaki, K. Tomioka, K. Koga, Tetrahedron Lett., 26, 3031 (1985).

27) S. H. Ahn, D. Kim, M. W. Chun, W. K. Chung, Tetrahedron Lett., 27, 943 (1986).

28) J. L. Herrmann, R. H. Schlessinger, Tetrahedron Lett., 2429 (1973).

29) D. Seebach, D. Wasmuth, Helv. Chim. Acta, 63, 197 (1980). M. Zuger, T. Weller, D. Seebach, Helv. Chim. Acta, 63, 2005 (1980). G. Frater, Helv. Chim. Acta, 62, 2825 (1979). G. Frater, U. Muller, W. Gunther, Tetrahedron, 40, 1269 (1984).

30) K. Narasaka, Y. Ukaji, Chem. Lett., 81 (1986).

31) K. K. Mahalanabis, M. Mumtaz, V. Snieckus, Tetrahedron Lett., 23, 3971 (1982).

32) S. Berrada, P. Metzner, Tetrahedron Lett., 28, 409 (1987).

32a) Y. Tamaru, Y. Furukawa, M. Mizutani, O. Kitao, Z. Yoshida, J. Org. Chem., 48, 3631 (1983).

32b) T. Nakai, K. Mikami, Chem. Rev., 86, 885 (1986).

33) a) H. O. House, T. M. Bare, J. Org. Chem., 33, 943 (1968). b) A. P. Krapcho, E. A. Dundulis, J. Org. Chem., 45, 3236 (1980).

34) a) H. O. House, B. A. Tefertiller, H. D. Olmstead, J. Org. Chem., 33, 935 (1968). b) B. J. L. Huff, F. N. Tuller, D. Caine, J. Org. Chem. 34, 3070 (1969). c) M. E. Kuehne, J. Org. Chem., 35, 171 (1970). d) C. M. Lentz, G. H. Posner, Tetrahedron Lett., 3769 (1978).

35) K. Tomioka, H. Kawasaki, K. Koga, Tetrahedron Lett., 26, 3027 (1985).

36) K. Tomioka, H. Kawasaki, Y. Iitaka, K. Koga, Tetrahedron Lett., 26, 903 (1985).

37) K. Tomioka, K. Yasuda, H. Kawasaki, K. Koga, Tetrahedron Lett., 27, 3247 (1986). K. Tomioka, H. Kawasaki, K. Yasuda, K. Koga, J. Am. Chem. Soc., 110, 3597 (1988).

38) a) H.-M. Shieh, G. D. Prestwich, J. Org. Chem., 46, 4319 (1981). b) A. R. Chamberlin, M. Dezube, Tetrahedron Lett., 23, 3055 (1982). c) S-Y. Chen, M. M. Joullie, J. Org. Chem., 49, 2168 (1984).

39) R. Naef, D. Seebach, Angew. Chem. Int. Ed. Engl., 20, 1030 (1981). J. D. Aebi, M. A. Sutter, D. Wasmuth, D. Seebach, Liebigs Ann. Chem., 2114 (1983).

40) W. Ladner, Angew. Chem. Int. Ed. Engl., 21, 449 (1982).

41) D. Seebach, J. D. Aebi, Tetrahedron Lett., 24, 3311 (1983).

42) J. Mulzer, A. Chucholowski, Angew. Chem. Int. Ed. Engl., 25, 655 (1986).

43) J. H. Horner, M. Vera, J. B. Grutzner, J. Org. Chem., 51, 4212 (1986).

44) T. Takahashi, M. Nisar, K. Shimizu, J. Tsuji, Tetrahedron Lett., 27, 5103 (1986). T. Takahashi, H. Ueno, M. Miyazawa, J. Tsuji, ibid., 26, 5139 (1985).

45) D. P. Curran, J-C. Chao, J. Am. Chem. Soc., 109, 3036 (1987).

45a) H-U. Reissig, I. Bohm, J. Am. Chem. Soc., 104, 1735 (1982).

46) H. O. House, M. Gall, H. D. Olmstead, J. Org. Chem., 36, 2361 (1971). H. O. House, L. J. Czuba, M. Gall, H. D. Olmstead, J. Org. Chem., 34, 2324 (1969).

47) E. J. Corey, A. W. Gross, Tetrahedron Lett., 25, 495 (1984).

48) For a review on the compatibility between organometallic reagents and Lewis acids, see Y. Yamamoto, Angew. Chem. Int. Ed. Engl., 25, 947 (1986).

49) R. D. Clark, C. H. Heathcock, Tetrahedron Lett., 2027 (1974); L. Fedor, R. C. Cavestri, J. Org. Chem., 41,1369 (1976).

50) G. H. Posner, C. M. Lentz, J. Am. Chem. Soc., 101, 934 (1979).

51) G. Stork, P. F. Hudrlik, J. Am. Chem. Soc., 90, 4462 (1968).

52) C. A. Brown, J. Org. Chem., 39, 3913 (1974).

53) E. Nakamura, T. Murofushi, M. Shimizu, I. Kuwajima, J. Am. Chem. Soc., 98, 2346 (1976).

54) R. D. Miller, D. R. McKean, Synthesis, 730 (1979).

55) M. E. Krafft, R. A. Holton, Tetrahedron Lett., 24, 1345 (1983).

56) C. W. Still, A. Galynker, Tetrahedron, 37, 3981 (1981).

57) A. I. Meyers, K. B. Kunnen, W. C. Still, J. Am. Chem. Soc., 109, 4405 (1987).

57a) C. L. Liotta, T. C. Caruso, Tetrahedron Lett., 26, 1599 (1985).

57b) M. E. Garst, J. N. Bonifiglio, D. A. Gruddoski, J. Marks, Tetrahedron Lett., 2671 (1978); J. Org. Chem., 45, 2307 (1980).

57c) T. R. Hoye, S. R. Duff, R. S. King, Tetrahedron Lett., 26, 3433 (1985).

58) R. E. Ireland, R. H. Mueller, A. K. Willard, J. Am. Chem. Soc., 98, 2868 (1976).

59) E. Nakamura, K. Hashimoto, I. Kuwajima, Tetrahedron Lett., 2079 (1978).

60) Z. A. Fataftah, I. E. Kopka, M. W. Rathke, J. Am. Chem. Soc., 102, 3959 (1980).

61) D. W. Moreland, W. G. Dauben, J. Am. Chem. Soc., 107, 2264 (1985).

61a) A. S. Kende, B. H. Toder, J. Org. Chem., 47, 163 (1982).

62) A. D. Adams, R. H. Schlessinger, J. R. Tata, J. J.Venit, J. Org. Chem., 51, 3068 (1986).

63) P. Beslin, P. Metzner, Y. Vallée, J. Vialle, Tetrahedron Lett., 24, 3617 (1983).

64) P. L. Block, D. J. Boschetto, J. R. Rasmussen, J. P. Demers, G. M.Whitesides, J. Am. Chem. Soc., **96**, 2814 (1974).

65) C. H. Heathcock, M. C. Pirrung, S. H. Montgomery, J. Lampe, Tetrahedron, **37**, 4087 (1981).

66) C. H. Heathcock, C. T. Buse, W. A. Kleschick, M. C. Pirrung, J. E. Sohn, J. Lampe, J. Org. Chem.,**45**, 1066 (1980).

67) K. G. Davenport, M. Newcomb, D. E. Bergbreiter, J. Org. Chem., **46**, 3143 (1981) and references cited therein.

68) D. A. Evans, L. R. McGee, Tetrahedron Lett., **21**, 3975 (1980).

69) a) C. Goasdoue, N. Goasdoue, M. Gaudemar, M. Mladenova, J. Organomet. Chem., **208**, 279 (1981); b) Y. Tamaru, T. Harada, S. Nishi, M. Mizutani, T. Hioki, Z. Yoshida, J. Am. Chem. Soc., **102**, 7806 (1980).

70) (a) D. A. Evans, J. V. Nelson, E. Vogel, T. R. Taber, J. Am. Chem. Soc., **103**, 3099 (1981). (b) D. E. Van Horn, S. Masamune, Tetrahedron Lett., **20**, 2229 (1979). (c) J. Hooz, J.Oudenes, Tetrahedron Lett., **24**, 5695 (1983). (d) S. Masamune, J. W. Ellingboe, W. Choy, J. Am. Chem. Soc., **104**, 5526 (1982).

71) Y. Tamaru, T. Hioki, S. Kawamura, H. Satomi, Z. Yoshida, J. Am. Chem. Soc., **106**, 3876 (1984).

72) a) T. T. Tidwell, Tetrahedron Lett., **20**, 4615 (1979). b) L. M. Baigrie, D. Lenoir, H. R. Seikaly, T. T. Tidwell, J. Org. Chem., **50**, 2105 (1985). c) D. Lenoir, H. R. Seikaly, T. T. Tidwell, Tetrahedron Lett., **23**, 4987 (1982). d) L. M. Baigrie, H. R. Seiklay, T. T. Tidwell, J. Am. Chem. Soc., **107**, 5391 (1985).

73) M. C. Pirrung. J. R. Hwu, Tetrahedron Lett., **24**, 565 (1983).

74) B.L.Chenard, E. D. Laganis, F. Davidson, T. V. RajanBabu, J. Org. Chem., **50**, 3666 (1985).

74a) D. Seebach, R. Amstutz, T. Laube, W. B. Schweizer, J. D. Dunitz, J. Am. Chem. Soc., **107**, 5403 (1985).

75) N. Kornblum, Angew. Chem. Int. Ed. Engl. **14**, 734 (1975).

76) G. A. Russell, B. Mudryk, M. Jawdosiuk, J. Am. Chem. Soc., **103**, 4610 (1981).

77) J. F. Bunnett, Acc. Chem. Res., **11**, 413 (1978). J. F. Bunnett, P. Singh, J. Org. Chem., **46**, 5022 (1981).

78) D. R. Carver, T. D. Greenwood, J. S. Hubbard, A. P. Komin, Y. P. Sachdeva, J. F. Wolfe, J. Org. Chem., **48**, 1180 (1983).

79) H. D. Zook, W. L. Kelly, I. Y. Posey, J. Org. Chem., **33**, 3477 (1968).

80) E. C. Ashby, J. N. Argyropoulos, Tetrahedron Lett., **25**, 7 (1984); J. Org. Chem., **50**, 3274 (1985).

86

81) E. C. Ashby, J. N. Argyropoulos, G. R. Meyer, A. B. Goel, J. Am. Chem. Soc., **104**, 6788 (1982).

82) a) H. O. House, V. Kramar, J. Org. Chem., **28**, 3362 (1963). b) H. O. House, B. A. Tefertiller, H. D. Olmstead, J. Org. Chem., **33**, 935 (1968). c) D. Seebach, T. Weller, G. Protschulk, A. K. Beck, M. S. Hoekstra, Helv. Chim. Acta, **64**, 716 (1981).

83) T. T. Tidwell, Tetrahedron **34**, 1855 (1978).

84) P. O'Neill, A. F. Hegarty, J. Org. Chem., **52**, 2113 (1987).

85) M. E. Jones, S. R. Kass, J. Filley, R. M. Barkley, G. B. Ellison, J. Am. Chem. Soc., **107**, 109 (1985).

86) K. N. Houk, M. N. Paddon-Row, J. Am. Chem. Soc., **108**, 2659 (1986).

87) L. M. Jackman, B. C. Lange, J. Org. Chem., **48**, 4789 (1983).

88) a) W. P. Weber, Silicon Reagents for Organic Synthesis, Springer-Verlag, Berlin (1983). b) E. W. Colvin, Silicon in Organic Synthesis, Butterworth, London (1981). c) M. T. Reetz, W. F. Maier, Angew. Chem. Int. Ed. Engl., **17**, 48 (1978). d) I. H. M. Wallace, T. H. Chan, Tetrahedron, **39**, 847 (1983). e) R. P. Alexander, I. Paterson, Tetrahedron Lett., **24**, 5911 (1983). f) S. L. Schreiber, T. Sammakia, W. E. Crowe, J. Am. Chem. Soc., **108**, 3128 (1986). S. L. Schreiber, M. T. Klimas, T. Sammakia, ibid., **109**, 5749 (1987).

89) T. A. Oster, T. M. Harris, Tetrahedron Lett., **24**, 1851 (1983).

90) L. M. Bairgrie, R. L.-Toung, T. T. Tidwell, Tetrahedron Lett., **29**, 1673 (1988).

91) D. A. Evans, Aldrichimica Acta, **15**, 21 (1982). D. A. Evans, J. Bartroli, T. L. Shih, J. Am. Chem. Soc., **103**, 2127 (1981).

92) R. M. DiPardo, M. G. Bock, Tetrahedron Lett., **24**, 4805 (1983).

93) D. A. Evans, M. D. Ennis, T. Le, N. Mandel, G. Mandel, J. Am. Chem. Soc., **106**, 1154 (1984).

94) G. M. P. Giblin, N. S. Simpkins, J. Chem. Soc. Chem. Commun., 207 (1987).

95) R. Tsang, B. Fraser-Reid, J. Chem. Soc. Chem. Commun., 60 (1984).

96) I. Kopka, M. W. Rathke, J. Org. Chem., **46**, 3771 (1981).

97) R. E. Tirpak, M. W. Rathke, J. Org. Chem., **47**, 5099 (1982).

98) E. Negishi, H. Matsushita, S. Chatterjee, R. A. John, J. Org. Chem., **47**, 3188 (1982).

99) E. Negishi, M. J. Idacavage, K.-W. Chiu, T. Yoshida, A. Abramovitch, M. E. Goettel, A. Silveira Jr., H. D. Bretherick, J. Chem. Soc. Perkin Trans, 2, 1225 (1978).

100) Y. Yamamoto, H. Yatagai, K. Maruyama, Tetrahedron Lett., **23**, 2387 (1982).

101) M. J. Idacavage, E. Negishi, C. A. Brown, J. Organomet. Chem., **186**, C55 (1980).

102) E. Negishi, M. J. Idacavage, F. DiPasquale, A. Silveira Jr., Tetrahedron Lett., **20**, 845 (1979).

103) B. M. Trost, Acc. Chem. Res., **13**, 385 (1980).

104) E. Negishi, R. A. John, J. Org. Chem., **48**, 4098 (1983).

105) T. Tsuda, Y. Chujo, S. Nishi, K. Tawara, T. Saegusa, J. Am. Chem. Soc., **102**, 6381 (1980).

106) I. Shimizu, T. Yamada, J. Tsuji, Tetrahedron Lett., **21**, 3199 (1980).

107) R. R. Schrock, J. D. Fellmann, J. Am. Chem. Soc., **100**, 3359 (1978).

108) J. R. Stille, R. H. Grubbs, J. Am. Chem. Soc., **105**, 1664 (1983).

109) L. Clawson, S. L. Buchwald, R. H. Grubbs, Tetrahedron Lett., **25**, 5733 (1984).

110) M. D. Curtis, S. Thanedar, W. M. Butler, Organometallics, **3**, 1855 (1984).

111) M. T. Reetz, R. Peter, Tetrahedron Lett., **22**, 4691 (1981).

112) E. Nakamura, J. Shimada, Y. Horiguchi, I. Kuwajima, Tetrahedron Lett., **24**, 3341 (1983).

113) C. P. Gibson, G. Dabbagh, S. H. Bertz, J. Chem. Soc. Chem. Commun., 603 (1988).

114) a) Y. Yamamoto, K. Maruyama, Tetrahedron Lett., **21**, 4607 (1980). b) D. A. Evans, L. R. McGee, ibid., **21**, 3975 (1980).

115) D. H. R. Barton, B. Charpiot, W. B. Motherwell, Tetrahedron Lett., **23**, 3365 (1982).

116) M. Pereyre, J.-P. Quintard, A. Rahm, Tin in Organic Synthesis, Butterworths, London (1987).

117) P. A. Tardella, Tetrahedron Lett., 1117 (1969).

118) Y. Yamamoto, H. Hatagai, K. Maruyama, Silicon, Germanium, Tin and Lead Compounds, **9**, 25 (1986).

119) Y. Yamamoto, H. Yatagai, K. Maruyama, J. Chem. Soc. Chem. Commun., 162 (1981).

120) S. Shenvi, J. K. Stille, Tetrahedron Lett., **23**, 627 (1982).

121) Y. Yamamoto, J. Yamada, J. Chem. Soc. Chem. Commun., 802 (1988).

122) a) G. J. Baird, S. G. Davies, R. H. Jones, K. Prout, P. Warner, J. Chem. Soc. Chem. Commun., 745 (1984). b) G. J. Baird, J. A. Bandy, S. G. Davies, K. Prout, ibid., 1202(1983).

123) a) P. J. Curtis, S. G. Davies, J. Chem. Soc. Chem. Commun., 747 (1984). b) S. G. Davies, I. M. Dordor, P. Warner, J. Chem. Soc. Chem. Commun., 956 (1984).

124) L. Liebeskind, M. E. Welker, Organometallics, 2, 194 (1983).

125) L. S. Liebeskind, M. E. Welker, R. W. Fengl, J. Am. Chem. Soc., 108, 6328 (1986). L. S. Liebeskind, M. E. Welker, V. Goedken, J. Am. Chem. Soc., 106, 441 (1984). L. S. Liebeskind, M. E. Welker, Tetrahedron Lett., 25, 4341 (1984). L. S. Liebeskind, R. W. Fengl, M. E. Welker, Tetrahedron Lett., 26, 3075 (1985).

126) S. G. Davies, I. M. Dordor, J. C. Walker, P. Warner, Tetrahedron Lett., 25, 2709 (1984).

127) S. G. Davies, J. C. Walker, J. Chem. Soc. Chem. Commun., 209 (1985).

128) S. G. Davies, I. M. D.-Hedgecock, P. Warner, Tetrahedron Lett., 26, 2125 (1985); P. W. Ambler, S. G. Davies, ibid., 26, 2129 (1985).

129) a) S. L. Brown, S. G. Davies, D. F. Foster, J. I. Seeman, P. Warner, Tetrahedron Lett., 27, 623 (1986). b) S. G. Davies, I. M. D.-Hedgecock, K. H. Sutton, J. C. Walker, R. H. Jones, K. Prout, Tetrahedron, 42, 5123 (1986).

130) J. W. Herndon, C. Wu, H. L. Ammon, J. Org. Chem., 53, 2873 (1988).

131) I. Weinstock, C. Floriani, A. C.-Villa, C. Guastini, J. Am. Chem. Soc., 108, 8298 (1986).

132) K. H. Theopold, P. N. Becker, R. G. Bergman, J. Am. Chem. Soc., 104, 5250 (1982).

133) C. A. Rusik, T. L. Tonker, J. L. Templeton, J. Am. Chem. Soc., 108, 4652 (1986).

134) T. Hirao, S. Nagata, Y. Yamana, T. Agawa, Tetrahedron Lett., 26, 5061 (1985). T. Hirao, Y. Harano, Y. Yamana, Y. Ohshiro, T. Agawa, ibid., 24, 1255 (1983).

135) E. R. Burkhardt, J. J. Doney, R. G. Bergman, C. H. Heathcock, J. Am. Chem. Soc., 109, 2022 (1987).

136) P. Dall'Antonia, M. Graziani, M. Lenarda, J. Organomet. Chem., 186, 131 (1980).

137) Y. Ito, M. Nakatsuka, N. Kise, T. Saegusa, Tetrahedron Lett., 21, 2873 (1980).

138) T. Hirao, Y. Fujihara, S.Tsuno, Y. Ohshiro, T. Agawa, Chem. Lett., 367 (1984).

139) J. M. Manriquez, P. J. Fagan, T. J. Marks, J. Am. Chem. Soc., 100, 7112 (1978). D. C. Sonnenberger, E. A. Mintz, T. J. Marks, ibid., 106, 3484 (1984).

140) E. J. Crawford, C. Lambert, K. P. Menard, A. R. Cutler, J. Am. Chem. Soc., **107**, 3130 (1985). M. Akaita, A. Kondo, Y. Moro-oka, J. Chem. Soc. Chem. Commun.,1296 (1986).

141) J. Egnelbrecht, T. Greiser, E. Weiss, J. Organomet. Chem., **204**, 79 (1981).

142) V. Galamb, G. Pályi, F. Cser, M. G. Furmanova, Y. T. Struchkov, J. Organomet. Chem., **209**, 183 (1981).

143) W. J. Sieber, M. Wolfgruber, F. R. Kreissl, O. Orama, J. Organomet. Chem., **270**, C41 (1984).

144) W. G. Hatton, J. A. Gladysz, J. Am. Chem. Soc., **105**, 6157 (1983).

145) R. Bertani, C. B. Castellani, B. Crociani, J. Organomet. Chem., **269**, C15 (1984). R. A. Wanat, D. B. Collum, Organometallics 5, 120 (1986). T. Hirao, N. Yamada, Y. Ohshiro, T. Agawa, Chem. Lett., 1997 (1982).

146) G. V. Nizova, M. V. Serdobov, A. T. Nikitaev, G. B. Shul'pin, J. Organomet. Chem., **275**, 139 (1984). T. Yoshida, T. Okano, S. Otsuka, J. Chem. Soc. Dalton Trans. 1, 993 (1976).

147) D. Milstein, Acc. Chem. Res., **17**, 221 (1984).

148) G. H. Posner, C. M. Lentz, J. Am. Chem. Soc., **101**, 934 (1979).

149) Y. Yamamoto, K. Maruyama, J. Am. Chem. Soc., **104**, 2323 (1982).

150) Y. Yamamoto, J. Yamada, T. Uyehara, J. Am. Chem. Soc., **109**, 5820 (1987).

151) G. Helmchen, A. Selim, D. Dorsch, I. Taufer, Tetrahedron Lett., **24**, 3213 (1983).

152) G. Helmchen. R. Schmierer, Tetrahedron Lett., **24**, 1235 (1983).

153) G. Helmchen, R. Wierzchowski, Angew. Chem. Int. Ed. Engl., **23**, 60 (1984).

154) W. Oppolzer, G. Poli, Tetrahedron Lett., **27**, 4717 (1986).

155) T. R. Kelly, A. Arvanitis, Tetrahedron Lett., **25**, 39 (1984).

156) D. Seebach, R. Naef, Helv. Chim. Acta, **64**, 2704 (1981). For diastereoselective alkylation of tartaric acid derivatives, see R. Naef, D. Seebach, Angew. Chem. Int. Ed. Engl., **20**, 1030 (1981). D. Seebach, R. Naef, G. Calderari, Tetrahedron, **40**, 1313 (1984). D. Seebach, T. Weber, Helv. Chim. Acta, **67**, 1650 (1984).

157) D. Seebach, A. Fadel, Helv. Chim. Acta, **68**, 1243 (1985).

158) J. D. Aebi, D. Seebach, Helv. Chim. Acta, **68**, 1507 (1985).

159) D. Seebach, J. D. Aebi, M. G. Coquoz, R. Naef, Helv. Chim. Acta, **70**, 1194 (1987). D. Seebach, J. D. Aebi, Tetrahedron Lett., **25**, 2545 (1984).

160) D. Seebach, J. Zimmermann, Helv. Chim. Acta, **69**, 1147 (1986).

161) D. Seebach, M. Boes, R. Naef, W. B. Schweizer, J. Am. Chem. Soc., **105**, 5390 (1983).

162) W. H. Pearson, M-C. Cheng, J. Org. Chem., **51**, 3746 (1986).

163) S. Karady, J. S. Amato, L. M. Weinstock, Tetrahedron Lett., **25**, 4337 (1984).

164) P. Duhamel, J. J. Eddine, J-Y. Valnot, Tetrahedron Lett., **25**, 2355 (1984).

165) I. Ojima, X. Qiu, J. Am. Chem. Soc., **109**, 6537 (1987).

166) A. Misumi, K. Iwanaga, K. Furuta, H. Yamamoto, J. Am. Chem. Soc., **107**, 3343 (1985).

167) M. Kobayashi, K. Koyabu, T. Shimizu, K. Umemura, H. Matsuyama, Chem. Lett., 2117 (1986).

168) D. A. Evans, M. D. Ennis, D. J. Mathre, J. Am. Chem. Soc., **104**, 1737 (1982).

169) R. P. Alexander, I. Paterson, Tetrahedron Lett., **26**, 5339 (1985).

170) E. R. Koft, M. D. Williams, Tetrahedron Lett., **27**, 2227 (1986).

171) Y. Ito, T. Katsuki, M. Yamaguchi, Tetrahedron Lett., **25**, 6015 (1984).

172) T. Hanamoto, T. Katsuki, M. Yamaguchi, Tetrahedron Lett., **27**, 2463 (1986).

173) K. Okano, T. Izawa, M. Ohno, Tetrahedron Lett., **24**, 217 (1983).

174) T. Hayashi, K. Kanehira, H. Tsuchiya, M. Kumada, J. Chem. Soc. Chem. Commun.,1162 (1982).

175) K. Saigo, A. Kasahara, S. Ogawa, H. Nohira, Tetrahedron Lett., **24**, 511 (1983).

176) C. Gluchowski, T. T.-Harding, J. K. Smith, D. E. Bergbreiter, M. Newcomb, J. Org. Chem., **49**, 2650 (1984).

177) J. K. Whitesell, S. W. Felman, J. Org. Chem., **45**, 755 (1980). T. Yamashita, H. Mitsui, H. Watanabe, N. Nakamura, Bull. Chem. Soc. Jpn.,**55**, 961 (1982). A. C. Lepan, J. Stannton, J. Chem. Soc. Chem. Commun., 764 (1982). M. Asami, Chem. Lett., 829 (1984). N. S. Simpkins, J. Chem. Soc. Chem. Commun., 88 (1986). S. Brandage, S. Josephson, L. Morch, S. Vallen, Acta Chim. Scand., 293 (1981).

178) H. Hogeveen, W. M. P. B. Menge, Tetrahedron Lett., **27**, 2767 (1986). M. B. Eleveld, H. Hogeveen, ibid., **27**, 631 (1986).

179) A. Ando, T. Shioiri, J. Chem. Soc. Chem. Commun.,656 (1987).

180) R. Shirai, M. Tanaka, K. Koga, J. Am. Chem. Soc., **108**, 543 (1986).

181) T. Laube, J. D. Dunitz, D. Seebach, Helv. Chim. Acta, **68**, 1373 (1985).

182) H. E. Zimmerman, Acc. Chem. Res., **20**, 263 (1987).

183) S. Takano, J. Kudo, M. Takahashi, K. Ogasawara, Tetrahedron Lett., **27**, 2405 (1986).

184) S. Takano, S. Yamada, H. Numata, K. Ogasawara, J. Chem. Soc. Chem. Commun., 760 (1983).

185) L. Duhamel, S. Fouquay, J-C. Plaquevent, Tetrahedron Lett., **27**, 4975 (1986). L. Duhamel, P. Duhamel, J. C. Launay, J. C. Plaquevent, Bull. Soc. Chim. Fr., II-421, 1984. L. Duhamel, J. C. Plaquevent, J. Am. Chem. Soc., **100**, 7415 (1978).

186) H. Hogeveen, L. Zwart, Tetrahedron Lett., **23**, 105 (1982).

187) C. Gennari, L. Colombo, G. Bertolini, J. Am. Chem. Soc., **108**, 6394 (1986).

187) a) W. Oppolzer, R. Moretti, Helv. Chim. Acta, **69**, 1923 (1986).

188) D. A. Evans, T. C. Britton, R. L. Dorow, J. F. Dellaria, J. Am. Chem. Soc., **108**, 6395 (1986).

189) L. A. Trimble, J. C. Vederas, J. Am. Chem. Soc., **108**, 6397 (1986).

190) E. J. Bailey, D. H. R. Barton, J. Elks, J. F. Templeton, J. Chem. Soc., 1578 (1962). J. N. Gardner, F. E. Carlon, O. Gnoj, J. Org. Chem., **33**, 3294 (1968). H. H. Wasserman, B. H. Lipshutz, Tetrahedron Lett., 1731 (1975).

191) E. Vedejs, J. Am. Chem. Soc., **96**, 5944 (1974). E. Vedejs, D. A. Engler, J. E. Telschow, J. Org. Chem., **43**, 188 (1978).

192) F. A. Davis, L. C. Vishwakarma, J. M. Billmers, J. Finn, J. Org. Chem., **49**, 3241 (1984).

193) D. A. Evans, M. M. Morrissey, R. L. Dorow, J. Am. Chem. Soc., **107**, 4346 (1985).

194) F. A. Davis, L. C. Vishwakarma, Tetrahedron Lett., **26**, 3539 (1985).

195) W. Oppolzer, P. Dudfield, Helv. Chim. Acta, **68**, 216 (1985).

196) a) F. A. Davis, M. S. Haque, J. Org. Chem., **51**, 4083 (1986): F. A. Davis, M. S. Haque, T. G. Ulatowski, J. C. Towson, ibid., **51**, 2402 (1986). b) F. A. Davis, T. G. Ulatowski, M. S. Haque, J. Org. Chem., **52**, 5288 (1987).

197) R. Gamboni, C. Tamm, Helv. Chim. Acta, **69**, 615 (1986); Tetrahedron Lett., **27**, 3999 (1986).

198) Y. Morizawa, A. Yasuda, K. Uchida, Tetrahedron Lett., **27**, 1833 (1986).

199) H. J. Gais, T. Lied, K. L. Lukas, Angew. Chem. Int. Ed. Engl., **23**, 511 (1984).

200) E. J. Corey, H. E. Ensley, J. Am. Chem. Soc., **97**, 6908 (1975).

201) a) G. M. Rubottom, J. M. Gruber, J. Org. Chem., **43**, 1599 (1978). b) N. K. Dunlap, M. R. Sabol, D. S. Watt, Tetrahedron Lett., **25**, 5839 (1984).

202) J. H. Hutchinson, T. Money, S. E. Piper, J. Chem. Soc. Chem. Commun., 455 (1984).

203) G. Mehta, K. S. Rao, J. Chem. Soc. Chem. Commun., 1464 (1985).

204) J. D. White, M. A. Avery, S. C. Choudhry, O. P. Dhingra, M. Kang, A. J. Whittle, J. Am. Chem. Soc., **105**, 6517 (1983).

205) M. A. Sutter, D. Seebach, Liebigs Ann. Chem., 939 (1983). D. Seebach, H.-F. Chow, R. F. W. Jackson, K. Lawson, M. A. Sutter, S. Thaisrivongs, J. Zimmermann, J. Am. Chem. Soc., 107, 5292 (1985); Liebigs Ann. Chem., 1281 (1986).

206) R. Naef, D. Seebach, Liebigs Ann. Chem., 1930 (1983).

207) D. A. Evans, J. Bartroli, Tetrahedron Lett., 23, 807 (1982).

208) a) D. A. Evans, D. J.Mathre, W. L. Scott, J. Org. Chem., 50, 1830 (1985). b) D. A. Evans, R. L. Dow, Tetrahedron Lett., 27, 1007 (1986). c) D. A. Evans, R. P. Polniaszek, ibid., 27, 5683 (1986).

209) K. Luthman, M. Orbe, T. Waglund, A. Claesson, J. Org. Chem., 52, 3777 (1987).

210) a) A. G. Schultz, J. P. Dittami, J. Org. Chem., 49, 2615 (1984). b) A. G. Schultz, J. P. Dittami, ibid., 48, 2318 (1983).

211) S. E. Denmark, J. P. Germanas, Tetrahedron Lett., 25, 1231 (1984).

212) L. A. Paquette, G. D. Annis, H. Schostarez, J. Am. Chem. Soc., 104, 6646 (1982).

213) D. N. Jones, M. R. Peel, J. Chem. Soc. Chem. Commun., 216 (1986).

214) G. Pattenden, S. J. Teague, Tetrahedron Lett., 23, 5471 (1982).

215) J. D. White, K. Takabe, M. P. Prisbylla, J. Org. Chem., 50, 5233 (1985).

216) H. Hirai, K. Sawada, M. Aratani, M. Hashimoto, Tetrahedron Lett., 25, 5075 (1984).

217) W. Ladner, Chem. Ber., 116, 3413 (1983).

218) T. L. Fevig, J. A. Katzenellenbogen, J. Org. Chem., 52, 247 (1987).

219) For example, R. A. Wanat, D.B. Collum, J. Am. Chem. Soc., 107, 2078 (1985) and references cited therein.

220) For example, R. J. K.Taylor, Synthesis, 364 (1985) and references cited therein.

221) D. Kahne, D. B. Collum, Tetrahedron Lett., 22, 5011 (1981).

222) T. Hiyama, K. Kobayashi, Tetrahedron Lett., 23, 1597 (1982). T. Hiyama, K. Nishide, K. Kobayashi, ibid., 25, 569 (1984); Chem. Lett., 361 (1984).

223) A. S. Kende, P. Fludzinski, J. H. Hill, W. Swenson, J. Clardy, J. Am. Chem. Soc., 106, 3551 (1984).

224) L. M. Fuentes, I. Shinkai, T. N. Salzmann, J. Am. Chem. Soc., 108, 4675 (1986). Y. Nagao, T. Kumagai, S. Tamai, T. Abe, Y. Kuramoto, T. Taga, S. Aoyagi, Y. Nagase, M. Ochiai, Y. Inoue, E. Fujita, ibid., 108, 4673 (1986).

TRANSITION-METAL COMPLEXES OF N_2, CO_2, AND SIMILAR SMALL MOLECULES. AB-INITIO STUDIES OF THEIR STEREOCHEMISTRY AND COORDINATE BONDING NATURE

S. Sakaki

TRANSITION-METAL COMPLEXES OF N_2, CO_2, AND SIMILAR SMALL MOLECULES. Ab-initio
MO STUDIES OF THEIR STEREOCHEMISTRY AND COORDINATE BONDING NATURE

S. SAKAKI

1. INTRODUCTION

In the last decade, non-Werner transition-metal complexes have received considerable attention because of their many interesting and useful properties; for instance, coordination and activation of inert molecules, variety of structure, various catalytic activities available to chemical industry and organic synthesis, complex formation of reactive chemical species considered as a model of reaction intermediates, and so on (ref. 1). Of these interesting properties, let us concentrate on transition-metal complexes of inert molecules such as dinitrogen, carbon dioxide, and similar small molecules (ref. 2). It is a matter of course that coordination of these molecules with transition metals is believed to be one of the powerful and universal ways of activating inert molecules and expected to be applied to N_2 and CO_2-fixation into useful substances (refs. 3 and 4), and NO_x and SO_x-treatment into innocuous substances (refs. 5 and 6). In this light, the reactivity of coordinated molecules is one of the central interests, from the viewpoint of applications of non-Werner transition-metal complexes. Moreover, not only is the above-mentioned application useful, but also basic interest is found in the complexes of inert molecules; for instance, the interaction between inert molecule and transition metal ion is an attractive subject of research in the field of molecular science of chemical bonds. Furthermore, structures of the complexes of inert molecules are full of variety; e.g., it is well known that transition metal complexes exhibit fluxionality of total structure and orientation of ligands (e.g., T-shape or Y-shape structure of three-coordinate complexes, Td or planar structure of four-coordinate complexes, etc. (ref. 7); rotation of ligand around its coordinate bond (ref. 8) etc.). Besides these general features of transition metal complexes, complexes of inert molecules exhibit several possibilities of coordination modes, as follows. In the transition-metal CO_2 complexes, three coordination modes have been proposed by Vol'pin et al (ref. 4a): The first is the η^2-side on coordination, <u>1,</u> the the second is the

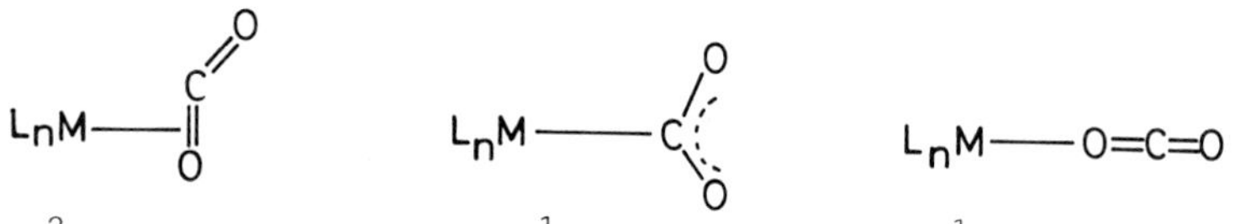

$$L_nM \underset{O}{\overset{O}{\parallel}} C{\overset{\displaystyle \parallel}{}} O \qquad \eta^2\text{-side on} \qquad \mathbf{1}$$

(structures shown)

η^2-side on

1

η^1-C

2

η^1-O end on

3

η^1-C coordination, $\underline{2}$, and the last is the η^1-O end on coordination $\underline{3}$. Transition-metal N_2 complexes also have two possibilities in the coordination modes (ref. 3); one is the η^1-end on coordination ($\underline{4}$ and $\underline{5}$), and the second is the η^2-side on coordination ($\underline{6}$ and $\underline{7}$). Which coordination mode is used depends on the electronic structure of the metal complex, in general. The reactivity of

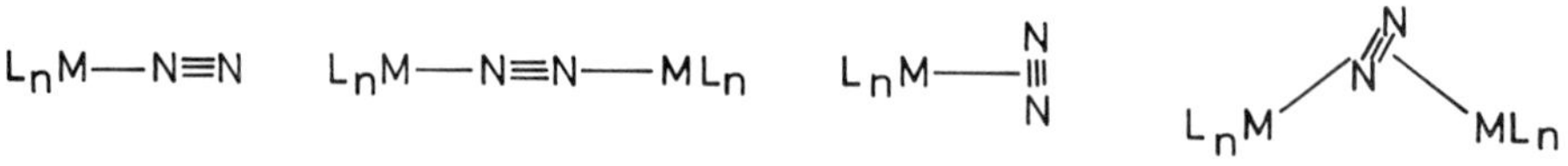

$L_nM-N{\equiv}N$ $L_nM-N{\equiv}N-ML_n$ $L_nM \underset{N}{\overset{N}{\parallel}} N$

4   **5**   **6**   **7**

η^1-end on η^2-side on

the coordinating molecule is also remarkably influenced by the coordination mode and the electronic structure of the metal complex. Thus, coordination modes, bonding nature, electronic structure, and reactivity of transition metal complexes are related to each other, and information about all these issues are indispensable for predicting structures of metal complexes, synthesizing a new metal complex, designing and finding a good catalyst applicable to various catalytic reactions, including CO_2 and N_2-fixation.

Theoretical investigations are expected to be effective in obtaining information on geometrical structure, electronic structure and bonding nature of transition-metal complexes. 10 or 15 Years ago, it was very difficult to apply quantitative or semi-quantitative theoretical methods to big molecules such as transition-metal complexes. However, methods and algorithms for molecular orbital calculations and ability of computers have made startling progress recently. As a result, ab-initio MO methods can be successfully applied now to many transition metal complexes (ref. 9). Even metalloporphyrins, which are big metal complexes, have been investigated with ab-initio MO methods (refs. 10 and 11). Furthermore, not only usual ab-initio MO methods based on Hartree-Fock-Roothaan LCAO MO method, but also more sophisticated methods have been applied to transition metal complexes (refs. 9b, 9c, and 12).

Such ab-initio MO methods have been applied to theoretical investigations of transition metal complexes of inert molecules. In this chapter, the author intends to review and summarize those theoretical investigations, and to offer information about stereochemistry, electronic structure, and bonding nature of transition metal complexes of inert molecules. One of the purposes of this review is to clarify how, and why, the coordination mode depends on the electronic structure of the central metal ion. The second is to present a way of predicting the structure of metal complexes. The third is to show a way of applying MO methods to transition-metal chemistry. Our goal is to understand theoretically transition metal complexes from a viewpoint of molecular science and to predict theoretically metal complexes useful for N_2 and CO_2-fixation, SO_x and NO_x-treatment, and various catalytic reactions. Although the present stage of the theoretical investigation is not far from the goal, theoretical methods are still imperfect. The author hopes that this review helps experimentalists and theoreticians to approach the goal cooperatively.

The outline of MO methods and related computational methods are first described, followed by general survey of the coordination bond. After that, theoretical investigations of transition-metal CO_2 and N_2 complexes are reviewed from the point of view of stereochemistry, coordinate bonding nature, and electronic structure.

2. BRIEF OUTLINE OF COMPUTATIONS
2.1 SEVERAL MO METHODS

MO methods of wide range, from semi-empirical to non-empirical methods, have been applied to theoretical investigation of transition-metal complexes. In this section, several MO methods are briefly described. The simplest method is the extended Hückel (EH) MO method. This method was proposed very early in theoretical investigation of MnO_4^- and CrO_4^{2-} by Wolfsberg and Helmholz (ref. 13), and has been popularized by Hoffmann with an object of applying it to various compounds including organic, inorganic, transition-metal and organo-metallic compounds (ref. 14). The greatest merit of this method is its simplicity; the computational time is very short, and therefore, it can be easily applied to big molecules. Nevertheless, results obtained are reliable qualitatively or semi-quantitatively in many cases. Furthermore, this method is very powerful in analyzing orbital interactions through the fragment orbital formalism (ref. 15), and very useful in investigating coordinate bonding nature, stereochemistry (especially, conformation of ligand), and various reactions of transition-metal complexes. Of course, this method has limitations; for instance, it is weak in examining problems related to spin-multiplicity because of neglect of electron-repulsion integrals, and also rather inadequate for electrostatic interaction. Anderson et al. improved this method (ref. 16), to yield better results on the structure and relative stability in energy (ref. 17). The MO method of Fenske-Hall is also one of the semi-empirical MO methods and has been applied to many transition-metal complexes (ref. 18). This method starts from a hamiltonian including electron-repulsion interaction and leads to a formalism very similar to the charge-consistent EH-MO method, after several approximations. In this regard, this method is considered to offer a theoretical base to the charge-consistent EH-MO method.

CNDO and INDO type MO methods have been proposed for organic molecules by Pople and his collaborators (ref. 19). In these methods, electron repulsion integrals are included in calculations, while many of them are neglected with the zero-differential overlap approximation. Thus, these methods take some parts of the electrostatic interaction into account, and is useful in examining the problem of spin-multiplicity. In this light, these are better than the EH-MO method. Several kinds of parametrizations have been proposed and their application to transition-metal complexes has been reported (ref. 20). However, these methods involve some weak points; for instance, the energetics of reaction is not well reproduced in many cases. MINDO and MNDO can offer reliable results about energetics (ref. 21) but their application is limited to compounds including first-row and several second-row elements, and no parameter has been proposed for transition-metal elements, with only one exception (ref. 22). Theoretical studies of transition-metal complexes using these methods seem to be decreasing at the moment.

Since Slater proposed the $X\alpha$ approximation of the exchange repulsion integral (ref. 23), $X\alpha$-MS (Multiple-Scattering; refs. 23 and 24) and $X\alpha$-DV (Differential Variational; ref. 25) methods have been actively applied to transition metal chemistry. This kind of MO methods require shorter computational times than the ab-initio MO method (vide infra) and gives better results than semi-empirical MO methods. These methods are especially powerful in investigating electronic spectra and ionization potentials of transition-metal complexes. Although these methods had a weak point in investigating energetics and structure, recent developments of $X\alpha$-DV methods can settle this problem and even energy gradient calculations are possible now (ref. 26). As a result, equilibrium geometry and reaction course can easily be investigated with this method.

The ab-initio MO method is based on the Hartree-Fock-Roothaan LCAO-SCF MO method and all the calculations are carried out from basis set functions which represent   atomic orbitals in many cases.   In this method, an atomic orbital is represented by one or several linear combinations of primitive gaussian functions because the use of gaussian functions decreases the computational time of two-electron integrals (ref. 27). Nevertheless, this method requires long computational times and its application to a big molecule is not easy. Results calculated with this method are considered more reliable than those of semi-empirical methods, especially in the cases of geometry, electron distribution, and energetics. However, it is noted that the reliability of this method significantly depends on the choice of basis set. Although the basis set effects on calculated results have been extensively investigated in organic molecules (ref. 28), only a little is known of the basis set effects in transition-metal complexes, unfortunately (ref. 29). Several basis sets of transition metal elements have been reported (ref. 30). Usually, these basis sets do not include p-primitive gaussians representing valence p orbital (the 4p orbital of the first transition series elements and the 5p orbital of the second transition series elements) with only one exception (ref. 30b). Thus, some diffuse p-primitive gaussians must be added to represent valence p orbitals. Furthermore, it has been recommended that a diffuse d-primitive gaussian is also added to the basis set, so as to describe accurately several electron configurations of the atom (refs. 30c and 30f). Not only basis sets of transition-metal elements but also basis sets of ligand atoms are important to obtain reliable results. Usually, a (9s 5p) set proposed by Huzinaga and Dunning (refs. 27 and 31), MINI and MIDI series basis sets proposed by Huzinaga's group (ref. 32), and M-31, M'-21G (M=4 or 6; M'=3), and STO-NG (N=3-6) basis sets proposed by Pople's group (ref. 33) are used for ligand atoms. In detailed investigations including electron correlation effects (vide infra), Wachter's basis set (ref. 30b), Goddard's basis set (ref. 30f), and Tatewaki-Huzinaga's DZC basis set (refs. 30d and 30g) are usually employed for metals, and either Huzinaga-Dunning, or MIDI's set, is used for ligand atoms. In geometry optimization with energy gradient techniques (ref. 34), Roos-Veillard-Vinot's basis set (ref. 30a) and Huzinaga's STD and DZC sets are used for transition metals, and either a 3-21G or a STO-3G set is used for ligand atoms, to reduce the computational times, since computational time of a gradient calculation increases significantly with increasing number of primitive gaussians. It has been found that a combination of Roos-Veillard-Vinot's basis set for a metal and a 4-31G set for a ligand atom gives good results on geometry, as well as a combination of a STD set for a metal and a 3-21G set for a ligand atom.

Although all the core orbitals are included in the above-described basis sets, these core orbitals are considered not to play an important role in chemical reactions, bond formation, and so on. Recently, attempts have been made to replace these core orbitals with some repulsive one-electron potentials. These potentials are called effective core potential (ECP; ref. 35). The ECP has several merits in MO calculations: The first is to incorporate effectively relativistic effects which are important in heavy elements such as 5d transition-metal elements. It is not easy to consider, directly, relativistic effects in the usual ab-initio MO calculations, but some parts of relativistic effects can be effectively incorporated in the ECP. The second is to reduce computational time in geometry optimization with energy gradient techniques: since gradient calculation of two-electron integrals is much more difficult than that of one-electron integrals, the use of ECP significantly reduces computational time of gradient calculation. The third is that the use of ECP decreases computational time in configuration interaction calculations (vide infra), because the large number of basis set functions requires very long computational time in integral transformation from an atomic orbital basis to a molecular orbital basis.

Although the ab-initio MO method is much more reliable than semi-empirical MO methods, the former includes an intrinsic flaw: even if the quality of the basis set is markedly improved and the basis set used approaches completeness, results calculated with this method are worse than the exact solution of the non-relativistic Schrödinger eq., and the total energy calculated is always higher than that obtained from a non-relativistic Schrödinger eq. This difference in energy is called correlation energy. The above-mentioned flaw results from the neglect of correlation between motion of an electron and the motions of other electrons. In other words, the Hartree-Fock method is based on the assumption that an electron in a molecule senses an average potential from nuclei and other electrons. This means that an electron moves on an orbital which is independent of the other electrons' motion. This difficulty can be improved upon by using not a single Slater determinant (or limited number of Slater determinants to represent spin eigen-function in the case of open-shell molecule) but a linear combination of Slater determinants. The most popular way of implementing this improvement is the configuration interaction (CI) method (ref. 36). In this method, a wave function is represented by a linear combination of many Slater determinants whose expansion coefficients are variationally determined. The MC SCF (Multi-Configurational SCF) method is also a popular and powerful way to calculate correlation energy (ref. 37), in which not only coefficients of Slater determinants but also LCAO coefficients are variationally determined at the same time. The CAS SCF (Complete Active Space SCF) method (ref. 38) is one kind of MC SCF methods in which all the electron configurations arising from chemically important orbitals are included in the MC SCF calculation. These two methods are specially powerful in calculating non-dynamical correlations which result from near-degeneracy of electron configurations (near-degeneracy means that several electron configurations lie close in energy). Thus, one of most attractive and efficient ways of introducing electron correlation effects is MC (or CAS) SCF Multi-Reference CI method (ref. 39), in which non-dynamical correlations are incorporated mainly by MC or CAS SCF methods and the dynamical contribution is incorporated by conventional CI methods. Recently, Moller-Plesset's perturbation theory (ref. 40) is often used in organic and transition metal compounds, since this method is incorporated in Gaussian 80, 82 and 86 programs by Pople's group (ref. 41). This method is easy to use and less time consuming than conventional CI and MC SCF methods. Furthermore, this method satisfies the size-consistency, whereas the single-double excitation CI method suffers from this problem and results must be corrected with Davidson's method (ref. 42).

2.2 ENERGY DECOMPOSITION ANALYSIS (EDA)

It is not easy, in general, to analyze results of ab-initio MO calculations in which diffuse basis sets are used. The energy decomposition analysis seems effective in analyzing such results of ab-initio MO calculations because this analysis offers a variety of information on chemical bonds, structure, and reactivity. In this section, the method of energy decomposition analysis is briefly outlined.

When two chemical species, A and B, approach each other, various types of interactions arise, and subsequently a new bond, A-B, is formed (with or

$$A \; + \; B \; \rightarrow \; A-B \tag{1a}$$
$$ML_n \; + \; L_0 \; \rightarrow \; L_n ML_0 \tag{1b}$$

without bond breaking). This type of phenomenon has been theoretically investigated in detail in order to elucidate chemical reactivity. Several methods have been proposed to analyze the interaction between two species and to clarify what type of interaction is important in the chemical reaction (refs. 43-45). Formation of a coordinate bond can be viewed quite the same as the chemical reaction, as shown in eq.'s (1a) and (1b). Thus, the methods used in analysis of chemical reactivity can be applied to investigations of the coordinate bond.

Let us briefly examine what type of interactions arise between two reacting species. First of all, an electrostatic (classical Coulombic) interaction arises between two species A and B (see ES in Figure 1). This interaction is important at rather long separation between two species, in comparison with the other interactions. Then, exchange interaction arises when occupied orbitals of two species start to overlap (see EX in Figure 1). This interaction results from Pauli's exclusion principle; at infinite separation,

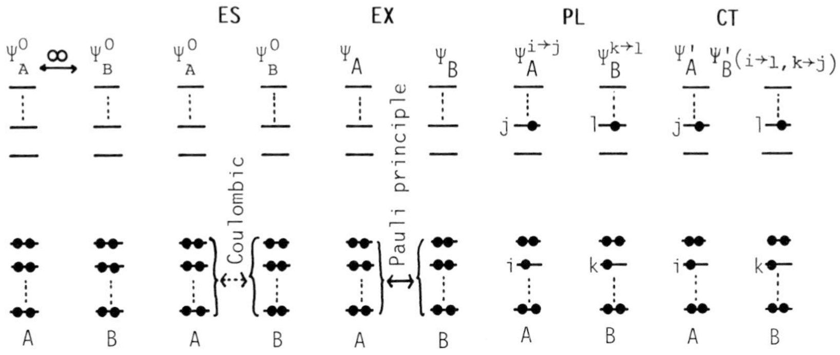

Fig. 1. Schematic pictures of various intermolecular interactions between molecules A and B
a) Transfer of many electrons, of course, take place in PL and CT. Here, only one electron transfer is pictured, for brevity.

electrons of A (or B) satisfy Pauli's exclusion principle in the A (or B) molecule. When two species start to interact with one another, electrons of A (or B) must satisfy the Pauli's exclusion principle in the super-molecule A-B, which requires a deformation of occupied orbitals on A and B. As a result, destabilization in energy arises from this term. At the same time, electrons on A (or B) perceive various potentials from the B (or A) molecule, and therefore, mixing occurs between occupied and virtual orbitals in A (or B), to stabilize the A (or B) molecule in the new situation (i.e., in the super-molecule). This type of interaction is called polarization (see PL in Figure 1). Electrons on A (or B) delocalize to B (or A), in which electrons transfer from occupied orbitals of A (or B) to vacant orbitals of B (or A) (see CT in Figure 1). Mutual charge-transfer interaction seems to correspond to covalent interaction.

The total energy of super-molecule can, thus, be represented as follows;

$$E_{AB} = E_0^A + E_0^B + E_{ES} + E_{EX} + E_{PL} + E_{CT} + E_{mix} \qquad (2a)$$

$$E_{PL} = E_{PL(A)} + E_{PL(B)} + E_{PL(mix)} \qquad (2b)$$

$$E_{CT} = E_{CT(A \to B)} + E_{CT(B \to A)} + E_{CT(mix)} \qquad (2c)$$

where E_0^A and E_0^B are the total energies of isolated molecules A and B, respectively, E_{ES}, E_{EX}, E_{PL}, and E_{CT} are electrostatic, exchange repulsion, polarization, and charge-transfer interaction energies, respectively. E_{mix} is the higher order term arising from couplings of exchange, polarization, and charge-transfer interactions. $E_{PL(A)}$ means the polarization of molecule A which is caused by influences from molecule B, and $E_{PL(mix)}$ is the higher order polarization term arising from coupling between the polarization in the A molecule and that in the B molecule. $E_{CT(A \to B)}$ indicates the charge-transfer interaction from occupied orbitals of A to vacant orbitals of B, and $E_{CT(mix)}$ is the higher order charge-transfer term arising from coupling between the charge transfer from A to B and that from B to A. These terms can be calculated with either the second-order perturbation theory (ref. 43) or diagonalization of model Hartree-Fock matrix elements in which only a part of MO space is included (ref. 45).

A very similar energy decomposition analysis has been proposed by Pullman et al. (ref. 46), Bagus et al. (ref. 47), and Stone and Erskine (ref. 48) in MO calculations, and also by Ziegler (ref. 49) and Baerend et al. (ref. 50) in the $X\alpha$-DV calculations.

Here, the method proposed by Morokuma and his collaborators (ref. 45) is described in slightly more detail. In this method, a total complex, $L_n ML^0$, is considered to consist of two fragments, a metal part ML_n and a ligand L^0, where we have an interest in the coordination of L^0. The binding energy (BE) is defined as a stabilization energy of $L_n ML^0$, relative to ML_n and L^0 which take their equilibrium structures:

$$BE = E_t(L_n ML^0) - E_t(ML_n)_{eq} - E_t(L^0)_{eq} = INT + DEF \qquad (3)$$

$$DEF = E_t(L^0)_{dist} - E_t(L^0)_{eq} + E_t(ML_n)_{dist} - E_t(ML_n)_{eq} \qquad (4)$$

$$INT = E_t(L_n ML^0) - E_t(ML_n)_{dist} - E_t(L^0)_{dist} \qquad (5)$$

The deformation energy (DEF) is defined as a destabilization energy required to distort L^0 and ML_n from their equilibrium structures to distorted ones present in the the total complex, as given in eq. 4, where subscripts "eq" and "dist" mean the equilibrium and distorted structures, respectively. Then, the interaction energy (INT) is defined as a stabilization energy of the total complex, $L_n ML^0$, with respect to ML_n and L^0, taking the same distorted structures as in the total complex. In general, INT is further divided into several chemically meaningful terms such as electrostatic (ES), exchange repulsion (EX), polarization (PL), and charge-transfer (CT) interactions, as shown in eq. 2a. In the case of transition-metal complexes, however, the coordinate bond is stronger than the weak intermolecular interactions such as van der Waals interaction. When an intermolecular interaction becomes strong, the

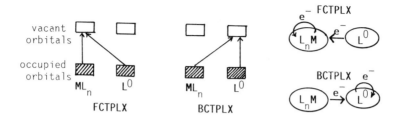

Fig. 2. Schematic pictures of FCTPLX and BCTPLX appearing in eq. (6).

coupling between the charge-transfer from A to B and the polarization of B is enhanced (ref. 51). Thus, the coupling is not divided into the charge-transfer and polarization in their new EDA scheme which has been applied to transition metal complexes (ref. 45c). As a result, INT is divided into the following terms, shown in eq. 6.

$$INT = ES + EX + FCTPLX + BCTPLX + R \qquad (6)$$

where FCTPLX includes a forward charge-transfer from L^0 to ML_n, a polarization of ML_n, and their coupling terms, and BCTPLX includes a back charge-transfer from ML_n to L^0, a polarization of L^0, and their coupling terms, as shown in the schematic picture in Figure 2. R represents the remaining higher order term.

An energy decomposition analysis of interactions between $Ni(PH_3)_2$ and C_2H_4 is given as an example in Table 1 and Figure 3 (note that negative values mean stabilization, from the definitions given in eq.'s. 3 - 5). The electrostatic stabilization (ES) is large, but at the same time, the exchange repulsion (EX) is also large. As a result, the static energy, which corresponds to a sum of ES and EX, is repulsive (positive value). The net stabilization of the C_2H_4 coordination results primarily from the BCTPLX stabilization, and it is noted that the BCTPLX stabilization is significantly larger than the FCTPLX stabilization. Mulliken populations are also analyzed according to the EDA scheme (see Table 1). The Ni atomic population is significantly decreased upon C_2H_4 coordination through the BCTPLX term, which mainly comes from the decrease in the Ni $3d\pi$ orbital population. Simultaneously, electron population of the C_2H_4 ligand increases significantly. These changes in electron density caused by the C_2H_4 coordination are displayed as difference density maps in Figure 3. Through the EX term, electron density decreases in the intermediate region between Ni and C_2H_4, which seems to correspond to the general picture of

TABLE 1

Component analysis of interaction energies between $Ni(PH_3)_2$ and C_2H_4 (kcal/mol unit) and Changes in Mulliken population[a]

	BE	DEF	INT	ES	EX	FCTPLX	BCTPLX	R
Energy Contribution[b]								
	-30	15	-45	-132	168	-16	-54	-11
Component analysis of Mulliken populations[c]								
Ni						-0.01	0.05	-0.27	-0.04
s						0	0	0	0.03
p						0.01	0.09	0	0.05
d						-0.01	-0.04	-0.27	-0.13
$3d\pi$						0	-0.03	-0.26	-0.29
C_2H_4						0	-0.06	0.34	0.21

a) Ref. 45c. b) Negative values mean stabilization in energy and positive values destabilization. c) Changes in Mulliken populations upon C_2H_4 coordination. Positive values mean an increase in population and negative values a decrease.

exchange repulsion. The FCTPLX term decreases electron density in the C_2H_4 region but slightly increases electron density in the $Ni(PH_3)_2$ region, which is in accordance with the general feature of donative interaction. The BCTPLX term, on the other hand, increases electron density in the C_2H_4 region but decreases electron density in the region of Ni $d\pi$ orbital. Thus, the BCTPLX term clearly corresponds to the π-back donative interaction. These results encouraged us to apply this EDA method to theoretical investigations of the coordinate bond.

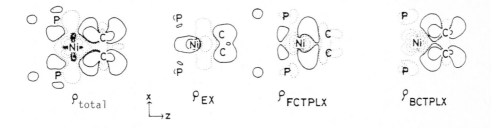

ρ_{total} ρ_{EX} ρ_{FCTPLX} ρ_{BCTPLX}

Fig. 3. Difference density maps of $Ni(PH_3)_2(C_2H_4)$. Solid lines show an increase in density and dotted lines a decrease upon C_2H_4 coordination. The contours are ±0.05, ±0.01, and ±0.001 e/a_0^3. (Reproduced with permission from ref. 45c. Copyright 1981 American Chemical Society).

3. GENERAL SURVEY OF COORDINATE BOND NATURE

Since the Dewar-Chatt-Duncanson model was proposed (ref. 52), the strength and nature of the coordinate bond, electron distribution, and reactivity of transition-metal complexes have very often been discussed in terms of σ-donation and π-back donation; in the former, a ligand donates its electrons to vacant orbitals of a central metal, and in the latter a central metal donates its d_π electrons to vacant orbitals of a ligand. It is not easy to extract information about the relative strengths of σ-donation and π-back donation from experimental results, in asmuchas many experimental results depend on both effects; for example, net charge of the central metal is determined as a sum of σ-donation and π-back donation. MO studies of transition-metal complexes are expected to offer a variety of information on coordinate bonds, including relative strength of σ-donation and π-back donation.

In this section, we will summarize MO studies of transition-metal carbonyl complexes as an useful example, since many theoretical analyses have been reported on these complexes. It is our intention to shed some light on how we can extract information on the nature of the coordinate bond and relative stabilities of σ-donation and π-back donation, from MO calculations.

3.1 VARIOUS MO STUDIES of METAL-CARBONYL COMPLEXES

A transition metal-monocarbonyl complex, M(CO), can be viewed as a good prototype of non-Werner transition-metal complexes and chemisorption on a metal surface. Specifically, Ni(CO) has been an attractive object of theoretical research, probably because this complex presents interesting problems in its electronic structure and, at the same time, belongs to the platinum group which is very common in transition-metal chemistry. As a result, various types of MO calculations have been reported on this complex. A comparison of them offers us a variety of interesting information about computational methods.

Before starting to discuss coordinate bond nature, let us briefly examine those computational results. The electronic state of this complex was previously believed to be $^3\Delta$ (refs. 53-56). Recently, however, several theoretical analyses have been carried out on Ni(CO) using ab-initio MO methods including electron correlation effect (refs. 57-59) and Xα-DV methods (ref. 60). In these studies, the ground state is suggested to be a singlet $^1\Sigma^+$, arising from the ^1S state of Ni (d^{10}). As shown in Table 2, the energy difference between them is not large. This result requires a careful comparison of two states, and, therefore, potential energy curves of these two states have been investigated with Hartree-Fock (HF) and CI methods, as shown in Figure 4 (ref. 57). Several important features are found in this figure. On the HF level, the $^1\Sigma^+$ state clearly exhibits an energy minimum in the potential curve but is less stable than the $^3\Delta$. On the other hand, the potential curve of $^3\Delta$ is more stable than that of $^1\Sigma^+$, but seems repulsive in the range of a coordinated bond distance. Introduction of electron correlation effects results in significant

TABLE 2

Various MO studies of Ni(CO) and Ni(CO)$_4$ and related complexes

Complex	$^3\Delta - {}^1\Sigma^+$ (eV)	R(M–C) (Å)	R(C–O) (Å)	BE[a] (kcal/mol)	Method	Reference
Ni(CO)		1.75		35	SCF	Rives and Fenske
	0.6 (1.2)	1.70		62	CI	(ref. 57)
Ni(CO)		1.65	1.15	71	Xα–DV	Dunlap et al.(ref. 60)
Ni(CO)		1.75		−58.7	SCF	Blomberg et al.(ref. 58)
		1.71		21.1	CAS–SCF	
	0.73(1.0)	1.71		29.9	CAS–SCF CI	
Ni(CO)	−2.12	1.844	1.134	30.0	SCF	Rohlfing and Hay
	2.38	1.711	1.208	93.3	MP2	(ref. 59)
Ni(CO)		1.726	1.145		SCF	Carsky and Dedieu
		1.670	1.215		MP2	(ref. 64)
expt.				68.4±15.0		Stevens et al.(ref. 61a)
Pd(CO)	1.64	2.056	1.130	13.0	SCF	Rohlfing and Hay
	4.98	1.882	1.185	37.3	MP2	(ref. 59)
Pt(CO)	2.22	2.073	1.129	19.8	SCF	
	5.82	1.977	1.184	44.5	MP2	
Ni(CO)$_4$		1.884	1.139		SCF	Schaefer et al.(ref. 62)
Ni(CO)$_4$		1.84	1.15		Xα–DV	Dunlap et al. (ref.60)
Ni(CO)$_4$		1.805	1.137		Xα–DV	Jorg and Rosch (ref. 63)
Ni(CO)$_4$		1.814	1.135		SCF	Carsky and Dedieu(ref. 64)
Ni(CO)$_4$		1.971	1.129	17.9	SCF	Rohlfing and Hay (ref.59)
		1.873	1.181	55.5	MP2	
expt.		1.82	1.12	40		refs.61b–d for distance
		∿ 1.84	∿1.15	∿45		ref. 61e for BE
Pd(CO)$_4$		2.169	1.128	9.5	SCF	Rohlfing and Hay(ref. 59)
		2.032	(1.178)	28.7	MP2	
Pt(CO)$_4$		2.202	1.128	12.8	SCF	Rohlfing and Hay(ref. 59)
		2.100	1.178	33.5	MP2	

a) Binding energy; BE = E_t(Ni) + E_t(CO) − E_t(Ni–CO).
b) in parentheses: an error included in atom calculation is corrected.

108

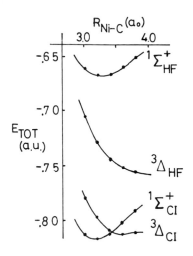

Fig. 4. Potential curves of Ni(CO)
Energies are relative to -151.0 a.u.
(Reproduced with permission from
ref. 57. Copyright 1981 American
Institute of Physics).

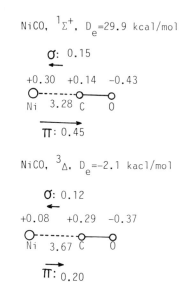

Fig. 5. Mulliken charges of Ni(CO)
(Reproduced with permission from
ref. 58. Copyright 1985 American
Chemical Society).

changes on the potential curves; (1) both potential curves are significantly stabilized in energy by CI calculations, (2) the $^1\Sigma^+$ gives a minimum slightly more stable than that of $^3\Delta$, (3) the potential curve of $^3\Delta$ is very shallow and carbonyl easily dissociates from Ni in this state, (4) on the other hand, the potential curve of $^1\Sigma^+$ is deep and the coordination of carbonyl seems stable. All these results imply that the ground state of Ni(CO) is the $^1\Sigma^+$. Furthermore, recent experimental results also suggest that the ground state is $^1\Sigma^+$ (ref. 61). Thus, it is reasonably concluded that the $^1\Sigma^+$ is the ground state and introduction of electron correlation effects is necessary for obtaining a correct ordering of these two states.

Computational results obtained for the $^1\Sigma^+$ state of Ni(CO) are summarized in Table 2, together with MO studies of $M(CO)_n$ (M = Ni, Pd, or Pt; n = 1 or 4). These are expected to offer a range of information about reliabilities of various methods, which is useful in understanding discussions of MO studies. The first to be noted is, again, electron correlation effect on the relative stabilities of $^1\Sigma^+$ and $^3\Delta$ states. Also in the ab-initio MO/MP2 study of Ni(CO) by Hay et al., the $^3\Delta$ is more stable than the $^1\Sigma^+$ state at Hartree-Fock level. However, introduction of electron correlation effects reverses the relative stabilities of these two states. In other words, electron correlation effects are indispensable in comparing two states which are close in energy. It is also noted that in Pd(CO), and Pt(CO), the $^1\Sigma^+$ is the ground state even at the Hartree-Fock level and, introduction of electron correlation effects stabilizes the $^1\Sigma^+$ more than the $^3\Delta$. The former result corresponds to the general trend that the Pd(0) and Pt(0) have a stable d^{10} electron configuration compared to the Ni(0) complexes. The latter result is also a general trend of electron correlation effects on the relative stabilities of singlet and triplet states. The second is electron correlation effects on the coordinate bond distance. The usual ab-initio MO method tends to overestimate the coordinate bond distance, but inclusion of electron correlation effects shortens the coordinate bond distance. It is noted this electron correlation effect is the reverse of the usual one found in organic compounds. The third is the binding energy (BE in Table 2) of carbonyl with metal. Again, the usual ab-initio MO method tends to underestimate the binding energy, but inclusion of electron correlation effects increases it. The fourth is that the MP2 method tends to overestimate the binding energy (see MP2 calculations of Ni(CO) and Ni(CO)$_4$).

Now, let us start to discuss the coordinate bond nature of the carbonyl ligand. We can obtain some information about σ-donation and π-back donation by examining Mulliken populations (of course, we must take notice that Mulliken

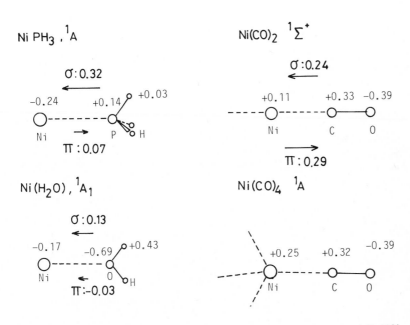

Fig. 6. Mulliken charges of Ni(PH$_3$), Ni(H$_2$O), Ni(CO)$_2$, and Ni(CO)$_4$. (Reproduced with permission from ref. 58. Copyright 1985 American Chemical Society).

110

populations only give qualitative information on electron distributions and, at the same time, that Mulliken populations have some ambiguity, probably owing to counter intuitive orbital mixing (ref. 65)). As shown in Figure 5, Mulliken populations of Ni(CO) are separated into σ-space and π-space. Apparently, the $^1\Sigma^+$ ground state exhibits stronger σ-donation and π-back donation than the $^3\Delta$ state does (ref. 58). The same result has been also suggested by Rives and Fenske (ref. 57). This is probably one of reasons that the potential curve for the Ni-CO distance is very shallow in the $^3\Delta$ state but rather deep in the $^1\Sigma^+$ state. The ground state of Ni(CO)$_4$ is of course the ^1A state arising from the ^1S (d^{10}) state of Ni which corresponds to the $^1\Sigma^+$ state of Ni(CO). Therefore, the coordinate bond of the $^1\Sigma^+$ state of Ni(CO) is investigated here in more detail. The charge-transfer in the σ-space, which arises from the σ-donation, is 0.15 e, as shown in Figure 5. The π-back donation causes charge-transfer of 0.45 e, in the σ-space. As a result, the Ni atom is positively charged but the carbonyl ligand is negatively charged. This electron distribution strongly suggests that the π-back donation is much stronger than the σ-donation. A similar analysis of electron distribution has been carried out on Ni(PH$_3$) and

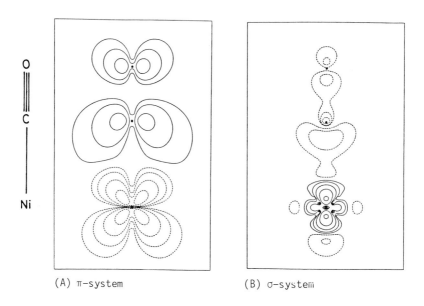

O
‖
C
|
Ni

(A) π-system   (B) σ-system

Fig. 7. Electron density maps comparing CI and HF calculations for the $^1\Sigma^+$ state of Ni(CO). Contour values are $\pm 0.001 \times 3^n$, n= 0,1,2,3,..... The Ni atoms are the atoms closest to the bottom of the figures. (Reproduced with permission from ref. 57. Copyright 1981 American Institute of Physics).

Ni(H_2O) complexes, as shown in Figure 6, in which the electronic states of these complexes have been taken to be either $^1A'$ for the former or $^1A_1$ for the latter, because these states arise from the ^1S (d^{10}) state of Ni. In the Ni(PH_3) complex, the σ-donation from PH_3 to Ni is much larger than that of Ni(CO), but the π-back donation from Ni to PH_3 is very weak. These interactions make the Ni atom negatively charged but the PH_3 ligand positively charged. In the Ni(H_2O) complex, the σ-donation from H_2O to Ni is much weaker than that of Ni(PH_3) and to a similar extent to that of Ni(CO). A critical difference of Ni(H_2O) from Ni(CO), and Ni(PH_3), is found in the charge-transfer interaction of the π-space; the π-back donation does not occur at all in Ni(H_2O), probably due to the absence of an acceptor orbital in this ligand. From these results, it is reasonably concluded that carbonyl is a good acceptor ligand but PH_3 is a good donor ligand, even in the Ni(0) complex.

The electron correlation effects on the bonding nature must be also examined here. Figure 7 shows differences in electron density between inclusion and exclusion of electron correlation effects (ref. 57). The solid lines indicate the increase in electron density, and the dashed lines the decrease upon including electron correlation effects. In the π-space that includes π-back donation, electron-correlation increases electron density in the region of carbonyl; but, in the σ-space that includes σ-donation, electron correlation increases electron density in the metal part. These changes in electron density suggest that both σ-donation and π-back donation strengthen upon introducing electron correlation effects. A similar enhancement of the π-back bonding interaction, by introducing electron correlation effects, has been reported by Hay et al, in their MP2 studies of various carbonyl complexes of platinum group (ref. 59).

Going to Ni$(CO)_2$ from Ni(CO), the charge transfer by σ-donation is enhanced to 0.24 e, but the charge-transfer by π-back donation is suppressed to 0.29 e (see Figure 6; ref. 58). This result seems reasonable, considering that CO is a π-acceptor ligand. In Ni$(CO)_2$, as a result, the σ-donation transfers electrons from CO to Ni, to almost the same degree that the π-back donation transfers electrons from Ni to CO. In Ni$(CO)_4$, a similar analysis has not been reported. However, electron populations of Ni and CO are very similar to those found in Ni$(CO)_2$. This electron distribution suggests that the relative contributions of σ-donation and π-back donation are similar in Ni$(CO)_2$ and Ni$(CO)_4$.

The coordinate bond is very often compared in the same group of the periodic table. In the platinum group, the coordinate bond of CO has been calculated to become stronger in the order Pd < Pt < Ni (ref. 59). In these calculations, relativistic effective core potentials, which include some part of relativistic effects, are used for core orbitals of Pt and Pd. The same trend is found in both M(CO) and M(CO)$_4$, and is independent of inclusion or exclusion of electron correlation effects. We are interested not only in the order of coordination strengths but also in the dependence of σ–donation and π–back donation on the kind of metal. This relative importance has been discussed on the basis of calculated C–O distance and ionization potential of metal (^{1}S) by Rohlfing and Hay (ref. 59), as follows. In Ni–carbonyl, the π–back donation is very important. In Pt– and Pd–carbonyls, however, the π–back donation is less important than in Ni–carbonyls, and σ–donation is still not dominant, but may be the factor that determines the relative stabilities of metal–carbonyls in the second– and third–row transition elements. The CO coordination of M(CO)$_6$ (M=Cr, Mo, W), M(CO)$_5$ (M=Fe, Ru, Os), and M(CO)$_4$ (M=Ni, Pd, Pt) has been investigated with Xα–DV method by Ziegler (ref. 66). When the relativistic effects are entered in the calculations, the same increasing order of coordination strength, the 4d metal < the 5d metal < the 3d metal, has been obtained. However, if the relativistic effects are not included in the calculations, the order is changed to a different one, the 5d metal < the 4d metal < the 3d metal. Thus, inclusion of relativistic effects in MO calculations is also important in order to obtain correct information on the coordinate bond as well as electron correlation effects.

3.2 ENERGY DECOMPOSITION ANALYSIS (EDA) between METAL and CARBONYL LIGAND

The energy decomposition analysis offers also a powerful means of investigating coordinate bonding nature. Several Ni(0) complexes have been examined with ab-initio MO and EDA methods (refs. 45c and 67). As compiled in Table 3 (note that positive values mean destabilization in energy and negative values mean stabilization), the ES term yields significantly large stabilization but the EX term is much more repulsive than the stabilization by ES term in all the

TABLE 3

Energy decomposition analysis of interaction between Ni(PH$_3$)$_2$ and several ligands[a] (kcal/mol)

L	BE	DEF	INT	ES	EX	BCTPLX	FCTPLX	R
CO[b]	−23	3	−26	−101	136	−43	−19	
H$_2$CO	−43	20	−62	−102	157	−74	−17	−26
C$_2$H$_2$	−37	40	−77	−148	189	−75	−24	−19

a) Ref. 67. b) EDA was carried out for interaction between Ni(PH$_3$)$_2$ and two CO's in Ni(PH$_3$)$_2$(CO)$_2$.

complexes examined. The static energy, corresponding to the sum of ES and EX, is therefore repulsive (positive value). The BCTPLX term, which includes π-back donation interaction as a main contributor, stabilizes the coordinate bond much more than the FCTPLX term which includes σ-donation as a main contributor. These results clearly indicate that the π-back donation is very important in Ni(0) complexes.

A similar EDA approach with $X\alpha$-DV method has been reported on $M(CO)_4$ (M = Ni, Pd, Pt), $M(CO)_5$ (M = Fe, Ru, Os), and $M(CO)_6$ (M = Cr, Mo, W) by Ziegler et al. (ref. 66), as shown in Table 4. In this report, energy decomposition analysis is based on symmetry considerations, because the π-back donation arises from the orbitals of $\underline{e}$ representation and the σ-donation comes from those of a_1 representation, in both T_d and O_h symmetries, as shown in Figure 8. Thus, the stabilization energy in the a_1 representation is produced by the σ-donation interaction and the stabilization energy in the $\underline{e}$ representation is the result of the π-back donation. In $M(CO)_4$ and $M(CO)_5$, the σ-donation is stronger than the π-back donation with only one exception, $Cr(CO)_6$, which has a stronger π-back donation than the σ-donation. However, the π-back donation is of much importance for determining the relative strength of the

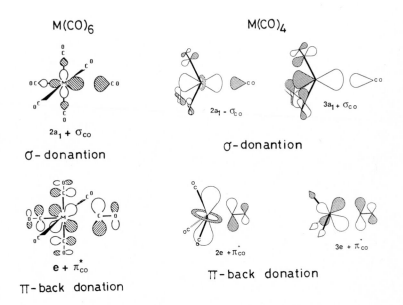

Fig. 8. Schematical pictures of orbital interactions concerning σ-donation and π-back donation in $M(CO)_6$ and $M(CO)_4$. (Reproduced with permission from ref. 66. Copyright 1987 American Chemical Society).

TABLE 4

Decomposition of the first ligand dissociation energy, ΔH, in $M(CO)_6$ (M = Cr, Mo, W) and $M(CO)_4$ (M = Ni, Pd, Pt)[a] (kJ/mol)

	static ΔE_0[b]	promotion ΔE_{prep}[b]	σ-donation $\Delta E(a_1)$	π-back donation $\Delta E(e)$	ΔE_R	ΔH[c]	exptl.		
$Cr(CO)_6$		193	-163	-177		147	162[d]	155[e]	154[f]
$Mo(CO)_6$		197	-165	-148	-3	119	126	142	169
$W(CO)_6$		197	-160	-154	-25	142	166	159	192
$Ni(CO)_4$	9.2	169.3	-145.3	-138.9		105.7	104[g]		
$Pd(CO)_4$	11.3	225.1	-143.7	-113.4	-7.1	27.3			
$Pt(CO)_4$	10.6	261.1	-145.6	-112.6	-51.5	38.0			

a) Ref. 66. b) ΔE_{prep}=the promotion energy, ΔE_0 = static energy.
c) $\Delta H = -[\Delta E_{prep} + \Delta E_0 + \Delta E(a_1) + \Delta E(e) + \Delta E_R]$. ΔH corresponds to binding energy. d) Ref. 68. e) Ref. 69. f) Ref. 70. g) Ref. 61e

coordination bond, while the strength of σ-donation is rather independent of the central metal. The increasing order of π-back donation is quite the same as the increasing order of occupied d-orbital energy, as expected (ref. 66).

A comparison of CO with isoelectronic CS, CSe, and CTe ligands is interesting. Ziegler carried out an Xα-DV and EDA study of $Ru(CO)_4L$ (L=CO, CS, CSe, or CTe) (ref. 71). In the CO, CS, CSe complexes, the σ-donation is stronger than the π-back donation, while both are almost the same in the CTe complex. Both σ-donation and π-back donation become strong, going to CTe from CO, leading to the increasing order of coordination strength CO < CS < CSe < CTe.

A different type of EDA study has been reported on $Ni(CO)_4$ and $Fe(CO)_5$ by Bauschlicher and Bagus (ref. 47b) (see Table 5; note positive values mean stabilization in energy). In $Ni(CO)_4$, the back donation is significantly stronger (about 30 times) than the donation. This result seems different from the result reported by Ziegler (vide supra). One of the reasons for this difference is probably the different EDA scheme; in the method of Bauschlicher, interaction between Ni and 4 CO's is analyzed, while in the method of Ziegler the interaction between $Ni(CO)_3$ and CO is analyzed. In $Fe(CO)_5$, the back donation is still stronger than the donation, but the relative importance of the back donation is smaller than in $Ni(CO)_4$. Because Fe(0) has a d^8 electron configuration with an empty d-orbital, the contribution from the donation is larger than that in $Ni(CO)_4$.

Although the electron correlation effects have not been directly included

TABLE 5

Energy decomposition analysis between metal and CO (eV unit)[a]

Complex	Ni(CO)$_4$	Fe(CO)$_5$	Ni(CO)		
Method	SCF	SCF	CAS-SCF		
FO(static)	-4.65	-6.08	-4.18		
V(Ni:Ni) (polarization)	3.48	3.69	1.74 (dσ-4s)	0.70 (others)	
V(Ni:full)(Ni→CO)	3.04	4.59	0.09 (σ)	1.82 (π)	
V(CO:CO) (polarization)	1.00	1.81	0.32		
V(CO:full)(CO→Ni)	0.49	2.50	0.07 (σ)	0.02 (π)	
E_{sup}(Ni)[b]	0.04	0.07	σ-covalent	0.30	
E_{sup}((CO)$_4$)[b]	0.30	0.39	others	0.21	

a) eV=23.045 kcal/mol. Ref. 47b and 72. b) Basis set superposition error.

in the above EDA investigations, Bauschlicher carried out an EDA study of
Ni(CO) on the CAS SCF level (ref. 72). As clearly shown in Table 5, the back
donation significantly contributes to the coordinate bond of carbonyl, but the
donation hardly contributes to it.   In this approach, the strength of covalent
interaction between Ni and CO is also estimated and its contribution in energy
is about 3 times larger than the donative interaction.   The sum of σ-donation
and   σ-covalent interactions is, however, still much smaller than the π-back
donation.   Compared with the discussion based on the population analysis, this
EDA study seems to emphasize the large contribution of π-back donation.   Of
course, electron flow is not directly related to the energy contribution.   So,
an important thing is to clarify the standpoint used in examining the coordinate bond; if electron distribution is investigated, a discussion must be based
on population analysis and/or density map. If the stabilization energy of the
coordination is studied, the discussion based on the EDA method offers better
results.

    At the end of this section, we had better summarize the above discussion:
(1) The relative importance of σ-donation and π-back donation can be successfully investigated with population analysis and EDA methods.
(2) Both analyses give qualitatively the same, but quantitatively different
results.
(3) Introduction of electron correlation effects strengthens both σ-donation
and π-back donation of Ni(CO), leading to shortening of M-CO distance and
lengthening C-O distance.
(4) The EDA results at the SCF level are not changed by introducing electron
correlation effects, in a qualitative sense.

(5) The π-back donation becomes strong in the order, the 4d-metal < 5d-metal < 3d-metal, in M(CO) and $M(CO)_4$ (M=Ni, Pd, Pt). Coordinate bond strengths of carbonyl in $M(CO)_6$ (M=Cr, Mo, W) and $M(CO)_5$ (M=Fe, Ru, Os) also becomes strong in the order, 4d-metal < 5d-metal < 3d-metal. The relativistic effects are important to obtain this correct order.

4. TRANSITION-METAL CO_2 COMPLEXES

4.1 EXPERIMENTAL RESULTS REGARDING STEREOCHEMISTRY of CO_2 COMPLEXES

Carbon dioxide is a very stable molecule and a thermodynamic end product of many reactions. It is not easy to use carbon dioxide as carbon source and to synthesize useful organic compounds from carbon dioxide. However, some biological systems can easily convert carbon dioxide into organic substances through light irradiation and formation of coordination complexes. This fact encourages us and suggests that coordination of carbon dioxide with metal complexes is one of the powerful and universal ways of activating carbon dioxide. In this regard, it is important and worthwhile to synthesize transition-metal CO_2 complexes and investigate their structure, coordinate bond nature, electronic structure, reactivity, and so on. Unfortunately, however, synthesis of transition-metal CO_2 complexes is rather difficult. For instance, only a limited number of transition-metal CO_2 complexes have been isolated, as shown in the following table

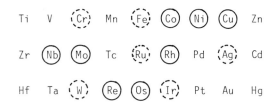

in which solid circles mean that the CO_2 complexes have been isolated but dashed circles mean that the presence of CO_2 complexes have been postulated in the reaction with CO_2. Molecular orbital studies of transition-metal CO_2 complexes are, therefore, expected to offer valuable informations about such labile and/or unstable transition-metal CO_2 complexes.

As described in the Introduction, one of the characteristic features of transition-metal CO_2 complexes is the presence of three possible coordination modes; η^2-side on, η^1-C coordination, and η^1-O end on coordination modes (see 1, 2, and 3 in the Introduction). The η^2-side on mode has been reported in $Ni(PCy_3)_2(CO_2)$ (Fig. 9; ref. 73), $Nb(\eta-C_5H_4Me)_2(CH_2SiMe_3)(CO_2)$ (Fig. 9; ref. 74), and $Mo(PR_3)_3(CN-R)(CO_2)_2$ (R = C_6H_5, Pr^i) (Fig. 10; ref. 75), and the η^1-C mode has been known in $K[Co(R-salen)(CO_2)](THF)$ (Fig. 11; ref. 76) and RhCl-

(diars)$_2$(CO$_2$) (diars = o-phenylene-bis(dimethylarsine)) (Fig. 12; ref. 77). In
the Ni(0)-CO$_2$ complex, the CO$_2$ ligand lies on the same plane of Ni(PR$_3$)$_2$, as in
many Ni(PR$_3$)$_2$(olefin) species. In the Mo(0)-CO$_2$ complex, two CO$_2$ ligands

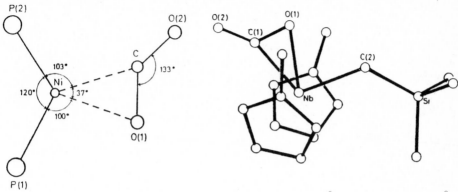

Ni–C=1.84 Å; Ni–O(1)=1.99 Å;
C–O(1)=1.22 Å; C–O(2)=1.17 Å.
(Reproduced with permission
from ref. 73. Copyright 1975
Royal Society of Chemistry).

Nb–C(1)=2.144(7) Å; Nb–O(1)=2.173(4) Å;
C(1)–O(1)=1.283(8) Å; C–O(1)=1.283(8)
Å; C(1)–O(2)=1.216(8) Å; O(1)–C(1)–O(2)=
132.4(7)°. (Reproduced with permission
from ref. 74. Copyright 1981 Royal
Society of Chemistry).

Fig. 9. Structures of [Ni(PCy$_3$)$_2$(CO$_2$)] 0.75-toluene and Nb(η-C$_5$H$_4$Me)$_2$-
(CH$_2$SiMe$_3$)(η^2-CO$_2$).

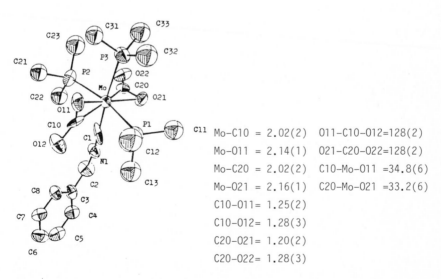

Mo–C10 = 2.02(2)   O11–C10–O12=128(2)
Mo–O11 = 2.14(1)   O21–C20–O22=128(2)
Mo–C20 = 2.02(2)   C10–Mo–O11 =34.8(6)
Mo–O21 = 2.16(1)   C20–Mo–O21 =33.2(6)
C10–O11= 1.25(2)
C10–O12= 1.28(3)
C20–O21= 1.20(2)
C20–O22= 1.28(3)

Fig. 10. Structure of Mo(CO$_2$)$_2$(PMe$_3$)$_3$(CN–CH$_2$–C$_6$H$_5$). (Reproduced with
permission from ref. 75. Copyright 1986 American Chemical Society).

118

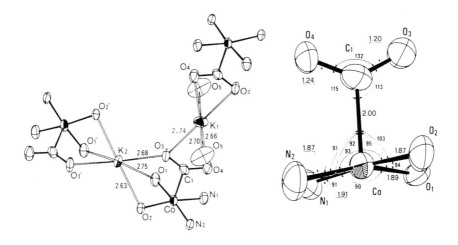

Fig. 11. View of the repetitive unit present in Co(n-Pr-salen)-K(CO$_2$)(THF), and the coordination sphere around cobalt, with the most relevant bond distances (Å) and angles (deg.). (reproduced with permission from ref. 76b. Copyright 1982 American Chemical Society).

coordinate with Mo, being staggered to each other, and at the same time, eclipsed to the Mo-P bond axis. It is very interesting that, although carbon dioxide coordination is difficult in general, this complex includes two CO$_2$ ligands coordinating a single Mo atom. The reason for this complex to be stable will be discussed later. The conformation of this complex is the same

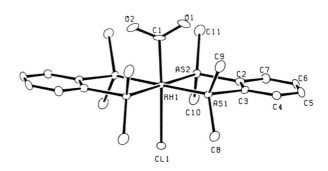

Fig. 12. View of the RhAs$_4$C$_{20}$H$_{32}$·Cl·CO$_2$ adduct. Rh-Cl=2.05(2) Å; C-O=1.20(2) and 1.25(2) Å; Rh-C-O=116(1)°; Rh-Cl=2.635(4) Å. (Reproduced with permission from ref.77. Copyright 1983 American Chemical Society).

as that found in trans-Mo(PR$_3$)$_4$(C$_2$H$_4$)$_2$ (ref. 78). These results suggest the
η^2-side on coordination of CO$_2$ resembles C$_2$H$_4$ coordination. An Os complex
includes a bridging type carbon dioxide (Fig. 13; ref. 79). The same type
bridging carbon dioxide has been reported for a Re-carbonyl complex (Fig. 14;
ref. 80). In these complexes, the bridging carbon dioxide coordinates with one
metal via the η^1-C mode, and with the other metals via the η^1-O end on mode.
Although pure examples including only the η^1-O end on coordination have not
been isolated, these complexes can be considered examples of the η^1-O end on
coordination mode. The CO$_2$ coordination in M[Co(R-salen)(CO$_2$)] (Figure 11)
resembles CO$_2$ adducts of Os and Re-carbonyl cluster compounds, because carbon
dioxide coordinates with Co(I) via the η^1-C mode and with the K$^+$ cation via the
η^1-O end on mode (Fig. 11). This type of coordination is called bi-functional
activation by Floriani and his collaborators (ref. 76). Certainly, the
stability of CO$_2$ coordination depends on the type of alkali cation, with the
Li$^+$ cation producing the most stable CO$_2$ adduct.

Besides the above-mentioned, well-characterized transition-metal CO$_2$
complexes, Ir(I) has been reported to coordinate with CO$_2$ through the η^1-C
mode; for instance, [Ir(dmpe)$_2$]Cl (dmpe = Me$_2$PCH$_2$CH$_2$PMe$_2$) reacts with carbon
dioxide at room temperature, yielding a white solid (ref. 81). The elemental

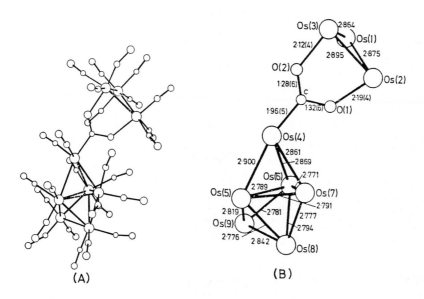

Fig. 13. Structure of [HOs$_3$(CO)$_{10}$·O$_2$C·Os$_6$(CO)$_{17}$]$^-$ (A), and significant
bond lengths (in Å) within the cluster, with CO groups omitted for clarity
(B). (Reproduced with permission from ref. 79. Copyright 1976 The Royal
Chemical Society).

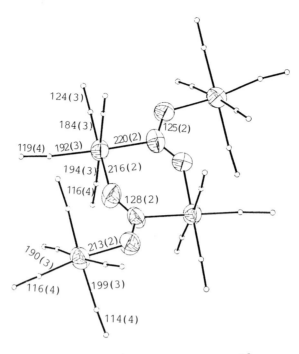

Fig. 14. Structure of $[(CO)_5Re(CO_2)Re(CO)_4]_2$ (10^{-2}Å unit). (Reproduced with permission from ref.80. Copyright 1982 Verlag Chemie GmbH).

analysis indicates the stoichiometry of this complex is $Ir(dmpe)_2Cl(CO_2)$. Also, $[Ir(diars)_2]Cl$ readily binds 1 equiv. of carbon dioxide. These $Ir-CO_2$ complexes have a d^8 electron configuration and are expected to have a structure similar to the isolated $Rh(I)-CO_2$ adduct, $RhCl(diars)_2(CO_2)$ (see Figure 12). An $Ir(I)$ complex, $IrCl(C_8H_{14})(PMe_3)_3$, reacts with carbon dioxide, yielding $Ir(C_2O_4)(PMe_3)_3$, 9 (ref. 82). This reaction has been postulated to proceed via stepwise uptake of two carbon dioxide molecules with a concomitant displacement

$$IrCl(C_8H_{14})(PMe_3)_3 \xrightarrow{CO_2} [IrCl(CO_2)(PMe_3)_3] \xrightarrow{CO_2}$$

8

9

of cyclooctene, in which an $Ir-CO_2$ adduct 8 is proposed as an intermediate. Although the coordination mode of 8 has not been described, the η^1-C mode seems most probable, as will be discussed later in this review. Four-coordinate Ir(I) complexes, $[Ir(OH)(CO)(PPh_3)_3]$ and $[Ir(L-L)_2]Cl$ (L-L = dmpe or $Et_2PCH_2CH_2PEt_2$) react with carbon dioxide under mild conditions, to form CO_2 adducts as well as $IrCl(C_8H_{14})(PMe_3)_3$ (ref. 83), but the coordination mode has not been reported. The η^1-C coordinated CO_2 complex, $NiL_3(CO_2)$ (L = PR_3), has also been proposed as an intermediate in the formation reaction of Ni-$(PR_3)_2(CO_2)$ from $Ni(PR_3)_4$, in asmuchas the i.r. band (either 1750 cm^{-1} for L = PEt_3 or 1760 cm^{-1} for L = PBu^n_3) differs from that of $Ni(PCy_3)_2(CO_2)$ (1728 cm^{-1} in solid and 1740 cm^{-1} in solution) (ref. 84). Several Rh(I) complexes have been reported to form CO_2 adducts, $RhCl(CO_2)L_2$ and $RhCl(CO_2)L_3$ (L = PBu^n_3, $PEtPh_2$, PEt_2Ph) (ref. 85). Of these two kinds of Rh-CO_2 adducts, the penta-coordinate ones are considered to have the η^1-C coordination mode and the four-coordinate ones the η^2-side coordination, given their i.r. spectra. A similar penta-coordinate CO_2 complex has been reported for $Fe(PMe_3)_4(CO_2)$ (ref. 86). The coordination mode, however, has been postulated to be the η^2-side on coordination. Several Co and Fe hydride complexes have been reported to form CO_2 adducts (ref. 87). While their CO_2 adducts have not been isolated, i.r. spectra and the CO_2 release from these complexes support the formation of CO_2 adducts.

Bifunctional activation of carbon dioxide, which slightly differs from that in $K[Co(R-salen)(CO_2)](THF)$ (Fig. 11), has been proposed in Co(I) complexes (ref. 88). A Co(I) complex, $(np_3)CoH$ (np_3 = tris(2-(diphenylphosphino)ethyl)amine) reacts with carbon dioxide, yielding not the

10

11

CO_2 adduct but a carbonyl complex, $[(np_3)Co(CO)](BPh_3)$. In this reaction the η^1-C coordinated CO_2 complex is postulated as an intermediate in which the sodium ion is considered to stabilize the CO_2 coordination, as shown in 10. The same type of bifunctional coordination of carbon dioxide has been proposed in $Li_2[W(CO)_5(CO_2)]$ 11 (ref. 89) and in $[Fe(Cp)(CO)_2(CO_2)]$ 12A, (ref. 90). In the former, the η^1-C mode has been supported by ^{13}C NMR and IR studies. The W-CO_2 bond is suggested to have a significant π-component from the J_{W-C} value (93 Hz), by considering this value as an intermediate between the value for W-C σ-single bond and that for the W-carbene bond. 12A is synthesized from M[Fe(η-C_5H_5)(CO)$_2$] and CO_2 at $-78°C$ and the stability of this CO_2 adduct increases in the order NBu$^t_4$ < K$^+$ < Na$^+$ < Li$^+$ (ref. 90). This stability dependence on the cation means the presence of an interaction between the Fe-CO_2 complex and these cations. In both 11 and 12A, an interesting oxide transfer reaction

12A   **12B**

proceeds very easily between the coordinated CO and the coordinated CO_2. This reaction can be considered a nucleophilic attack of the coordinated CO_2 to CO, as shown in 12B, which suggests that CO_2 is activated by the η^1-C coordination with metal. It seems worthwhile to predict the coordination mode of all these, not-isolated CO_2 adducts from a theoretical viewpoint. The author intends to try this prediction after discussing the coordinate bond nature and structure of isolated CO_2 complexes.

The second characteristic feature in the stereochemistry of transition metal CO_2 complexes is the significant distortion of coordinated carbon dioxide; in general, the OCO angle is around 130°, as clearly shown in Fig.'s 9 - 14. Corresponding to this significant distortion, the C=O bond distance is lengthened by the coordination. Such distortions are interesting in relation to the electronic structure and coordinate bond nature of CO_2 ligands. The driving force to distort the coordinated CO_2 ligand, as well as the coordination modes, will be theoretically investigated in the following section from the viewpoint of coordinate bond nature.

4.2 COORDINATION MODES and COORDINATE BOND NATURE

Prior to starting a detailed discussion, let us briefly examine several molecular orbitals of carbon dioxide which are important for interactions with transition metals. They are non-bonding π (HOMO of CO_2; abbreviated as nπ), π^* (LUMO of CO_2), a lone pair and π orbitals. Three π-type orbitals are doubly degenerate, but the lone pair is non-degenerate. The difference between carbon

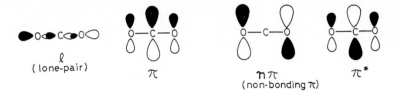

dioxide and the usual η^2-ligand, such as ethylene, is that carbon dioxide has a non-bonding π orbital, which induces some different feature to the coordinate bond. When carbon dioxide becomes distorted (i.e. the OCO angle becomes bent), these doubly-degenerate orbitals become non-degenerate and can be classified into two groups; π, $n\pi$, and π^* orbitals located on the CO_2 plane are classified as the first group and denoted as $\pi_{/\!/}$, $n\pi_{/\!/}$, and $\pi^*_{/\!/}$. π, $n\pi$, and π^* orbitals perpendicular to this plane are classified as the second, and denoted as $\pi_\perp$, $n\pi_\perp$, and $\pi^*_\perp$. The distortion of carbon dioxide slightly

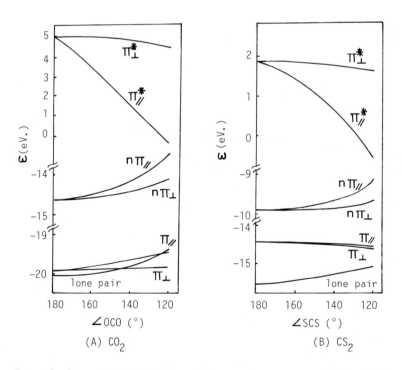

Fig. 15. Schematical pictures of lone pair, π, $n\pi$, and π^* orbitals of CX_2 (X = O or S) and changes of those orbitals by CX_2 bending (ab-initio MO calculation with 4-31G. R(C-O)=1.182, R(C-S)=1.55 Å from experimental value).

destabilizes the $\pi_{//}$ and $n\pi_{//}$ orbitals in energy, as shown in Figure 15, whereas the distortion changes little the energy level of $\pi_{\perp}$ and $n\pi_{\perp}$ orbitals. On the other hand, the distortion significantly stabilizes the $\pi^{*}_{//}$ orbital in energy but causes little change to the energy of the $\pi^{*}_{\perp}$ orbital. These changes suggest that the distortion of CO_2 strengthens significantly the back donation from metal to CO_2 but little affects donation from CO_2 to metal.

In the η^2-side on mode, the $\pi_{//}$ and $n\pi_{//}$ orbitals can participate in the donor interaction from CO_2 to metal since they are doubly occupied. A donor interaction is strong, when these orbitals overlap well with metal unoccupied orbitals and these orbitals lie higher in energy. Therefore, the $\pi_{//}$ and $n\pi_{//}$ orbitals in the first group are expected to contribute to the coordinate bond of carbon dioxide to a greater extent than the $\pi_{\perp}$ and $n\pi_{\perp}$ orbitals in the second group. Because two π^* orbitals are unoccupied, they can form back donor interactions with the metal occupied orbitals. Of these two orbitals, the $\pi^{*}_{//}$ orbital contributes more to the back donation than the $\pi^{*}_{//}$ orbital, since the former lies lower in energy than the latter. Besides these bonding interactions, the occupied π and $n\pi$ orbitals cause four-electron destabilizing interactions with the metal occupied orbitals.

In the η^1-C coordination mode, the $n\pi$ orbital is expected not to contribute as much to the CO_2 coordination because this orbital cannot overlap well with the metal orbitals. On the other hand, π and π^* orbitals should make a greater contribution to the donor interaction from CO_2 to metal and the back-donor interaction from metal to CO_2, respectively, since these orbitals have a large C $p\pi$ contribution which can directly interact with a metal. In the η^1-O end on mode, the lone pair can contribute to the donor interaction

M=Ni(0) or Cu(I)

13

14

M=Ni(0) or Cu(I)

15

and the π^* orbital can participate in the π-back donation. In the following discussion, let us observe these orbitals and examine electrostatic, donation, back-donation, and four-electron destabilizing interactions.

The coordinate bond nature of the η^2-side on and η^1-O end on modes has been examined in $Ni(PH_3)_2(CO_2)$ 13 with ab-initio MO and energy decomposition analysis, as shown in Table 6 (ref. 91). A similar analysis has been carried out on the η^2-side on coordinated $Ru(CO)_4(CO_2)$ 14 using the Xα-DV method (see Table 7; ref. 71). Note that negative values of Table 6 mean stabilization in energy as defined in eq. 5, whereas the negative values of Table 7 mean destabilization in energy. Both ab-initio MO and Xα-DV methods indicate that

TABLE 6

Energy decomposition analysis of interaction between $M(PH_3)_2$ and CO_2 (M = Ni or Cu$^+$) (kcal/mol) (negative values mean stabilization)

| | | $Ni(PH_3)_2(CO_2)$ | | | | $Cu(PH_3)_2(CO_2)^+$ | | |
| | | η^2-side on | η^1-O-end on | | | η^2-side on | | η^1-O end on |
		180^a 139^a	2.0	2.1	1.7^b	139^a	139+0.4^c	2.06^d
Binding energy	(BE)	19 -27	1	1	17	57	38	-14
Deformation energy	(DEF)	4 34	4	4	4	34	34	4
Interaction energy	(INT)	15 -61	-3	-3	13	23	4	-18
Electrostatic	(ES)	-71 -76	-21	-14	-75	-31	6	-22
Exchange	(EX)	148 131	36	24	131	105	20	20
Donative	(FCTPLX)	-12 -16	-5	-4	-13	-13	-12	-4
Back-donative	(BCTPLX)	-41 -76	-9	-7	-27	-35	-10	-8
Remaining	(R)	-9 -30	-4	-3	-3	-3		-3

a) The OCO angle. b) The Ni-O distance. c) CO_2 is displaced away by 0.4 Å. keeping the other geometrical parameters fixed. d) The optimized Cu-O distance. (Reproduced with permission from ref. 91. Copyright 1982 American Chemical Society)

the η^2-side on CO_2 coordination receives significantly larger stabilization from a back-donor interaction than that from a donor interaction. In the η^2-side on coordinated $Ni(PH_3)_2(CO_2)$, the contribution from the back-donor interaction is about 4 to 5 times larger than that from the donor interaction. In $Ru(CO)_4(CO_2)$, the back donor interaction contributes to the stabilization about 2.5 times more than the donor interaction. Although the back donor interaction is stronger than the donor interaction in both Ni(O) and Ru(O) complexes, the relative importance of the back-donor interaction is much larger in the Ni(O) complex than in the Ru(O) complex. Because different methods have

TABLE 7

Energy decomposition analysis of interaction between $Ru(CO)_4$ and several ligands including CO_2 (KJ/mol) (positive values mean stabilization)

	CO_2	CS_2	H_2CO	H_2CS	CO	CS
Stabilization energy(ΔE)	102.3	157.9	181.3	227.9	180	237
Preparation energy(ΔE_{prep})	-186.3	-201.5	-169.9	-158.2	-16	-16
Static energy (ΔE^0)	-141.2	-128.1	-155.2	-163.8	-199	-204
Donative energy (ΔE_D)	118.9	132.4	123.7	159.1	220	241
Back-donative (ΔE_{BD})	271.0	318.3	345.6	359.3	175	216
mixing (ΔE_{mix})	39.9	36.8	37.1	31.5	–	–

(Reproduced with permission from ref. 71. Copyright 1986 American Chemical Society).

been employed in investigating these complexes, we cannot reach a definitive conclusion. However, the large difference between Ni and Ru systems suggests that $Ni(PH_3)_2$ tends to form stronger π-back donor interaction with η^2-ligands than does $Ru(CO)_4$. The electron distribution of transition-metal CO_2 complexes musc, then, be examined in detail since it should be related to the coordinate bond nature. Mulliken populations of the η^2-side on coordinated $Ni(PH_3)_2(CO_2)$, given in Table 8, clearly indicates the Ni atomic population significantly decreases upon coordination by carbon dioxide; and, at the same time, the electron population of carbon dioxide increases remarkably. The decrease in the Ni atomic population mainly results from the decrease in the Ni d_{xz} orbital population (see 13 for x and z axes). Since the Ni d_{xz} orbital participates in the π-back bonding with the CO_2 $\pi^*_{//}$ orbital, all these results in electron distribution apparently show the presence of a strong π-back donation in the η^2-side on $Ni(PH_3)_2(CO_2)$. This result agrees well with the results of EDA described above.

The coordinate bond of carbon dioxide is compared with that of the analogous carbon disulfide in Table 7 (ref. 71). CS_2 coordination receives greater stabilization from both donor and back-donor interactions and, simultaneously, smaller destabilization from the static interaction than in CO_2 coordination. The greater distortion energy of carbon disulfide is sufficiently overwhelmed by the above-mentioned, stronger, bonding interaction and smaller static interaction, to produce a stronger coordinate bond in carbon disulfide than in carbon dioxide. A similar trend is also found in comparisons of H_2CO and H_2CS coordinations and of CO and CS coordinations. In these

TABLE 8

Changes in atomic and atomic orbital populations of $M(PH_3)_2(CO_2)$ (M = Ni or Cu^+) caused by CO_2 coordination

	$Ni(PH_3)_2$	$Ni(PH_3)_2(CO_2)$		$Cu(PH_3)_2$	$Cu(PH_3)_2(CO_2)^+$	
		η^2-side on	η^1-O end on		η^2-side on	η^1-O end on
M	27.75	27.43	27.70	28.16	28.22	28.15
s	6.03	6.10	6.03	6.13	6.15	6.12
p	12.01	12.21	12.03	12.11	12.25	12.13
d	9.71	9.12	9.64	9.93	9.82	9.91
d		-0.69	$-0.02(d_{xz})$		-0.12	-0.0
			$-0.01(d_{yz})$			
PH_3	18.13	17.92^a	18.14	17.92	17.82^a	17.93
		17.96^b			17.88^b	
CO_2		22.69	22.02		22.08	21.98

a) Trans to C of CO_2. b) Cis to C of CO_2. (Reproduced with permission from ref. 91. Copyright 1982 American Chemical Society).

ligands, substitution of oxygen by sulfur pushes the π and nπ orbitals to higher energy but lowers the π* orbital energies, as shown in Figure 15B. This is probably the principal reason why CS_2, CS, and H_2CS coordinations are stronger than those of CO_2, CO, and H_2CO.

The η^2-side on coordination mode, **13**, is now compared with the η^1-O end on coordination mode **15** of $Ni(PH_3)_2(CO_2)$ and its isoelectronic complex, $[Cu(PH_3)_2(CO_2)]^+$ (ref. 91), as shown in Table 6. In the $Ni(PH_3)_2(CO_2)$, the η^2-side on coordination is stable and its binding energy is 27 kcal/mol, while the η^1-O end on coordination yields little stabilization. In $[Cu(PH_3)_2(CO_2)]^+$, on the other hand, the η^1-O end on coordination produces small but non-negligible binding energy, whereas the η^2-side on coordination is very unstable. Corresponding to these differences in the CO_2 binding energy, several critical contrasts are found in electron distributions (Table 8). In the η^1-O end on coordinated $Ni(PH_3)_2(CO_2)$, the Ni d_{xz} and d_{yz} orbital populations decrease only slightly upon CO_2 coordination. In the η^2-side on coordinated $[Cu(PH_3)_2(CO_2)]^+$, again the Cu d_{xz} orbital population decreases to a much lesser extent than in the Ni d_{xz} orbital, upon CO_2 coordination. These results suggest the absence of strong π-back bonding in the η^1-O end on $Ni(PH_3)_2(CO_2)$ and η^2-side on $[Cu(PH_3)_2(CO_2)]^+$, whereas strong π-back bonding occurs in the η^2-side on $Ni(PH_3)_2(CO_2)$. Therefore, a comparison of $Ni(PH_3)_2(CO_2)$ with $[Cu(PH_3)_2(CO_2)]^+$ offers clear differences between the η^1-O end on and η^2-side on coordinations.

The coordinate bond nature of these complexes will be compared from an energetical viewpoint, based on the EDA investigation. In the equilibrium structure, the η^2-side on coordinated $Ni(PH_3)_2(CO_2)$ suffers considerable destabilization from the exchange repulsion interaction but receives great stabilization from electrostatic and back-donation interactions. In the η^1-O end on, the exchange repulsion gives rise to small destabilization and the electrostatic, donation, and back-donation interactions also offer small stabilization. Thus, a comparison of the η^2-side on coordination with the η^1-O end on mode of the equilibrium structure would not clarify the difference between two coordination modes. In order to ascertain the intrinsic difference between these two modes, a comparison of energy components at the same inter-fragment distance would be helpful. It is not easy, however, to define the same inter-fragment separation, since these two coordination modes have significantly different structures. In this case, the exchange repulsion, EX, might be considered a measure of inter-fragment distance, since it depends on the contact of electron clouds between two fragments. Thus, two coordination modes have been compared, when their EX terms have the same value. We choose somewhat arbitrarily, 131 kcal/mol; e.g., the EX value for the optimized η^2-side on coordinated $Ni(PH_3)_2(CO_2)$, as a reference in comparing the two modes of Ni(0) complex. Calculations of the η^1-O end on coordinated Ni(0) complexes were carried out at various R_{Ni-O} distance, and the $R_{Ni-O}= 1.7$ Å was found to exhibit the EX value of 131 kcal/mol. Values of energy components of this geometry are shown in Table 6, and compared with η^2-side on $Ni(PH_3)_2(CO_2)$. These two coordination modes have nearly the same ES and FCTPLX stabilization. The principal difference between two modes comes from the BCTPLX stabilization, supplemented in part by R, which also includes the back-bonding coupling term. As shown in Table 6, a related comparison is made for the Cu(I) complexes, at EX = 20 kcal/mol, between the most stable η^1-O end on mode and the η^2-side on mode, in which energy component values of the latter are obtained by moving

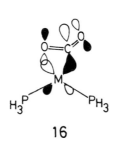

16

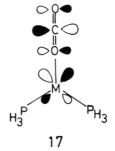

17

CO_2 away from Cu by 0.4 Å. Although the stabilization in energy from BCTPLX and FCTPLX+R terms is nearly the same in these two modes, the ES stabilization strongly favors the η^1-O end on mode. Thus, the η^2-side on coordination is stable when the back-donation interaction results in a large stabilization, as in $Ni(PH_3)_2(CO_2)$. The η^1-O end on coordination is stable when the electrostatic interaction yields a large stabilization. As shown in **16**, the η^2-side on coordination mode has a favorable overlap between the Ni $d\pi$ and the CO_2 $\pi^*_{\!/\!/}$ orbital, whereas the η^1-O end on mode has a small overlap between d_{xz} and d_{yz} orbitals of Ni and π^* orbital of CO_2 (see 17). This is probably the main reason why the π-back bonding is very weak in the η^1-O end on $Ni(PH_3)_2(CO_2)$ but very strong in the η^2-side on $Ni(PH_3)_2(CO_2)$ (vide supra). When the metal ion has a sufficiently large positive charge, the η^2-side on mode is unfavorable due to a repulsive electrostatic interaction between the $M^{\delta+}$ and $C^{\delta+}$, while the η^1-O end on mode is electrostatically favorable. Thus, the η^1-O end on mode is stabilized in the Cu(I) complexes since the Cu atom has +0.84 Mulliken charge in $[Cu(PH_3)_2]^+$ but the Ni atom has only +0.25 Mulliken charge in $Ni(PH_3)_2$ (see Table 8). If the metal has a sufficiently large Lewis basicity, as in $Ni(PH_3)_2$, strong π-back bonding would be induced so as to stabilize the η^2-side on mode. In such a case, however, the end-on mode would not be stable, because the metal atom would have a considerably larger electron population, as shown in Ni(0) complex (see Table 8); thus only a weak electrostatic interaction develops between $Ni^{\delta+}$ and $O^{\delta-}$ atoms. Therefore, the η^2-side on mode

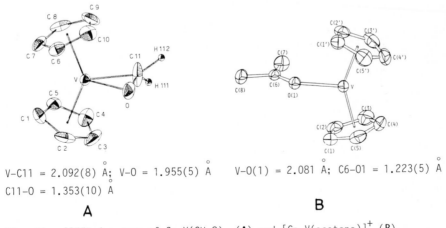

V-C11 = 2.092(8) Å; V-O = 1.955(5) Å   V-O(1) = 2.081 Å; C6-O1 = 1.223(5) Å
C11-O = 1.353(10) Å

A   **B**

Fig. 16. ORTEP drawings of $Cp_2V(CH_2O)$ (**A**) and $[Cp_2V(acetone)]^+$ (**B**). (Reproduced with permission from ref. 92 and 93. Copyright 1982, 1983 American Chemical Society).

is stable in the Ni(0) complex. In $[Cu(PH_3)_2]^+$, the Cu $3d_{xz}$ orbital lies at −15.8 eV (d_{xz}-PH$_3$ anti-bonding orbital) and at −20.2 eV (d_{xz}-PH$_3$ bonding orbital), which are much lower in energy than the Ni d_{xz} orbital in Ni(PH$_3$)$_2$ (−6.9 eV for the d_{xz}-PH$_3$ anti-bonding orbital). Because of stable d orbitals, $[Cu(PH_3)_2]^+$ cannot effectively cause back-donation between the Cu $3d_{xz}$ and the CO_2 π^* orbitals, as has been suggested above by the electron distribution. Therefore, the η^2-side on coordination mode would be unstable in the Cu(I) complex, because this mode requires considerable stabilization from the back-donating interaction.

Some experimental results of (Cp)$_2$V(H$_2$CO) and $[(Cp)_2V(acetone)]^+$ support the above discussion. When the central V atom takes on the low oxidation state of +2, H$_2$CO coordinates with V via the η^2-side on mode (see Figure 16A; ref. 92). However, when the V atom is oxidized to the higher oxidation state of +3, coordination of acetone is changed to the η^1-O end on mode (see Figure 16B; ref. 93). The former result means the η^2-side on mode needs strong π-back donation and the latter suggests the η^1-O end on mode requires an electrostatic interaction with the positively charged metal ion.

Three possible coordination modes have recently been compared in detail for $[Co(alcn)_2(CO_2)]^-$ (alcn = HNCHCHCHO$^-$) using the ab-initio MO method (ref. 94). This complex has been investigated as a model of M[Co(R-salen)(CO$_2$)]. The real complex has been isolated by Floriani et al., and is known to have the η^1-C coordination mode (ref. 76; Figure 11), as described above. Structures of the coordination modes examined are schematically shown in Figure 17. Of three possible coordination modes, only the η^1-C coordination structure is stable.

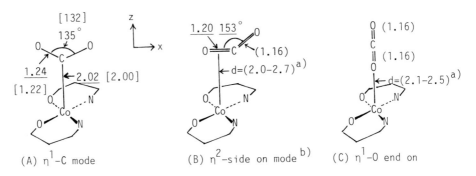

Fig. 17. Geometries of $[Co(alcn)_2(CO_2)]^-$ examined in ref. 94.
underline = optimized value; [ ] = experimental value; ( ) = assumed value
a) No local minimum was found in this region. b) The geometry of CO$_2$ was optimized at d=2.0 Å.
(Reproduced with permission from ref. 94. Copyright 1987 American Chemical Society).

Several important geometrical parameters have been optimized in this mode, which agree well with experimental values (see Figure 17). The η^1-O end on mode is, on the other hand, destabilized in energy as carbon dioxide approaches Co(I), as shown in Figure 18. This seems rather strange, considering that Co(I) is positively charged as suggested by its +1 oxidation state and that the η^1-O end on mode is expected to be stable when the central metal is positively charged. A detailed inspection of orbital interactions, however, clarifies the reason why this mode is unstable in the Co(I) complex. Corresponding to the destabilization in the total energy, all the occupied d-orbitals rise in energy upon carbon dioxide approaching Co(I), as shown in a lower part of Figure 18 (ref. 94b). This is probably a reason why the η^1-O end on mode is unstable in the Co(I) complex. The HOMO of $[Co(alcn)_2]^-$, mainly composed of the Co d_{z^2} orbital, lies considerably high in energy. The Co d_{xz} and d_{yz} orbitals lie lower in energy than the HOMO. The empty Co 4s and $4p_z$ orbitals, which might be important for the charge-transfer interaction from CO_2 to Co, are found remarkably high in energy, above some π^* orbitals of the alcn ligand, thus

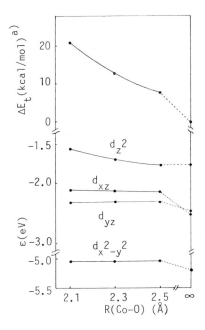

Fig. 18 Total energy and d-orbital energies vs. R(Co-O) in the η^1-O end on $[Co(alcn)_2(CO_2)]^-$. a) ΔE_t= change in total energy caused by CO_2 approach. (Reproduced with permission from ref. 94b. Copyright 1987 American Chemical Society).

132

suggesting a weak Lewis acidity of $[Co(alcn)]^-$. From symmetry considerations, the η^1-O end on mode is expected to involve several σ-type interactions between the CO_2 lone pair and Co $3d_{z^2}$, 4s, and $4p_z$ orbitals and also π-type interactions between CO_2 π, $n\pi$, π^* and Co d_{xz} and d_{yz} orbitals. Stabilizing interactions arise from the CO_2 lone pair-Co 4s, CO_2 lone pair-Co $4p_z$, and CO_2 π^*-Co d_{xz}, d_{yz} orbital pairs, whereas destabilizing interactions arise from the CO_2 lone pair-Co $3d_{z^2}$, CO_2 π-Co d_{xz} d_{yz}, and CO_2 $n\pi$-Co d_{xz}, d_{yz} pairs. The d_{xz}, d_{yz}, and d_{z^2} orbitals of cobalt are calculated to be raised in energy by the coordination of CO_2, as described above. Also, CO_2 approaching Co hardly stabilizes the lone pair and $n\pi$ orbitals. Furthermore, anti-bonding interactions are found in the Co d_{z^2} and d_{xz}, d_{yz} orbitals, as shown in right-half of Figure 19. All these results suggest that the destabilizing interactions are predominant, which are well explained by the schematical orbital interaction diagram pictured in the left-half of Figure 19. A characteristic feature of this scheme is two pairs of four-electron destabilizing interactions, one between the Co d_{z^2} and the CO_2 lone pair and the other between the Co $d\pi$ and CO_2 $n\pi$ orbitals. Such unfavorable situation results from the presence of high-lying occupied d_{z^2}, d_{xz}, and d_{yz} orbitals and the absence of a good accepting orbital in $[Co(alcn)_2]^-$. From these results, it is concluded that the η^1-O end on mode is repulsive owing to four-electron destabilizing interaction when the metal part has a σ-type HOMO lying at a high energy level.

The next topic is a comparison of the η^1-C coordination mode with the η^2-side on mode (ref. 94). In the Co(I)-CO_2 complex, the η^1-C mode exhibits a stabilization energy of 6.2 kcal/mol. Although this value is not sufficient

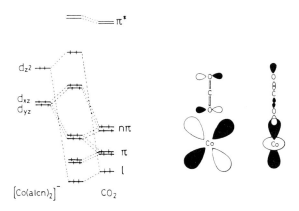

Fig. 19. Schematical orbital interaction diagram and important four-electron destabilizing interaction in $[Co(alcn)_2(CO_2)]^-$.

TABLE 9

Energy decomposition analysis of the interaction between $[Co(alcn)_2]^-$ and CO_2 (alcn = $HNCHCHCHO^-$) (kcal/mol) and Mulliken population's changes caused by CO_2 coordination (negative values mean stabilization in energy)

	η^2-side on	η^1-C mode	
		$\angle OCO=135°$	$180°$
Binding energy (BE)	27.0	-6.2	54.8
Deformation energy (DEF)	16.0	49.2	13.1
Interaction energy (INT)	11.0	-55.4	41.5
Electrostatic energy (ES)	-85.2	-67.9	-83.7
Exchange repulsion (EX)	147.6	103.8	170.2
donative interaction (FCTPLX)	-14.9	-12.1	-10.8
back-donative interaction (BCTPLX)	-34.1	-53.0	-29.1
Remaining (R)	-2.4	-26.2	-5.1
Changes in Electron populations[a]			
Co	-0.18	-0.40	-0.24
d_{z^2}	-0.07	-0.63	-0.24
d_{xz}	-0.19	0.08	0.04
CO_2	0.28	0.71	0.31

a) Positive values mean an increase in electron population by CO_2 coordination. (Reproduced with permission from ref. 94b. Copyright 1987 American Chemical Society).

for the usual coordinate bond, the presence of an alkali cation significantly increases the binding energy to 20 kcal/mol, which seems sufficient for the coordination bond. The η^2-side on coordination mode, on the other hand, exhibits a repulsive potential curve in the range of d = 2.0 ∿ 2.7 Å (see Fig. 19 for d). Conversion to the η^1-C mode from the η^2-side on mode leads to a continuous decrease of total energy, again suggesting that the η^2-side on mode is less stable than the η^1-C mode. Thus, it is reasonably concluded that only the η^1-C mode is stable and the other two modes are not stable in the Co(I) complex. Results of energy decomposition analysis, given in Table 9 (ref. 94b), show several contrasts between the η^1-C and η^2-side on modes; (1) the η^2-side on mode suffers from very large EX repulsion, (2) the ES stabilization of the η^2-side on mode is larger than that of the η^1-C mode, (3) the η^1-C mode receives greater stabilization from the BCTPLX term than the η^2-side on mode does. The fact that the EX repulsion is stronger in the η^2-side on mode than in the η^1-C mode can be easily explained by the schematic pictures on Figure 20. The η^2-side on mode involves four pairs of four-electron destabilizing interaction, such as the Co d_{z^2}-CO_2 $n\pi_{/\!/}$, Co d_{z^2}-CO_2 $\pi_{/\!/}$, Co d_{xz}-CO_2 $n\pi_{/\!/}$, and Co d_{xz}-CO_2 $\pi_{/\!/}$ orbital pairs, which play a role of main contributor to the EX

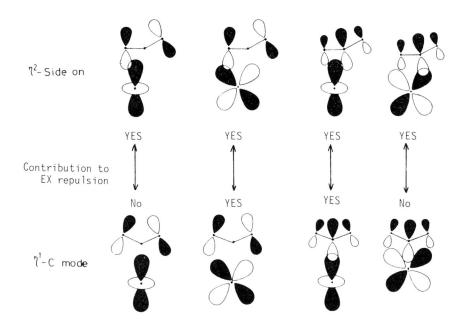

η^2- Side on

YES   YES   YES   YES
↑   ↑   ↑   ↑
Contribution to
EX repulsion
↓   ↓   ↓   ↓
No   YES   YES   No

η^1-C mode

Fig. 20. Four-electron destabilizing interaction between Co and CO_2 in η^1-C and η^2-side on coordination modes. (Reproduced with permission from ref. 94b. Copyright 1987 American Chemical Society).

repulsion. In the η^1-C mode, on the other hand, two of them, the Co d_z2-CO_2 $n\pi_{//}$ and Co d_{xz}-CO_2 $\pi_{//}$, cannot contribute to the EX repulsion, due to symmetry properties of interacting orbitals (note that orbitals belonging to the same symmetry can contribute to the EX repulsion). As a result, the η^1-C mode suffers less from the EX repulsion than the η^2-side on mode.

In addition to the energy decomposition analysis, changes in Mulliken populations caused by the CO_2 coordination can also provide meaningful information concerning the charge-transfer interaction (Table 9). In the η^1-C mode, the electron population of Co is significantly decreased by the coordination of carbon dioxide, mainly due to a substantial decrease in the Co d_z2 orbital population. Corresponding to these decreases, the electron population of CO_2 is increased markedly by the coordination of carbon dioxide. In the case of the η^2-side on mode, both Co d_z2 and d_{xz} orbital populations are decreased but the electron population of carbon dioxide is increased by the coordination of carbon dioxide. However, these changes in electron population are smaller than those found in the η^1-C mode. All these results of energy decomposition analysis, and Mulliken populations, imply the following results about the nature of the bonding: (1) the η^1-C mode has stronger back-donation interactions than the η^2-side on mode does, and (2) in the η^1-C mode, only the Co d_z2 orbital can participate in back-donation interactions, whereas in the η^2-side on mode both Co d_z2 and d_{xz} orbitals can contribute to back-donation interactions. A stronger back-donation interaction of the η^1-C mode (than that of the η^2-side on mode) can be easily understood by examining the energy levels of the cobalt d orbitals and their overlap with the CO_2 $\pi^*_{\parallel}$ orbital. In the η^1-C mode, the cobalt d_z2 orbital, a main component of HOMO of $[Co(alcn)_2]^-$, overlaps well with the $\pi^*_{\parallel}$ orbital of carbon dioxide, as schematically pictured in Figure 21-A, yielding a strong back-donation interaction. In the

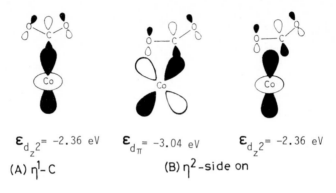

$$\varepsilon_{d_z2}= -2.36 \text{ eV} \qquad \varepsilon_{d_\pi}= -3.04 \text{ eV} \qquad \varepsilon_{d_z2}= -2.36 \text{ eV}$$

(A) η^1-C   (B) η^2-side on

Fig. 21. Back-donation interaction in η^1-C and η^2-side on coordination modes. (Reproduced with permission from ref. 94b. Copyright 1987 American Chemical Society).

η^2-side on mode, on the other hand, the Co d_z2 orbital cannot overlap well with the $\pi^*_{\parallel}$ orbital of carbon dioxide unlike in the η^1-C mode, because of the nodal plane of the $\pi^*_{\parallel}$ orbital; as shown in Figure 21-B, the positive overlap between the d_z2 and carbon p_π orbitals is decreased significantly by the negative overlap between the d_z2 and oxygen p_π orbitals. The Co d_{xz} orbital can, in this case, effectively overlap with the $\pi^*_{\parallel}$ orbital of carbon dioxide, producing a back-donation interaction (see Figure 21-B). But, the d_{xz} orbital lies lower in energy by ca. 0.6 eV than the d_z2 orbital; and, as a result, the back donation interaction of the η^2-side on mode is weaker than that of the η^1-C mode.

The electrostatic interaction, on the other hand, disfavors the η^1-C mode relative to the η^2-side on mode. This result is easily explained in terms of the electron distribution of interacting species; the carbon atom is positively charged (+0.73 e Mulliken charge), but the oxygen atom is negatively charged (-0.37 e) in the distorted carbon dioxide. The cobalt atom is also

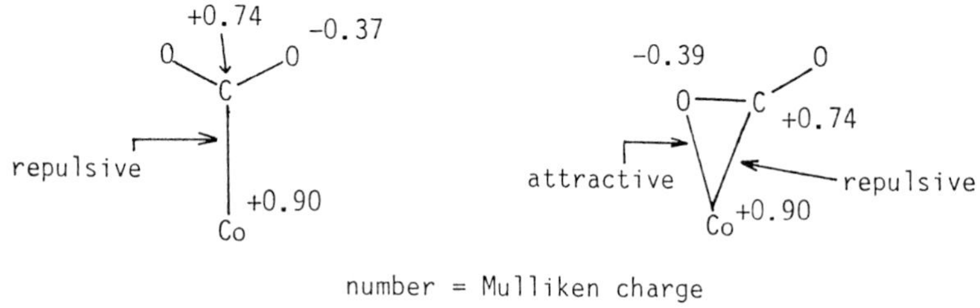

number = Mulliken charge

Fig. 22. Schematical pictures of electrostatic interactions in η^1-C and η^2-side on coordination modes of $[Co(alcn)_2(CO_2)]^-$. (Reproduced with permission from ref. 94b. Copyright 1987 American Chemical Society).

positively charged (+0.90 e Mulliken charge), in spite of the effective negative charge of $[Co(alcn)_2]^-$. In the η^1-C mode, the carbon atom is placed rather near to the cobalt atom, as shown in Figure 22, which gives rise to an electrostatic repulsion between them. In the η^2-side on mode, however, the electrostatic repulsion between cobalt and carbon atoms is compensated, to some extent, by the electrostatic attraction between cobalt and oxygen atoms, because both the carbon and oxygen atoms interact with the cobalt atom (see Figure 22). Consequently, the η^1-C mode receives less stabilization from the ES term than the η^2-side on mode does.

In conclusion, the η^1-C mode is stable due to the strong back-donation

TABLE 10

Summary of coordinate bonding nature in three coordination modes

Coordination mode	Interaction		
	back-bonding	electrostatic	exchange
η^2-side on	strong when HOMO is dπ.	medium	strong when dσ is occupied.
η^1-C	strong when HOMO is dσ.	unfavorable	strong when dπ is occupied.
η^1-O end on	weak in general	favorable for $M^{\delta+}$	strong when dσ is occupied.

interaction and the weak EX repulsion, in spite of the unfavorable situation for the electrostatic interaction. The η^2-side on mode is, on the other hand, unstable in $[Co(alcn)_2(CO_2)]^-$, because the great EX repulsion can not be over-whelmed by the weak back-donation interaction and rather strong electrostatic interaction.

The above-mentioned results, concerning of coordinate bonding nature of three coordination modes, are summarized in Table 10.

4.3 A WAY of PREDICTING the COORDINATION MODE of CARBON DIOXIDE

The discussion presented above provides us with a way of predicting the coordination mode of carbon dioxide. When the metal part has a d_{z^2} orbital as its HOMO, the η^1-C mode can yield a stronger charge-transfer interaction from metal to CO_2 $\pi^*_{//}$, and suffers less from the four-electron destabilizing interaction than the η^2-side on mode. From the point of view of the electro-static interaction, the η^1-C mode is inferior to the η^2-side on mode when the central metal is positively charged. In this regard, the most desirable situation for this mode is the presence of a HOMO mainly composed of a $d\sigma$ orbital, and, also, a low oxidation state of the metal. $[Co(R-salen)-(CO_2)]^-$ (Figure 11; ref. 76) and $RhCl(diars)_2(CO_2)$ (Figure 12; ref. 77) have a low spin d^8 electron configuration of low oxidation state (+1) and the HOMOs of $RhCl(diars)_2$ and $[Co(R-salen)]^-$ are considered to mainly consist of the d_{z^2} orbital, as shown in Figure 23-A (d-orbital splitting and the shape of $d\sigma$-orbital shown in Figure 23-A are given in ref. 95). Thus, both satisfy the above-mentioned conditions, and these two complexes are well-known as η^1-C coordinated CO_2 complexes.

When the metal part has a HOMO mainly composed of a $d\pi$ orbital as is the case with $Ni(PR_3)_2L$ (L = η^2-ligand such as olefin, acetylene, etc.), the η^2-

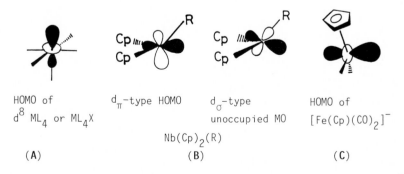

HOMO of d^8 ML_4 or ML_4X	d_π-type HOMO	d_σ-type unoccupied MO $Nb(Cp)_2(R)$	HOMO of $[Fe(Cp)(CO)_2]^-$
(A)	(B)		(C)

Fig. 23. Schematical pictures of HOMO of metal fragments interacting CO_2

side on mode is considerably stable, owing to a strong π-type back donating interaction. However, if the metal has a doubly occupied $d\sigma$ orbital, a four-electron destabilizing interaction arises from the overlap of the $d\sigma$ orbital with the π and $n\pi$ orbitals of CO_2, as described above (see Figure 20 and Table 9). Thus, the best condition for this coordination mode is the presence of a $d\pi$ orbital as a HOMO of the metal fragment and an empty $d\sigma$ orbital pointing to the CO_2 ligand. $Nb(\eta-C_5H_4Me)_2(CH_2SiMe_3)(CO_2)$ and $Mo(PR_3)_3(CN-Pr^i)(CO_2)_2$ satisfy these two conditions, as follows; $Nb(\eta-C_5H_4Me)_2(CH_2SiMe_3)$ has a $d\pi$ type orbital as its HOMO and an empty $d\sigma$ type orbital, as shown in Figure 23-B (these schematical pictures are supported by the orbital inter-action diagram between Cp_2NbR and CO; ref. 96). Because the Mo atom in $Mo(PR_3)_3(CN-Pr^i)$ has a low-spin d^6 electron configuration, simple crystal field considerations indicates the $Mo(PR_3)_3(CN-Pr^i)$ fragment takes the $(d_{xy})^2$-$(d_{xz})^2(d_{yz})^2$ electron configuration. Again, this complex has a $d\pi$ orbital as a HOMO and an empty $d\sigma$ orbital. Unfortunately, $Ni(PR_3)_2(CO_2)$ has not only a $d\pi$ type orbital as HOMO but also a doubly occupied $d\sigma$ orbital pointing to the CO_2 ligand. Probably, this is one of the reasons why the η^2-side on coordi-nation slightly deviates to the η^1-C coordination mode, i.e., the Ni-C distance is shorter than the Ni-O distance by 0.15 Å (see Figure 9).

Although three coordination modes of carbon dioxide have been theoreti-cally investigated in limited cases, such as in $Ni(PH_3)_2(CO_2)$ and $[Co(salen)_2(CO_2)]^-$, the results summarized in Table 10 can easily explain the coordination mode of the isolated transition metal CO_2 complexes. This encourages us to predict the coordination mode of several transition metal CO_2 complexes whose existence have been postulated, but not isolated. $IrCl(dmpe)_2(CO_2)$ has been isolated but its coordination mode has not been determined (ref. 81), as described above. In this Ir(I)-CO_2 adduct, the Ir atom has a +1 oxidation state, with a low spin d^8 electron configuration, and the HOMO of the $IrCl(dmpe)_2$ fragment is the $d\sigma$ type orbital. Thus, not only is the stoichiometry, but also the electron configuration of this complex, similar to $RhCl(diars)_2(CO_2)$. A consideration of these conditions strongly suggests this complex has the η^1-C coordination mode. $IrCl(CO_2)(PMe_3)_3$, **8**, is postulated as an intermediate in the reaction yielding $[Ir(C_2O_4)(PMe_3)_3]$, **9**, from $IrCl(C_8H_{14})(PMe_3)_3$ (ref. 82). The very similar Rh(I) complex, $RhClL_3$ (L = PBu^n_3, $PEtPh_2$, PEt_2Ph) has been reported to form a CO_2 adduct, $RhClL_3(CO_2)$ (ref. 85). Both Ir(I) and Rh(I) complexes include a central metal ion of +1 oxidation state possessing a d^8 electron configuration. If the $MClL_3$ fragment is square-planar, its HOMO is the $d\sigma$ type orbital, as in $RhCl(diars)_2$ and $[Co(salen)_2]^-$ (see Figure 23 A). From the above-mentioned discussion and the results of Table 10, the η^1-C coordination mode is considered the most

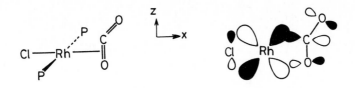

Fig. 24. Predicted structure of trans-$RhCl(PR_3)_2(CO_2)$ and important back donating interaction between Rh and CO_2.

favorable in these complexes. In $[W(CO)_5(CO_2)]^{2-}$ **11** and $[Fe(\eta-C_5H_5)(CO)_2-(CO_2)]^-$ **12A**, the η^1-C coordination mode has been proposed (**11** and **12A**; refs. 80 and 90), as described above. Simple crystal field examination suggests that $[W(CO)_5(CO_2)]^{2-}$ has a $(d_{xz})^2(d_{yz})^2(d_{xy})^2(d_z2)^2$ electron configuration, if this complex is an orbital singlet. The HOMO is the $d\sigma$ type orbital and this electron configuration is quite the same as that of $RhCl(diars)_2(CO_2)$ (Figure 23-A). Theoretical investigation of the $[Fe(\eta-C_5H_5)(CO)_2]^-$ fragment suggests that this complex has a $d\sigma$ type orbital as its HOMO (see Figure 23C; although the HOMO of $[Fe(Cp)(CO)_2]^-$ has not been reported, the LUMO of $[Fe(Cp)(CO)_2]^+$ has been investigated in ref. 97). The η^1-C coordination mode proposed for these complexes is reasonable, considering their electron configurations. However, not the η^1-C mode but the η^2-side on mode is proposed in $RhCl(CO_2)L_2$ (L = tertiary phosphine), although this complex includes a d^8 Rh(I) ion. In this case, a $d\sigma$ orbital expanding toward the CO_2 ligand is unoccupied but a $d\pi$ orbital is doubly occupied, if this complex has a square planar structure. Of d_{xz} and d_{xy} orbitals, the d_{xz} orbital is the HOMO and lies slightly higher in energy than the d_{xy} orbital, probably because of the absence of back-donating interaction between d_{xy} and PR_3 (ref. 98). This electron configura-

18 A   **18 B**   **19**

tion favors the η^2-side on coordination mode in which the CO_2 ligand is perpendicular to molecular plane because, in this orientation, the π^* orbital of CO_2 can interact well with the Rh d_{xz} orbital (HOMO), as shown in Figure 24. The above discussion offers theoretical support to the η^2-side on mode proposed by Aresta and Nobile (ref. 85).

Although the η^2-side on coordination of CO_2 has been accepted for $Ni(PCy_3)_2(CO_2)$ (ref. 73), the η^1-C mode has been proposed for $Ni(PR_3)_3(CO_2)$ (ref. 84). A previous ab initio MO calculation indicates the HOMO of $Ni(PH_3)_3$ is not a dσ orbital but a dπ orbital (ref. 99). The same ordering of d orbitals have been reported for $M(CO)_3$ in EH-MO studies (refs. 7a, 8b, 100). However, the HOMO of $Ni(PH_3)_3(\eta^1-SO_2)$ 18A is calculated to be a d_z2 orbital which strongly interacts with the π^* orbital of SO_2, as shown in 18B (ref. 99). This result is easily explained by considering that the d_z2 orbital is destabilized in energy by the approach of the forth ligand, and becomes the HOMO in

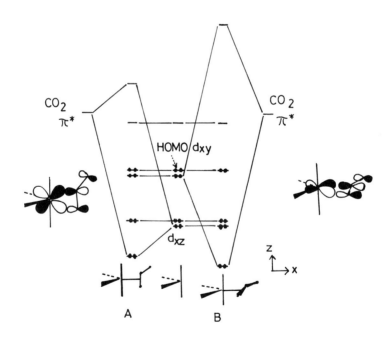

Fig. 25. Orbital interaction diagrams for $Fe(PR_3)_4(CO_2)$ in which CO_2 lies at an equatorial position. (A) CO_2 is perpendicular to the equatorial plane, and (B) CO_2 is on the equatorial plane. (this diagram is pictured by reference to the orbital interaction diagram which is presented for common penta-coordinate complex with pseudo-trigonal bipyramid structure (ref.7a)).

$Ni(PH_3)_3(SO_2)$. Although $Ni(PH_3)_3(CO_2)$ has not been theoretically examined, the above calculation suggests the d_z2 orbital of $Ni(PR_3)_3$ can strongly interact with the η^1-C coordinated CO_2 ligand, which reasonably supports $Ni(PH_3)_3(CO_2)$ acquiring the η^1-C mode, as shown in 19 (note the rotational isomer around the Ni-C bond can not be suggested from the present discussion and the orientation of CO_2 is arbitrarily shown in 19).

$Fe(PR_3)_4(CO_2)$ is a d^8 penta-coordinate complex of the first-row transition elements. Such complexes tend to take a psuedo-trigonal bipyramid structure. Orbital energy diagrams of penta-coordinate complexes have been qualitatively described by EH-MO calculations (refs. 7a, 101), as shown in Figure 25. The HOMO is a $d\pi$ orbital in the equatorial plane. Thus, the CO_2 ligand is expected to coordinate with Fe via a η^2-side on mode in the equatorial plane (see Fig. 25B). If the CO_2 ligand is perpendicular to the equatorial plane, the CO_2 π^* orbital cannot interact with the HOMO but with the $d\pi$ orbital lying lower in energy, as shown in Figure 25A; and, therefore, this conformation is less stable than the in-plane conformation. Of course, the situation becomes different if this complex takes a square pyramidal structure. In such a case, we cannot neglect the possibility of the η^1-C mode, as found in [Co(R-salen)-(CO_2)]$^-$ and $Rh(diars)_2(CO_2)$.

As discussed above, the results obtained in Section 4.2 successfully provide us with the way of predicting and explaining the coordination modes of CO_2 complexes. This kind of discussion is expected to be useful in the chemistry of transition metal CO_2 complexes.

4.4 DISTORTION of COORDINATED CARBON DIOXIDE and MISCELLANEOUS ISSUES of STRUCTURE

In all the transition metal carbon dioxide complexes whose structures have been verified experimentally, the carbon dioxide ligand is significantly bent from its equilibrium linear structure (see Figures 9 - 14). The driving force to cause such bending has been investigated in $Pt(PH_3)_2(CO_2)$ with a CNDO-type MO method (ref. 102) and in $Ni(PH_3)_2(CO_2)$ with ab initio MO methods and energy decomposition analysis (ref. 91). The former calculation indicates that the bending of CO_2 strengthens the interaction between Pt and CO_2, which makes the bent structure stable. This strengthening of CO_2 coordination has been discussed in terms of enhancement of the back-donating interaction. In the latter, a clearer analysis has been given. As shown in Table 5, the bent structure receives a greater stabilization from the BCTPLX term and suffers less from the EX repulsion than the non-distorted structure. The decrease in the EX repulsion caused by the CO_2 bending is small and about half of the increase in the BCTPLX stabilization. This means that the driving force to distort CO_2 is the BCTPLX stabilization, supplemented by the small decrease in the EX repulsion. As shown in Figure 15, the $\pi^*_{/\!/}$ orbital is remarkably stabilized in energy as the bending of CO_2 increases. Because of this lowering of the CO_2 $\pi^*_{/\!/}$ orbital energy, the π-back donation strengthens, leading to a large BCTPLX stabilization of the distorted structure. Changes in Mulliken populations, shown in Table 8, are consistent with the strengthening of back-bonding in the distorted structure; as the CO_2 bending increases, the CO_2 population increases but the Ni atomic population decreases owing to a decrease in the 3d, in particular $3d_{xz}$, orbital population. Corresponding to these changes, the electron population of PH_3 decreases with increasing bending of CO_2. This decrease probably results from an increase in the donation from PH_3 to Ni, which is strengthened by the increasing back-donation from Ni to the bending CO_2.

In $[Co(alcn)_2(CO_2)]^-$, as well as in $Ni(PH_3)_2(CO_2)$, the bending of CO_2 makes the CO_2 coordination stable while the non-distorted structure yields considerable destabilization in energy upon CO_2 coordination (ref. 94). Results of energy decomposition analysis are compared for the linear and bent structures of CO_2, in Table 9. The bent structure receives greater stabilization from the BCTPLX term and suffers less from the EX repulsion than the linear structure does, leading to the greater stabilization of the bent structure. Corresponding to the greater BCTPLX stabilization of the bent structure, the electron population of carbon dioxide is indeed increased by the bending, compared to that of the linear structure. This stronger back-donation interaction in the bent structure can be easily explained in terms of the lowering of CO_2 π^* orbital energy, as in $Ni(PH_3)_2(CO_2)$. On the other hand, a difference is found in the EX repulsion between this complex and $Ni(PH_3)_2(CO_2)$. The EX repulsion of $[Co(alcn)_2(CO_2)]^-$ is remarkably decreased by the CO_2 bending, while the EX repulsion in $Ni(PH_3)_2(CO_2)$ is decreased a little by the CO_2 bending. In the former, the bending of CO_2 weakens two pairs of four-

TABLE 10A

The driving force to bend CO_2 in transition metal-CO_2 complexes

Coordination mode	η^1-C mode	η^2-side on mode
Electrostatic	No (linear > bent)	No (little changes)
Exchange repulsion	Yes (linear > bent)	Yes but small
Back-donation	Yes (linear < bent)	Yes (linear < bent)

electron destabilizing interaction, Co d_{xz}–CO_2 $n\pi_{//}$ and the Co d_{z^2}–CO_2 $\pi_{//}$ pairs, as is easily understood by referring to Figure 20. In the case of $Ni(PH_3)_2(CO_2)$, the CO_2 bending hardly influences overlaps between metal d_{xz} and CO_2 $n\pi_{//}$ orbitals and between metal d_{z^2} and CO_2 $\pi_{//}$ orbitals. Thus, as the CO_2 distortion increases, the EX repulsion hardly decreases in the η^2–side on mode, but significantly decreases in the η^1–C mode. The ES stabilization of the linear structure is greater than that of the bent one, probably owing to the shorter distance between cobalt and oxygen atoms. This greater ES stabilization, however, cannot compensate the greater destabilization of EX and the smaller stabilization of BCTPLX in the linear structure. Thus, the bent structure of CO_2 is stable in $[Co(alcn)_2(CO_2)]^-$.

The driving force to bend CO_2 is summarized in Table 10A. It is noted that although the coordinated CO_2 has a bending structure in both η^1–C and η^2–side on modes, the driving force is slightly different between two coordination modes.

Besides coordination modes and distortion of carbon dioxide, the orien-

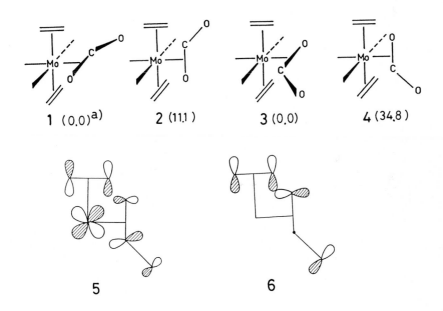

1 $(0.0)^{a)}$ 2 (11.1) 3 (0.0) 4 (34.8)

5   6

Fig. 26. Several rotational isomers of trans, mer–$Mo(PH_3)_3(C_2H_4)_2(CO_2)$, and back–donation and four electron destabilizing interactions of the isomer 4. a) in parentheses: relative stabilities in energy (kcal/mol), using isomer 1 as a standard (energy 0). (Reproduced with permission from ref. 103. Copyright 1988 American Chemical Society).

144

tation of the CO_2 ligand has been discussed in trans, mer-$Mo(PH_3)_3(C_2H_4)_2(CO_2)$ (ref. 103). Several rotational isomers, shown in Figure 26, have been examined and isomer 1 is calculated to be the most stable owing to the best situation for back–bonding interactions between Mo and two ethylene ligands and between Mo and carbon dioxide. Isomer 3 is much less stable than 1 and 2 (see Figure 26 for 1- 6), which results from weak back–bonding interaction between Mo and carbon dioxide (see 5 in Figure 26) and larger four-electron destabilizing interaction between C_2H_4 π and CO_2 nπ orbitals (see 6 in Figure 26). The latter interaction seems characteristic in CO_2 coordination, because an nπ orbital does not exist in the ethylene ligand but is present in carbon dioxide ligands.

A formation reaction of the transition metal η^2-CO_2 adducts has been investigated with the EH-MO method (ref. 104). In an early stage of the reaction, CO_2 approaches the metal from its 0-end, yielding the end-on conformer. In a later step of the reaction, the η^1-0 end on conformer changes to the η^2-side on conformer and then starts to cause the bending of CO_2, as shown in Figure 27. This postulated approach seems very reasonable, because

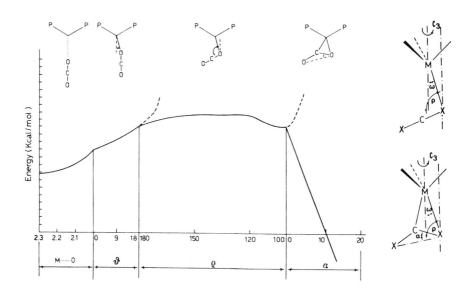

Fig. 27. Total energy variation along the pathway that brings a linear CO_2 and a $Ni(PH_3)_2$ fragment, non-interacting, to the η^1-0 end on conformer, and hence to the η^2-side on complex. The angular parameters are defined in pictures on the right hand side. (Reproduced with permission from ref. 102. Copyright 1984 American Chemical Society).

the coulombic interaction is important at rather long inter-fragment distance, and the charge-transfer interaction becomes important at rather short inter-fragment distance. It is noted that, although the final structure is the η^2-side on CO_2 complex, the energetically accessible reaction path starts from the η^1-O end on structure.

In conclusion, the stereochemistry of transition-metal CO_2 complexes has been successfully explained by considering important interactions between transition metal and carbon dioxide (see Tables 10 and 10A). The discussion about coordinate bond nature and stereochemistry can be easily applied to predicting structures of transition metal complexes which are postulated or proposed in several reactions of carbon dioxide.

5. STRUCTURES AND BONDING NATURE OF TRANSITION-METAL DINITROGEN COMPLEXES
5.1 STRUCTURES of TRANSITION-METAL DINITROGEN COMPLEXES: EXPERIMENTAL STUDIES

Compared to transition-metal CO_2 complexes, a larger number of stable transition-metal dinitrogen complexes have been reported. For instance, the solid circle, in the following table, means the metal forms stable dinitrogen complexes (ref. 105):

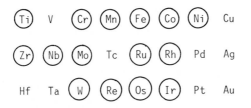

As described in the Introduction, the structures of coordinated dinitrogen are grouped into two categories, the η^1-end on and η^2-side on modes. In each, N_2 coordination has two possibilities; in the first, the dinitrogen ligand coordinates with only one metal, and in the second, the dinitrogen ligand coordinates with two or more metal atoms. X-ray studies of those dinitrogen complexes studied before about 1978 are summarized in ref. 105 and X-ray studies after 1979 are compiled in Table 11. From this table, and ref. 105, it is apparent that the η^1-end on coordination mode is common in mononuclear transition-metal dinitrogen complexes. A typical example of the η^1-end on dinitrogen complex is shown in Figure 28. Several characteristic features are found in their structures; (1) the M-N-N moiety is nearly linear, in asmuchas very small deviations are found in almost all mononuclear dinitrogen complexes. (2) Although the C-O bond is significantly lengthened in transition-metal CO_2 complexes, the N-N distance is only slightly lengthened in many transition metal dinitrogen complexes except for $[Ta(CHCMe_3)_3]_2(\mu\text{-}N_2)$ (ref. 108). It is

TABLE 11

Experimental structures of transition-metal dinitrogen complexes

Complexes	R(M-N) (Å)	R(N-N) (Å)	∠M-N-N (degree)	ν(N-N) (cm^{-1})	ref
trans-RhCl(PPr$^i_3$)$_2$(N$_2$)	1.885(4)	0.958(5)	177.3		106
[RhH(PPr$^i_3$)$_2$]$_2$(μ-N$_2$)	1.977(6)	1.134(5)	179.3	1927	107
[Ta(CHCMe$_3$)$_3$]$_2$(μ-N$_2$)	1.837(8)	1.298(12)	171.4(7)		108
	1.842(8)				
Mo(PMe$_3$)$_5$(N$_2$)	2.02(3)	1.12(3)			109
cis-Mo(PMe$_3$)$_4$(N$_2$)$_2$	1.97(1)	1.14	179(1)	2010	110
			177(1)	1965	
trans-MoCl(PMe$_3$)$_4$(N$_2$)	2.08(1)	1.14(2)	180.0		111
trans-WCl(PMe$_3$)$_4$(N$_2$)	2.04(2)	1.19(3)	180.0		111
trans-Mo(PMePh$_2$)$_2$(Ph$_2$PCH$_2$CH$_2$SMe)(N$_2$)$_2$					
	1.98(2)	1.04(3)	168(3)	2014w	112
	2.00(3)	1.10(3)	174(3)	1942s	
trans-Mo(N$_2$)$_2$(Me$_6$[16]aneS$_4$)	2.008	1.108	176.7		143
	1.991	1.105	176.2		
W(η^6-C$_6$H$_5$PPr$^n_2$)(PPr$^n_2$Ph)$_2$(N$_2$)	1.980(8)	1.126(0)	175.4(8)	1970	113
trans-[W(PEt$_2$Ph)$_4$(N$_2$)$_2$]thf	1.986(6)	1.146(7)	179.3(5)		113
	1.994(6)	1.139(7)	179.5(5)		
[(η^5:η^5-C$_{10}$H$_8$)(η-C$_5$H$_5$)$_2$Ti$_2$][η^1:η^5-C$_5$H$_4$)(η-C$_5$H$_5$)$_3$Ti$_2$](μ$_3$-N$_2$)					114
Ti-η^1-N$_2$	1.953(11)	1.301(12)			
	1.857(11)				
Ti-η^2-N$_2$	2.181(10)				
	2.097(11)				
N$_2$ (gas)		1.0976			115
				2359.6	116
Ph-N=N-Ph		1.23			115
				1441	117
H$_2$N-NH$_2$		1.449			115
				1111	118

surprising that even a decrease in the N-N bond distance is found in several transition-metal dinitrogen complexes, such as trans-Mo(CO)(N$_2$)(dppe)$_2$ (R$_{N-N}$=1.087 Å; ref. 119), trans-ReCl(N$_2$)(PMe$_2$Ph)$_4$ (R$_{N-N}$=1.06(3) Å; ref. 120), trans-RhH(N$_2$)(PBu$^t_2$Ph)$_2$ (R$_{N-N}$=1.074 (7) Å; ref. 121), and trans-RhCl(PPr$^i_3$)$_2$(N$_2$) (R$_{N-N}$=0.958(5) Å; ref. 106). This shortening has been discussed in some detail in terms of large thermal vibration (refs. 106 and

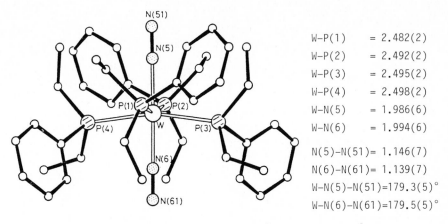

W–P(1)	= 2.482(2)
W–P(2)	= 2.492(2)
W–P(3)	= 2.495(2)
W–P(4)	= 2.498(2)
W–N(5)	= 1.986(6)
W–N(6)	= 1.994(6)
N(5)–N(51)	= 1.146(7)
N(6)–N(61)	= 1.139(7)
W–N(5)–N(51)	=179.3(5)°
W–N(6)–N(61)	=179.5(5)°

Fig. 28 Experimental structure of trans-$[W(N_2)_2(PEt_2Ph)_4]$ thf
(Reproduced with permission from ref. 113. Copyright 1986 Royal
Society of Chemistry).

121). If the error caused by this thermal vibration is corrected, the N–N
distance would become longer (ref. 121). Such correction was carried out in
$RhH(N_2)(PBu^t_2Ph)_2$, which lengthens the N–N distance from 1.074 Å to 1.18 Å;
but, this corrected value was considered to be overestimated. (3) The N–N
distance does not correlate with the N–N vibration (ref. 121). Although the
lower shift of the N–N vibration is rather large, the lengthening of the N–N
distance is small. (4) In the η^1-end on bridging type dinitrogen complexes,
the N–N distance is considerably lengthened; for instance, R(N–N)=1.134(5) Å
in $[RhH(PPr^i_3)_2]_2(\mu-N_2)$ (see Table 11; ref. 107), R(N–N)=1.298(12) Å in $[Ta-
(CHCH_3)_3]_2(\mu-N_2)$ (see Table 11; ref. 108), and R(N–N)=1.12 Å in $[(PCy_3)_2Ni]_2-
(\mu-N_2)$ (ref. 122). Especially in $[Ta(CHCMe_3)_3]_2(\mu-N_2)$, the N–N distance is
intermediate between the N–N single and N=N double bonds, but on the other
hand, the Ta–N distance is rather short. Thus, the Ta–N_2 interaction is described as being Ta=N–N=Ta (ref. 108).

The pure η^2-side on N_2 coordination is found in Ni(0) clusters, $\{C_6H_5[Na-
O(C_2H_5)_2]_2[(C_6H_5)_2Ni]_2(N_2)NaLi_6(OC_2H_5)_4 \cdot O(C_2H_5)_2\}_2$, which includes a dinitrogen
ligand interacting with two Ni atoms via the η^2-side on coordination, as shown
in Figure 29 (ref. 123). The other type of η^2-side on coordination has been
reported in $[(\eta-C_5H_4)(C_5H_5)_3Ti_2]_2(\mu^3-N_2)$. In this complex, a η^3-N_2 ligand
coordinates with one Ti atom via the η^2-side on mode and with the other two Ti
atoms via the η^1-end on mode, as shown in Figure 30 (ref. 114). This type of
coordination structure can be considered bifunctional coordination and
resembles the bifunctional CO_2 coordination in Re and Os carbonyl clusters

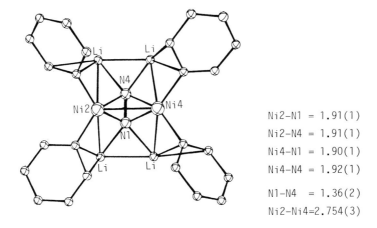

Ni2–N1 = 1.91(1)

Ni2–N4 = 1.91(1)

Ni4–N1 = 1.90(1)

Ni4–N4 = 1.92(1)

N1–N4 = 1.36(2)

Ni2–Ni4=2.754(3)

Fig. 29 Structure of a main part of {C₆H₅[Na O(C₂H₅)₂]₂[(C₆H₅)₂Ni₂–N₂NaLi₆(OC₂H₅)₄ O(C₂H₅)₂}₂   (Reproduced with permission from ref. 112c. Copyright 1976 American Chemical Society).

(see Figures 13 and 14; refs. 79 and 80) and K[Co(R-salen)(CO₂)] (Figure 11; ref. 76). Although the lengthening of the N–N distance is small in mononuclear transition metal dinitrogen complexes, the N–N distance is significantly lengthened in η²-side on bridging type N₂ complexes. Not only the lengthening of N–N distance, but also the lowering shift of the N–N stretching, is remarkably large. These two results indicate that the N–N bond order of the coordinated dinitrogen is intermediate between those of azo (–N=N–) and hydrazo (>N–N<) compounds, as compared in Table 11.

Besides the above-mentioned transition-metal dinitrogen complexes, which include only one-type of dinitrogen ligand, two types of dinitrogen ligands are found in [W(N₂)₂(PEt₂Ph)₃]₂(μ-N₂) (ref. 113) and [(η⁵-C₅Me₅)₂Zr(N₂)]₂(μ-N₂). As shown in Figure 31, these complexes have both terminal type η¹-end on N₂ and bridging type η¹-end on N₂ ligands. A common picture of these two complexes is that the coordination structure around the metals are rotated about the M-N=N-M axis by 90°, i.e., a terminal η¹-N₂ ligand on one metal is staggered to a terminal η¹-N₂ ligand on the other metal. This configuration is preferred to the structure in which the terminal η¹-N₂ ligand on one metal is eclipsed to a terminal η¹-N₂ ligand on the other metal, owing to both steric and electronic factors. From consideration of the steric factors, the staggered form has a less crowded arrangement of ligands than the eclipsed form. At the same time, in the eclipsed form   two π* orbitals of the μ-N₂ ligand can overlap well with the π-type orbitals of each metal, which does not form π-type interaction

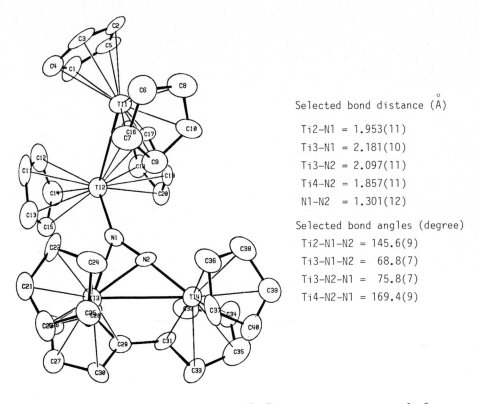

Selected bond distance (Å)

Ti2–N1 = 1.953(11)
Ti3–N1 = 2.181(10)
Ti3–N2 = 2.097(11)
Ti4–N2 = 1.857(11)
N1–N2 = 1.301(12)

Selected bond angles (degree)
Ti2–N1–N2 = 145.6(9)
Ti3–N1–N2 = 68.8(7)
Ti3–N2–N1 = 75.8(7)
Ti4–N2–N1 = 169.4(9)

Fig. 30.   Ortep view of the $(\mu_3\text{-}N_2)[(\eta^5:\eta^5\text{-}C_{10}H_8)(\eta\text{-}C_5H_5)_2Ti_2][(\eta^1:\eta^5\text{-}C_5H_4)\text{-}$ $(\eta\text{-}C_5H_5)_3Ti_2]$ unit.   (Reproduced with permission from ref. 114.   Copyright 1982 American Chemical Society).

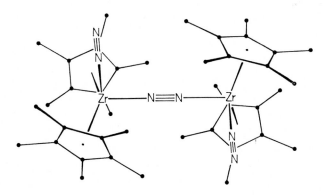

Fig. 31.   Experimental structure of $[(\eta^5\text{-}C_5Me_5)_2ZrN_2]_2N_2$ (Reproduced with permission from ref. 124. Copyright 1978 American Chemical Society).

with the terminal N_2 ligands (ref. 113).

Although the η^2-side on coordination has not been isolated in mononuclear transition-metal complexes, the presence of this coordination mode has been proposed in several cases. $Ti(Cp)_2(N_2)$ has been investigated with ^{1}H, ^{13}C, and ^{15}N NMR, in which the presence of the equilibrium "η^1-end on $N_2 \rightleftarrows \eta^2$-side on N_2" has been suggested (ref. 125). The same type equilibrium has been proposed in $[Ru(NH_3)_5(N_2)]^{2+}$ (see eq. 7; ref. 126). The activation enthalpy $\Delta H^{\ddagger}$ of this equilibrium has been reported to be 21 kcal/mol, and the $\Delta H^{\ddagger}$ for the N_2 loss is 28 kcal/mol. Some indications for the presence of the η^2-side

$$M-N_a\equiv N_b \quad \rightleftarrows \quad \left[M-\overset{N_a}{\underset{N_b}{\lVert\rVert}}\right] \quad \rightleftarrows \quad M-N_b\equiv N_a \qquad (7)$$

on coordinated N_2 complexes have also been reported in $(\eta^1-C_5H_5)_2-ZrCl[(Me_3Si)_2CH](N_2)$ (ref. 127).

In conclusion, the η^2-side on coordination is not impossible, but the η^1-end on coordination is common in transition-metal dinitrogen complexes. The N_2 molecule has π and π^* orbitals as in acetylene. A difference between dinitrogen and acetylene is that the former has a lone-pair orbital but the latter does not. Considering the existence of many transition-metal acetylene complexes, it seems strange that the η^2-side on coordination of dinitrogen has not been found in the case of simple mononuclear transition-metal complexes. In the following section, the coordinate bonding nature of the dinitrogen ligand is discussed in relation to the electronic structure of transition-metal dinitrogen complexes, and then the reason why the η^1-end on coordination mode is so common will be investigated in the following section, based on MO studies and energy decomposition analysis of an interaction between transition-metal and dinitrogen.

5.2 ELECTRON DISTRIBUTION and COORDINATE BONDING NATURE

The coordinate bonding nature of transition-metal dinitrogen complexes is one of the important subjects in complexes of inert molecules. In general, the dinitrogen coordination is discussed in terms of σ-donation from N_2 to metal and π-back donation from metal to N_2, as shown in the following scheme.

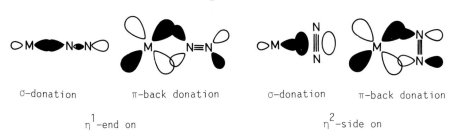

σ-donation   π-back donation   σ-donation   π-back donation

η^1-end on η^2-side on

Thus, the relative importance of σ-donation and π-back donation has been an attractive issue for long; both, in experimental and theoretical fields (refs. 105 and 128).

Early on, the relative importance of the σ-donation and π-back donation was investigated in $Os(N_2)X_2(PR_3)_3$ and its carbonyl analogue, using infrared spectral measurement, in which the coordinate bond of the dinitrogen ligand was compared with that of carbonyl ligands (ref. 129). By examining group dipole moment derivatives, it was concluded that the π* orbital of the dinitrogen ligand has a much lesser acceptor ability than the carbonyl ligand; and, at the same time, the σ-donor ability of the dinitrogen ligand is weaker than that of carbonyl ligands.

TABLE 12

N 1s ionization potentials of transition-metal dinitrogen complexes

Complexes	N 1s binding energy(eV)		Separation(eV)	Ref.
trans-RuCl(diars)$_2$(N$_2$)	402.3	400.7	1.6	130
ReCl(diphos)$_2$(N$_2$)	399.9	397.9	2.0	130
(C$_5$H$_5$)Mn(CO)$_2$(N$_2$)	403.0	401.8	1.2	131
ReCl(PMe$_2$Ph)$_4$(N$_2$)	400.1	398.4	1.6	132
ReBr(PMe$_2$Ph)$_4$(N$_2$)	400.1	398.5	1.5	132
ReBr(Py)(PMe$_2$Ph)$_3$(N$_2$)	399.6	398.0	1.6	132
ReCl(Py)(PMe$_2$Ph)$_3$(N$_2$)	399.8	398.3	1.5	132
ReCl(Py)(PMePh$_2$)$_3$(N$_2$)	399.9	398.2	1.7	132
ReCl(Py)(Ph$_2$PCH$_2$CH$_2$PPh$_2$)$_2$(N$_2$)	400.9	398.8	2.1	132
(PhMe$_2$P)$_4$ClRe(N$_2$)MoCl$_4$(OMe)	398.6			132

The relative importance of σ-donation and π-back donation significantly influences the electron distribution of dinitrogen complexes. ESCA studies are expected to offer reliable information of electron distribution, and several investigations were reported on dinitrogen complexes. A characteristic feature of ESCA chemical shifts is that two kinds of N 1s ionization, which are separated by about 1.5 to 2 eV, are observed, as displayed in Table 12 (ref. 130 - 132). The peak of lower energy has been assigned to be the 1s ionization of the terminal nitrogen atom, and the higher peak to be the 1s ionization of the coordinating nitrogen atom (ref. 130 - 131). This assignment was based on the idea that the transition-metal dinitrogen complex has two major resonance forms (eq. 8); however, no direct evidence has been described for validity of this assignment.

$$M\overset{-}{\text{———}}\overset{+}{N}\text{≡}N \quad \longleftrightarrow \quad M\text{═}\overset{+}{N}\text{═}\overset{-}{N} \tag{8}$$

The next interesting result derived from ESCA studies is that nitrogen atoms appear to be absolutely negatively charged (ref. 132). Furthermore, the oxidation state of the rhenium in $ReCl(N_2)(PMe_2Ph)_4$ has been estimated to be almost the same as that in $ReCl_2(PMe_2Ph)$. This means that the dinitrogen ligand is as electron-withdrawing as chlorine is in the Re(I) complex. By examining the $Re(4f_{\frac{7}{2}})$ binding energy, the electron-withdrawing ability of NO, CO and N_2 ligands are estimated to decrease in the order; NO > CO ∿ N_2.

A theoretical study with MO methods is expected to offer a variety of information about the coordinate bonding nature of dinitrogen complexes. A pioneering theoretical examination of $Cr(CO)_6$ and $Cr(N_2)_6$ was carried out with the Fenske-Hall MO method, and the coordinate bond of a dinitrogen ligand was discussed in terms of σ-donation and π-back donation (ref. 133). In this study, the following interesting results were reported; (1) The smaller t_{2g} and larger e_g metal orbital populations of $Cr(CO)_6$, compared to $Cr(N_2)_6$, indicate that a carbonyl ligand is a better σ-donor and π-acceptor ligand than a dinitrogen ligand. (2) The lone-pair orbital of dinitrogen and carbonyl ligands is anti-bonding in the N-N and C-O regions; and, as a result, the σ-

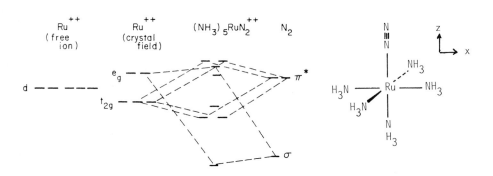

Fig. 32. Schematical molecular orbital splitting diagram of $[Ru(NH_3)_5(N_2)]^{2+}$. The N_2 orbitals have been shifted down from their free molecule values in view of the 2+ charge of Ru. The lower two t_{2g} orbitals are d_{xz} and d_{yz}. (the left half of this picture is reproduced with permission from ref. 134. Copyright 1981 American Chemical Society).

donation slightly strengthens the internal ligand σ-bonding. The second result suggests that the bond weakening of the coordinating CO and N_2 ligands cannot be related to the strength of π-back donation only. The first result accords with the infrared study of Os-dinitrogen complexes (ref. 129). In an Xα-DV study of $[Ru(NH_3)_5(N_2)]^{2+}$, the importance of π-back bonding has been deduced from the electron distribution (ref. 134). The nitrogen ligand exhibits −0.27e of the overall charge. On the other hand, the net charge of Ru is +2.09e, which is higher than that in $[Ru(NH_3)_6]^{2+}$ by about 0.7 e and higher even than that in $[Ru(NH_3)_6]^{3+}$. Furthermore, we can extract meaningful information about the contribution of π-back donation from the orbital interaction diagram shown in Figure 32. Of three t_{2g} orbitals, d_{xy}, d_{yz}, and d_{xz}, of Ru, the d_{xz} and d_{yz} orbitals can interact with dinitrogen π* orbitals, but the d_{xy} orbital cannot interact with the π* orbital. Thus, the π-back donation stabilizes the d_{xz} and d_{yz} orbitals in energy but does not change that of the d_{xy} orbital. The energy difference between these two orbitals is about 0.3 eV (see Figure 32), which means the π-back donation contributes to the dinitrogen coordination in this complex. All these results strongly suggest the importance of π-back bonding interaction in the dinitrogen coordination.

The relative importance of σ-donation and π-back donation has also been investigated with energy decomposition analysis (ref. 49c, 135). As shown in Table 13, the η^1-end on trans-RhCl(PH₃)₂(N₂), 20, receives more stabilization

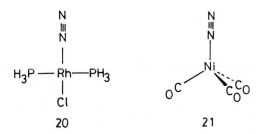

20   21

from the BCTPLX term than from the FCTPLX term. The π-back bonding interaction is included as a main contributor in the BCTPLX term and the σ-donation is in the FCTPLX term. Thus, the results of EDA suggest that the π-back donation is stronger than the σ-donation (ref. 135). A similar EDA study was carried out on the dinitrogen coordination of Ni(CO)₃, 21 (see Table 14; ref. 49c). In this complex, the σ-donation contributes to the stabilization of dinitrogen coordination to a similar extent to the π-back donation. This is probably because the π-back donation between Ni and dinitrogen ligand would be weakened by strong π-back donation between Ni and CO. In these studies, the coordinate bond of a dinitrogen ligand is compared with a variety of similar ligands, such

TABLE 13

Energy decomposition analysis of the interaction between $RhCl(PH_3)_2$ and various ligands with ab-initio MO method (kcal/mol; Note negative value means the stabilization in energy)

ligand	C_2H_4	η^1-N_2			η^1-N_2	CO	HNC	HCN
		2.246^a	2.70^b	2.250^c				
BE	−24.0	1.9	−5.7	0	−21.5	−35.8	−42.5	−26.5
DEF	3.2	0.1	0.0	0.1	0	0	0	0
INT	−27.2	1.8	−5.8	−0.1	−21.5	−35.8	−42.5	−26.5
ES	−52.6	−26.7	−1.2	−21.4	−39.2	−90.7	−94.5	−47.9
EX	66.0	66.0	7.1	54.1	54.1	121.5	111.5	52.7
FCTPLX	−16.2	−11.1	−4.1	−9.8	−13.0	−27.9	−26.2	−13.8
BCTPLX	−17.6	−20.1	−4.6	−17.3	−17.7	−32.7	−25.5	−14.0
R	−6.8	−6.3	−3.1	−5.7	−5.7	−6.1	−7.8	−10.0

a) The distance between Rh and the center of the N-N bond (A) yielding the same EX value as that of the C_2H_4 complex. b) The equilibrium structure. c) The Rh-N_2 distance yielding the same EX value as that of the η^1-N_2 complex. (Reproduced with permission from ref. 135. Copyright 1985 American Chemical Society).

as CO, HNC, HCN, CS, $CNCH_3$ which have a lone pair, π and π^* orbitals like those in dinitrogen ligands. In both EDA studies of trans-$RhCl(PH_3)_2L$ and $Ni(CO)_3L$, the carbonyl coordination is estimated to yield stronger σ-donation and π-back donation than the dinitrogen coordination, which agrees with the experimental conclusions of the infrared study (ref. 129) and the theoretical results of Fenske-Hall MO calculations (ref. 133). This conclusion is readily understood by considering that the carbonyl ligand has a lone pair orbital at a

TABLE 14
Energy decomposition analysis of the interaction between $Ni(CO)_3$ and various ligands with Xα-DV method (kcal/mol; Note negative value mean stabilization in energy).

ligands	CO	η^1-N_2	CS	PF_3	$CNCH_3$
ΔE (total)	−62.7	−48.3	−70.9	−82.8	−85.6
ΔE (static)	0	3.1	3.8	−16.3	−9.4
ΔE (donative)	−25.1	−23.8	−31.4	−32.0	−38.9
ΔE (back donative)	−35.1	−23.8	−38.2	−35.1	−37.0
ΔE (remaining)	−2.5	−3.8	−5.1	+0.6	−0.3

(Reproduced with permission from ref. 49c. Copyright 1979 American Chemical Society).

TABLE 15
Ligand properties: Mulliken charge of the coordinating atom, π, π^*, and lone pair orbital energy

Ligands	HNC	CO	HCN	η^1-N_2
Mulliken charge of coordinating atom	+0.30	+0.40	−0.34	0
Orbital energies (eV) π	−14.1	−17.4	−13.6	−17.1
π^*	5.6	4.0	5.4	4.4
lone pair	−13.0	−14.9	−15.6	−17.1

4-31 G basis set was used for calculations.

higher energy level and the π^* orbital at a lower energy level than the dinitrogen ligand does (table 15).

In both theoretical EDA studies, the dinitrogen coordination is calculated to be the weakest. The σ-donation of the dinitrogen ligand is the weakest, probably because the lone pair orbital is the most stable (Table 15). Also, the dinitrogen ligand forms a weaker π-back bonding than related ligands, except for HCN which has a slightly weaker ability for π-back bonding. The π^* orbital energy of a dinitrogen ligand is not low, very much, compared to the other ligands. However, the π^* orbital of CO, CS, HNC, and CH_3NC ligands places a larger p_π contribution on the coordinating atom than the dinitrogen ligand does, as schematically shown in the following picture. Thus, the π^*

orbital of the dinitrogen ligand overlaps with the metal $d\pi$ orbitals to a lesser extent than the π^* orbital of the other ligands, yielding weaker π-back donating interactions in the coordination of a dinitrogen ligand. Nevertheless, the π-back bonding is primarily important in the η^1-N_2 coordination and in the carbonyl coordination, because the dinitrogen coordinate bond receives a larger stabilization from the BCTPLX term (back bonding) than from the FCTPLX term (donation); this, of course, is not the case with HCN and HNC ligands, whose coordinate bonds receive almost the same stabilization from these two interactions (see Table 13).

Not only are the σ-donation and π-back donation weakest in dinitrogen coordination, but the same is true of the electrostatic interaction. The

TABLE 16

Constrained Space Orbital Variation (CSOV) analysis for
$Ni(N_2)$,[a] with reference[b] to Ni and N_2. ΔE_{INT} means
stabilization in energy by each interaction (eV unit).

Components	E_{INT}[c]	ΔE_{INT}[d]
frozen orbital	-2.90	
mix Ni 3d and 4s	-1.85	1.05
Ni polarization	-1.45	0.40
Ni σ donation to N_2	-1.42	0.03
Ni π donation to N_2	-0.68	0.74
N_2 polarization	-0.45	0.23
N_2 σ donation to Ni	-0.34	0.03
N_2 π donation to Ni	-0.32	0.03
N_2 covalent σ bond to Ni	-0.01	0.31
remaining bonding effects	0.14	0.15

1 eV=23.045 kcal/mol. a) R(Ni-N)=3.3 bohr and R(N-N)=2.07 bohr.
b) Both the orbitals and CI coefficients are taken from infinite
separation. c) SCF energy with the limited orbital space. d) Stabilization in energy at each step of calculation inducing additional
component to the bonding. (Reproduced with permission from ref. 72.
Copyright 1985 Elsevier).

coordinating atom of CO and HNC ligands is positively charged but these two
molecules have a good lone pair orbital that gives C^-O^+ and HN^+C^- polarity to
σ-electrons. This polarity of σ-electrons is expected to yield a significant
electrostatic stabilization in CO and HNC coordinations. The HCN ligand has a
negatively charged coordinating atom, which is a more favorable situation for
electrostatic interaction than in the η^1-end on N_2 ligand. Thus, the
dinitrogen ligand is the weakest ligand insofar as electrostatic interactions.

Since the above-described investigations are based on the Hartree-Fock ab
initio MO method, a more refined method including electron correlation effects
seems necessary for quantitative discussions. Recently, ab initio CAS SCF and
SD-CI calculations have been carried out on a model complex, $Ni(N_2)$ (ref. 72).
The ground state of $Ni(N_2)$ is taken as the $^1\Sigma^+$ state, since this complex is
considered a model. The CSOV analysis (Constrained Space Orbital Variation
method), with which partition of the interaction energy was performed, was
applied to the investigation of dinitrogen coordination. As shown in Table 16,
the π-back donation from Ni to dinitrogen is remarkably large, but the σ-
donation from the dinitrogen ligand to metal is very small. Besides the
donative σ-interaction, the dinitrogen coordination receives some stabilization from σ-type covalent interactions. This covalent interaction is about
3 times stronger than the donative σ-interaction. Nevertheless, the sum of
σ-interactions is still smaller than the π-back donation. The above results

indicate that the π-back donation is a main contributor to the dinitrogen coordination. Additional investigations of this type are desirable for more realistic complexes.

Now, let us start to discuss the electron distribution that was the second issue to be examined. Mulliken charges, which was not exact charges on an atom, but a useful measure of atomic charge, are summarized in Table 17. A clear contrast between semi-empirical (EH-MO) and ab initio MO (including $X\alpha$-DV) calculations is found in this table; In the former, the terminal nitrogen atom is more negatively charged than the coordinating atom; but, in the latter, the coordinating nitrogen atom is more negatively charged than the terminal atom with two unrealistic exceptions, $Cr(PH_3)_4(N_2)$ and $Mo(NH_3)_5(N_2)$. A simple back-bonding from metal $d\pi$ orbital to dinitrogen π^* orbital increases the electron population on both nitrogen atoms, because both nitrogen p_π orbitals contribute equally to the π^* orbital. The unequivalent electron distribution on the coordinating and terminal nitrogen atoms means the presence of some polarization in the dinitrogen ligand. Hoffmann and his collaborators reported an orbital mixing that explains the unequivalent electron distribution in such ligand as CN^-, N_2, and CO (ref. 140). As shown in Figure 33, the metal $d\pi$ orbital mixes with the π^* orbital of the ligand in a bonding way, but at the same time, in an anti-bonding way with π orbital of the ligand (this kind of orbital mixing is derived with second-order perturbation theory (ref. 141)). Such orbital mixing increases the electron density on

TABLE 17

Mulliken charges of various transition-metal dinitrogen complexes

Complexes	Method	M	N_c	N_t	ref.
trans-$Mo(PH_3)_4(N_2)_2$	EH	−0.59	0.05	−0.67	136
cis-$Mo(PH_3)_4(N_2)_2$	EH	0.15	−0.11	−0.88	136
$Ti(Cp)_2(N_2)$	ab-initio	1.53	−0.24	0.0	137
$[Ru(NH_3)_5(N_2)]^{2+}$	X −DV	2.09	−0.38	0.21	134
$[Ru(NH_3)_5(N_2)]^{2+}$	ab-initio	0.96	−0.12	0.15	138
$[Fe(NH_3)_5(N_2)]^{2+}$	ab-initio	1.50	−0.17	0.14	138
$Mo(PH_3)_5(N_2)$	ab-initio	0.40	−0.28	−0.37	138
trans-$Mo(NH_3)_4(N_2)_2$	ab-initio	0.59	−0.24	−0.23	138
trans-$Mo(PH_3)_4(N_2)_2$	ab-initio	0.65	−0.11	−0.03	138
$Cr(PH_3)_4(N_2)$	ab-initio	−0.048	−0.103	−0.222	139
trans-$RhCl(PH_3)_2(N_2)$	ab-initio	−0.077	−0.051	−0.027	135

a) N_c is a coordinating nitrogen atom and N_t is a terminal nitrogen atom.

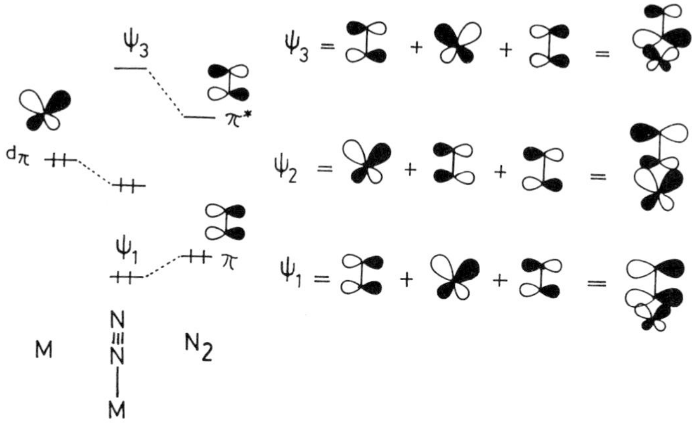

Fig. 33 Orbital mixing between metal dπ, dinitrogen π and π* orbitals.

the terminal atom but decreases the electron density on the coordinating atom. Consequently the terminal atom is more negatively charged than the coordinating atom. In many ab initio MO calculations, on the other hand, the coordinating atoms is more negatively charged, as described in Table 17. Often, the Mulliken charge includes some ambiguity in electron distribution. Thus, difference density maps of $[Ru(NH_3)_5(N_2)]^{2+}$ and trans-$RhCl(PH_3)_2(N_2)$ are examined, to obtain changes in electron density caused by dinitrogen coordination, as shown in Figure 34. In both complexes, the electron density increases around the coordinating nitrogen atom but decreases around the terminal nitrogen atom. The interesting feature is that the electron density appears to increase in the σ space near the coordinating N atom. Thus, the orbital mixing described in Figure 33 is not the origin of this electron distribution; rather, the σ polarization due to a positively charged metal atom seems to be the origin of the electron distribution found in these complexes. This explanation is supported by the electron distribution of Mo dinitrogen complexes given in Table 17; When the Mo atom is less positively charged, as in $Mo(NH_3)_5(N_2)$, the terminal nitrogen atom is more negatively charged than the coordinating nitrogen atom; but, when the positive charge of the Mo atom increases, the coordinating nitrogen atom becomes more negatively charged than the terminal nitrogen atom, as found in $Mo(NH_3)_4(N_2)_2$ and $Mo(PH_3)_4(N_2)_2$. Summarizing the above discussion, two factors must be considered, at least, in discussing the electron distribution of dinitrogen ligand; one is the mixing of the metal dπ orbital with dinitrogen π and π* orbitals, which makes the terminal nitrogen atom more negatively charged.

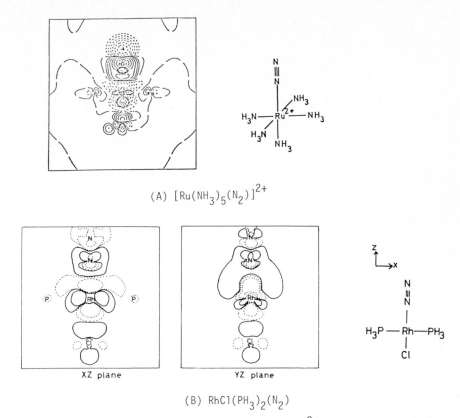

(A) $[Ru(NH_3)_5(N_2)]^{2+}$

XZ plane   YZ plane

(B) $RhCl(PH_3)_2(N_2)$

Fig. 34. Difference density maps of $[Ru(NH_3)_5(N_2)]^{2+}$ and $RhCl(PH_3)_2(N_2)$. Difference density $= \rho[ML_n(N_2)] - \rho(ML_n) - \rho(N_2)$; the solid lines indicate an increase in the density, and the dashed lines a decrease. (Reproduced with permission from ref. 135 and 138. Copyright 1980 Royal Society of Chemistry and 1985 American Chemical Society).

The other is the polarization of the dinitrogen ligand induced by the positive charge on the metal atom, which makes the coordinating atom more negatively charged. These two factors cause opposite results in the electron distribution; e.g., if the π-back bonding is sufficiently strong, the orbital mixing of Figure 33 is predominant, and as a result, the terminal nitrogen atom is more negatively charged. When the positive charge of the central metal atom is considerably large, the polarization due to the positive charge is predominant; and consequently, the coordinating nitrogen atom is more negatively charged. This seems the main reason that the electron distribution of transition-metal dinitrogen complexes changes considerably from complex to complex.

The last issue to be examined here is the assignment of ESCA chemical shifts of N 1s ionization. The previously proposed assignment suggested that the chemical shift at higher energy comes from the coordinating nitrogen 1s ionization and the chemical shift of lower energy arises from the terminal nitrogen 1s ionization (refs. 130 - 132). This assignment implies that the terminal nitrogen atom is more negatively charged than the coordinating nitrogen atom, which accords with the electron distribution obtained by semi-empirical MO calculations but does not agree with the results of ab initio MO calculations. The N 1s ionization in trans-$RhCl(PH_3)_2(N_2)$ has been investi-

TABLE 18

Calculated N 1s ionization energy (eV) of $RhCl(PH_3)_2(N_2)$

	$RhCl(PH_3)_2(N_2)$		free-N_2	
	Koopmans	ΔSCF	Koopmans	ΔSCF
N_c	428.19	412.20	426.66	412.60
N_t	427.88	411.21		
ΔI_p	0.31	0.91		

N_c means the coordinating nitrogen atom and N_t means the terminal nitrogen atom. (Reproduced with permission from ref. 135. Copyright 1985 American Chemical Society).

gated with ab initio MO methods (ref. 135). In this complex, the coordinating nitrogen atom is calculated to be more negatively charged than the terminal nitrogen atom, as described above. Nevertheless, the N 1s ionization of the coordinating nitrogen atom was estimated to be higher than that of the terminal nitrogen atom, in both calculations by Koopmans theorem and ΔSCF method, as shown in Table 18. The estimated energy difference between two N 1s ionization potentials is about 0.31 eV, by the Koopmans theorem, but increases to 0.99 eV in the ΔSCF calculation. In many transition-metal dinitrogen complexes, the energy difference between two 1s peaks is about 1.5 - 2 eV. and the estimated value is slightly smaller than the experimental value. Thus, while the estimation with ΔSCF calculation is better than with the Koopmans theorem, the improvement of the theoretical method by taking electron correlation into account is still necessary to get quantitatively correct values. However, the reason that the coordinating N 1s ionization is higher than that of the terminal N 1s ionization seems to be successfully discussed at the Hartree-Fock level. Not only the atomic charge, but the electrostatic potential of the metal, is expected to play an important role in the N 1s ionization potential since the ionized state has a positive hole localized on the nitrogen atom. The electrostatic potential of $RhCl(PH_3)_2$ is shown in Figure 35. The 1s

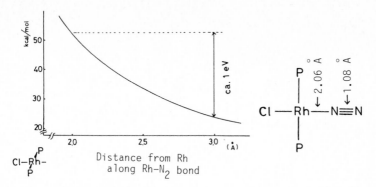

Fig. 35 Electrostatic potential of RhCl(PH$_3$)$_2$
(Reproduced with permission from ref. 135. Copyright 1985 American
Chemical Society).

ionization of the coordinating nitrogen atom yields a positive charge at about
a 2 Å distance from the Rh atom; but, the 1s ionization of the terminal
nitrogen atom yields a positive charge at about a 3.0 Å distance from the Rh
atom (see the right-half of Figure 35).   The difference in electrostatic
potential is about 1 eV, as shown in Figure 35, which roughly corresponds to
the difference in N 1s ionization between coordinating nitrogen and terminal
nitrogen atoms.   These results suggest several interesting features; (1)
Although the coordinating nitrogen atom is calculated to be more negatively
charged than the terminal nitrogen atom, the 1s ionization energy of the
coordinating nitrogen atom is calculated to be higher than that of the terminal
nitrogen atom.   (2) Not only atomic charge but also electrostatic potential due
to the metal are important in determining the nitrogen 1s ionization potential.
(3) Because the 1s ionization of the coordinating nitrogen atom yields a
positive charge near the positively charged Rh atom, this ionization requires
higher energy.   And, (4) the relaxation energy is not neglected and a more
refined theoretical method,   including electron correlation effects,   is
necessary for a quantitative discussion.

5.3 WHY the η^1-END ON COORDINATION IS VERY COMMON in TRANSITION-METAL
DINITROGEN COMPLEXES
     As described in section 5.1, we wondered why the η^1-end on mode is common
but the η^2-side on mode is very rare   in transition-metal dinitrogen
complexes.   This interesting issue has been investigated with EH-MO and ab
initio MO methods.

22   **23**

Two coordination modes of $Cp_2Ti(N_2)$, <u>22</u> and <u>23</u>, were initially compared in the EH-MO study (ref. 96). The orbital interaction diagrams for these two coordination modes are given in Figure 36. In the η^1-end on coordination, one π^* orbital (b_2 symmetry) interacts with the Ti $d\pi$ orbital, to form the π-back bonding interaction (note the π^* orbital in the b_1 symmetry cannot form a π-back donatimg interaction with the Ti atom since the Ti $d\pi$ orbital of this symmetry is empty). The lone pair orbital of the dinitrogen ligand also interacts with the Ti σ-type orbital (of a_1 symmetry), to form the σ-donating interaction. In the η^2-side on mode, only one π^* orbital interacts with the Ti $d\pi$ orbital (b_2 symmetry), yielding the π-back bonding interaction. In this mode, the lone pair and σ orbitals of dinitrogen belonging to the same symmetry as the $2a_1$ orbitals on Cp_2Ti orbital give rise to two kinds of donating interaction. However, the σ-donating interaction between the N_2

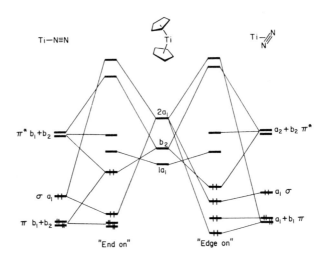

Fig. 36 Interaction diagram for η^1-end on (left) and η^2-side on (right) in $Cp_2Ti(N_2)$. (Reproduced with permisssion from ref. 96. Copyright 1976 American Chemical Society).

lone pair and Cp_2Ti $2a_1$ orbitals would be very weak due to the small overlap between these orbitals. Thus, the σ-donating interaction contributes to a similar extent to both η^1-end on and η^2-side on N_2 coordinations. On the other hand, the contribution of the π-back bonding is different in the η^1-end on and η^2-side on modes, if the central metal has more than three d electrons; in this case, the η^1-end on mode can form two types of π-back bonding interactions, which would make the η^1-end on mode more stable than the η^2-side on mode. If the metal has a d^2 electron configuration, only one π-back bonding interaction is formed, even in the η^1-end on mode, which has the ability to form two kinds of π-back bonding. Thus, it is suggested from this EH-MO study that if the η^2-side on dinitrogen coordination occurs, it should be found in a d^2 system. Certainly, the presence of the equilibrium between "η^1-end on $\rightleftarrows$ η^2-side on modes" has been experimentally reported in $Cp_2Ti(N_2)$ (ref. 125).

$Cp_2Ti(N_2)$ has been also investigated with ab initio MO methods, in which the η^1-end on N_2 mode has been calculated to be more stable than the η^2-side on mode(ref. 137). This means that the most stable structure is the η^1-end on coordination, even in the d^2 system, and probably the η^2-side on mode is possible only as an intermediate, being less stable than the η^1-end on mode. This ab initio MO study suggests an interesting difference in bonding nature between two coordination modes; the σ-donation favors the η^1-end on mode, but the π-back donation favors the η^2-side on mode.

The possibility of the η^2-side on bridging type coordination has been examined in a dinuclear complex, $[Co(CO)_3]_2(\mu-N_2)$, with the EH-MO method (ref. 142). Two bridging structures are possible; one is perpendicular to the Co-Co axis, 24, and the other is parallel to the Co-Co axis, 25. These dinitrogen

24   25

coordinations are discussed on the basis of the orbital interaction diagram given in Figure 37. In the perpendicular structure, two π^* orbitals of the dinitrogen ligand overlap well with the HOMO of the $[Co(CO)_3]_2$ part, yielding a strong π-back bonding interaction. In the parallel structure, on the other hand, one π^* orbital of dinitrogen overlaps with an empty d orbital, $3b_2$, of the $[Co(CO)_3]_2$ part (Figure 37), which makes the parallel structure less stable. Thus, the η^2-side on the bridging type dinitrogen coordination is

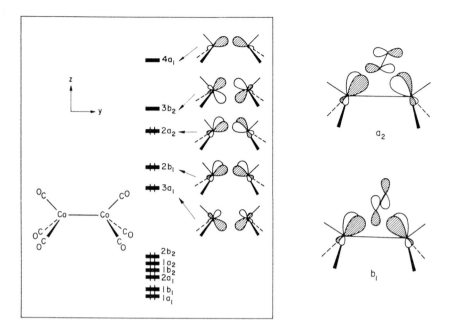

Fig. 37 Several important MO's near HOMO and LUMO of $Co_2(CO)_6$
(Reproduced with permission from ref. 142. Copyright 1982 American
Chemical Society).

predicted to be possible, while the parallel coordination is difficult.

Two coordination modes, η^1-end on and η^2-side on modes, have been compared
in trans-$RhCl(PH_3)_2(N_2)$, **20**, with ab initio MO and EDA methods (ref. 135). As
clearly shown in Table 13, the η^1-end on coordination is more stable than the
η^2-side on coordination by ca. 16kcal/mol. The η^2-side on mode exhibits a very
small binding energy, and at the same time, a very long Rh–N distance. The
difference in binding energy between two coordination modes is comparable with
the activation enthalpy ($\Delta H^{\ddagger}$= 21 kcal/mol) estimated for the intramolecular
linkage isomerization of N_2 in $[Ru(NH_3)_5(N_2)]^{2+}$ (ref. 126). In the equilibrium
structure, the η^2-side on complex suffers a small EX destabilization and
receives a small stabilization from ES, FCTPLX, and BCTPLX terms which
correspond to electrostatic, donating, and π-back donating interactions,
respectively. The η^1-end on coordination suffers significant destabilization
from EX repulsion but receives large stabilization from ES, FCTPLX, and BCTPLX
terms. To find an intrinsic difference between these two modes, we had better
compare these two modes at a similar Rh–N distance. However, it is not easy to

define the same Rh–N$_2$ distance for these different coordination modes. Here, we propose the EX repulsion as a measure of inter-fragment distance between RhCl(PH$_3$)$_2$ and N$_2$ in view of the following; the EX repulsion depends on the contact of electron clouds between two fragments, RhCl(PH$_3$)$_2$ and N$_2$; therefore, the same EX value corresponds to a comparable inter-fragment distance. The EX value of the η^1-end on N$_2$ complex (54.1 kcal/mol) is chosen as a standard for a comparison between the η^1-end on and η^2-side on modes, and the Rh–N$_2$ distance of the η^2-side on complex is shortened to acquire this EX value. Such comparison offers us a clear contrast between two modes, as shown in Table 13. Although both coordination modes receive a similar stabilization from the BCTPLX term, the η^1-end on mode receives a much larger ES stabilization and a slightly larger FCTPLX stabilization than the η^2-side on mode does. It is noted that, although the η^1-end on N$_2$ coordination receives a greater stabilization from the BCTPLX term than from the FCTPLX term (vide supra), the BCTPLX term does not influence the relative stability of the two coordination modes.

It is interesting to compare the η^2-side on dinitrogen complex with the ethylene complex since ethylene can ligate to Rh(I) by way of an η^2-side on coordination and form a stable complex, unlike N$_2$. Again, the comparison was carried out at the inter-fragment distance giving the same EX value (66.0 kcal/mol). As shown in Table 13, the Rh–C$_2$H$_4$ complex is more stable than the η^2-side on N$_2$ complex because of a much larger ES and slightly larger FCTPLX stabilizations. These results suggest that the ES and FCTPLX interactions seem

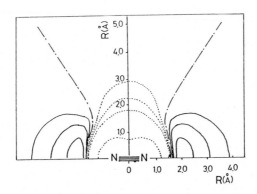

Fig. 38. Electrostatic potentials of N$_2$. ±0.02, ±0.01, ±0.005, 0.0 hartree. —— Negative value (stabilization of e$^+$); ----- Positive value (destabilization of e$^+$); —·— 0.0. (Reproduced with permission from ref. 135. Copyright 1985 American Chemical Society).

166

to play a key role in determining the relative stabilities of η^1-end on and η^2-side on N_2 coordinations. The ES potential map of N_2 was examined and shown in Figure 38, where a positive ES potential (given by dashed lines) mean destabilization of a positive charge and a negative ES potential (given by solid lines) means stabilization. Apparently, the positive ES region expands perpendicularly to the N≡N bond, but the negative region expands toward the outside of an N_2 bond, along the N≡N axis. This feature of the ES potential, corresponding to the negative quadrupole moment of N_2, results from a sufficient electron accumulation on the lone pair region, enough to compensate the ES repulsion due to the charge of an N nucleus, but insufficient electron accumulation on the π-orbital region, which is not enough to compensate the ES repulsion due to the charges of two N nuclei. This ES potential map indicates that ES stabilization can be obtained when a positive metal ion, such as Rh(I) ion, approaches N_2 along the N≡N bond axis, i.e., in the manner of η^1-end on coordination, but the ES destabilization would arise when a positive chemical species approaches to N_2 perpendicularly to the N≡N bond, i.e., in the way of the η^2-side on coordination. The ES potential must be compared between N_2 and C_2H_4. ES potentials of these ligands are given as a function of distance between Rh and the center of the C=C or N≡N bond, in Figure 39. In contrast to the positive ES potential of N_2, the ES potential of C_2H_4 is found to be negative around the coordinate bond distance. This ES potential of C_2H_4 seems to correspond to a positive quadrupole moment of C_2H_4, which probably results from the negative charge on the C atom and a sufficient electron accumulation in the π orbital region. The negative ES potential of C_2H_4 offers a large ES

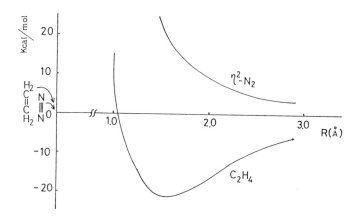

Fig. 39. Electrostatic potentials of C_2H_4 and N_2 in the direction perpendicular to the C=C and N≡N bonds. (Reproduced with permission from ref. 135. Copyright 1985 American Chemical Society).

stabilization to $RhCl(PH_3)_2(C_2H_4)$, unlike its η^2-side on N_2 analogue. In other words, the ES potential favors the η^1-end on N_2 and C_2H_4 coordination but disfavors the η^2-side on N_2 coordination.

The FCTPLX stabilization, corresponding to a donating interaction, is different for the η^1-end on and η^2-side on modes. This interaction depends on the energy level and expanse of the donor orbital. In the η^1-end on N_2 coordination, the lone pair orbital of N_2 contributes to this interaction, but in the η^2-side on and C_2H_4 coordinations, the π-orbital forms this donating interaction. Both the lone pair and the π-orbitals of N_2 lie similarly in energy, as shown in Table 15, but the lone pair orbital extends toward Rh(I) more than the π-orbital does, due to the hybridization in lone pair orbital. Thus, the η^1-end on coordination can receive larger stabilization from the FCTPLX term than the η^2-side on coordination. The π-orbital of C_2H_4 lies higher in energy than the π-orbital of N_2, leading to larger stabilization of FCTPLX in the C_2H_4 complex. A comparison of these interactions are summarized in Table 19.

In conclusion, the η^2-side on dinitrogen coordination is less stable than the η^1-end on coordination, mainly due to the smaller ES and FCTPLX stabilization. This means the neutral metal without acceptor ligands is desired for the η^2-side on coordination, since a neutral metal causes no ES repulsion with the

TABLE 19

Qualitative comparison of various interactions between the η^1-end on and η^2-side on N_2 coordinations

	η^1-end on	η^2-side on
Electrostatic	strong	weak
donating($N_2 \rightarrow$ Metal)	strong	weak
back donating (Metal $\rightarrow N_2$)	nearly equal	

positive quadrupole moment of N_2, and does not form strong donating interaction which favors the η^1-end on mode. Certainly, the dinitrogen coordination has been observed in a Ni(0) cluster compound which contains a Ni(0) atom and does not have any acceptor ligand. In this compound, the π-back donation is expected to be important and the ES interaction is considered not to disfavor the η^2-side on mode.

6. CONCLUDING REMARKS

The coordinate bond of non-Werner transition metal complexes has been discussed in terms of σ-donating and π-back donating interactions. Attempts have been actively made to estimate the strength of each interactions and their energetic contribution to the coordinate bond in both experimental and theoretical fields. Electron distribution and bond strength are susceptible to influences from both σ-donating and π-back donating interactions. So, it is not easy to assess how much each interaction contributes to the coordinate bond by examining electron distribution and bond strength. However, the energy decomposition analysis (EDA) method is expected to be effective in estimating the relative contribution of σ-donation and π-back donation, as described in Section 2. The EDA method has been successfully applied to investigations of many transition-metal complexes. Some of examples are given in Section 3, in which coordinate bonds of carbonyl and similar ligands are investigated with the EDA methods. The successful results obtained using EDA methods encourage us to apply this method to theoretical studies of transition-metal complexes of inert molecules, such as carbon dioxide and dinitrogen molecules.

In Section 4, discussions based on various MO and EDA studies are presented which examined coordinate bond nature, coordination modes, and ligand distortion of transition-metal carbon dioxide complexes. Relations between coordination modes and metal-ligand interactions are summarized in Table 9. The best condition for the η^2-side on mode are that the $d\pi$ orbital is HOMO and the $d\sigma$ orbital is empty, and the best conditions for the η^1-C mode are that the $d\sigma$ orbital is HOMO and the central metal is low valent. From these results, we can successfully predict which coordination mode is expected in a transition metal carbon dioxide complex, as described in Section 4.3. The significant distortion of coordinated carbon dioxide is one of the characteristic features of transition-metal carbon dioxide complexes. The driving force to cause the distortion is successfully discussed, based on the EDA calculations, in Section 4.4; as summarized in Table 10, strengthening of π-back bonding is a main driving force in the η^2-side on mode, and not only strengthening of π-back bonding but also a decrease in EX repulsion is a main driving force in the η^1-C mode.

In Section 5, the coordinate bond nature, electron distribution, and coordination modes of transition-metal dinitrogen complexes are examined, using ab-initio MO and EDA methods. Two coordination modes, η^1-end on and η^2-side on modes, are compared with each other in Section 5.3; although the π-back bonding is more important than the σ-donation in both η^1-end on and η^2-side on modes, the π-back bonding interaction does not play an important role in determining the coordination mode. On the other hand, the electrostatic interaction and σ-donating interaction make the η^1-end on mode more stable than the η^2-side on mode, when the metal is positively charged. From this discussion, we can suggest conditions to synthesize η^2-side on N_2 complexes, as described in the later half of Section 5.3. The electron distribution of transition-metal dinitrogen complexes is determined by two factors; one is the orbital mixing given in Figure 33, and the other is the polarization caused by the positive charge on the metal atom. These two factors result in different electron distribution relative to each other. Thus, we can easily understand the reason that the electron distribution changes considerably from complex to complex.

Many theoretical investigations presented here are carried out at the Hartree-Fock level. We have seen the importance of electron correlation effects on the structure and bonding nature of transition metal complexes in Section 3. Thus, the discussion given in Sections 4 and 5 must be considered to be semi-quantitative, in some cases. More sophisticated theoretical approaches including electron correlation effects have been applied to small model compounds. However, such methods are much more time-consuming than the usual ab-initio MO method, and their application to realistic complexes is still difficult. The author hopes that such methods will be applied to more realistic complexes in a near future and theoretical methods will be more and more powerful tools in investigating transition-metal chemistry.

Finally, several important and interesting papers, which have been published after completing this review, will be briefly described below. An excelent CAS SCF CI study of $Ni(CO)_4$ was carried out by Siegbharn group to elucidate electron correlation effects on carbonyl coordination with Ni(0) (ref. 144). Structures and coordinate bond nature of several iron(0) complexes of carbon dioxide and similar small molecule, $Fe(CO)_2(PH_3)_2(\eta^2-OCX)$ and $Fe(CO)_2(PH_3)_2(\eta^2-SCX)$ (X=O, S, NH, or CH$_2$), were investigated with ab initio MO method (ref. 145). An ab initio MO/MP2 study was performed on the η^1-C coordinate CO_2 complex, $RhCl(AsH_3)_4(CO_2)$ (ref. 146), in which the enhanced reactivity of the coordinated carbon dioxide and its origin were discussed in the basis of electronic structure of Rh(I)-CO_2 complex. Two ab initio MO studies have been reported on a CO_2 reaction with transition-metal complexes. In one, the CO_2 insertion into Cu(I)-H bond was investigated and several characteristic features of this reaction were elucidated (ref. 147). In the second, the CO_2 reaction with $[CrH(CO)_5]^-$ was studied and an interesting CO_2 adduct was proposed as an intermediate (ref. 148). Electron correlation effects on two coordination modes of dinitrogen complex, $Ni(PH_3)_2(N_2)$, was recently investigated using ab initio MO/SD-CI method (ref. 149). It was shown in this work that introduction of electron correlation effects, especially non-dynamical correlation effects, is indispensable for discussing the relative stabilities of two coordination modes in Ni(0) complex.

170

REFERENCES

1 For example, (a) G. Wilkinson, F. G. A. Stone, and E. W. Abel (Ed.),
   Comprehensive Organometallic Chemistry, Pergamon Press., Oxford, 1982.
   (b) F. R. Hartley and S. Patai (Ed.), The Chemistry of the Metal-Carbon
   Bond, John Wiley & Sons, Chichester, 1982, vol. 1; 1985, vol's. 2 and 3.
   (c) I. Wender and P. Pino (Ed.), Organic Synthesis via Metal Carbonyls,
   Vol. 2, A Wiley-Interscience, New York, 1977.
   (d) H. Alper (Ed.), Transition Metal Organometallics in Organic Synthesis,
   Vol. 2, Academic Press, New York, 1978.
   (e) R. P. Houghton, Metal Complexes in Organic Chemistry, Cambridge Univ.
   Press., Cambridge, 1979.
   (f) L. H. Pignolet, Homogeneous Catalysis with Metal Phosphine Complexes,
   Plenum, New York, 1983.
2 For example, (a) M. M. Taqui Khan and A. E. Martell, Homogeneous Catalysis
   by Metal Complexes, Activation of Small Inorganic Molecules, Academic
   Press. New York, 1974.
   (b) G. Henrici-Olive, S. Olive, Coordination and Catalysis, Verlag Chemie,
   Weinheim, 1977.
   (c) P. C. Ford (Ed.), Catalytic Activation of Carbon Monoxide, ACS Sympo-
   sium series 152, American Chemical Society, Washington, D. C., 1981.
   (d) A. E. Shilov, Activation of Saturated Hydrocarbons by Transition Metal
   Complexes, D. Reidel Publ., Dordrecht, 1984.
3 For example, (a) J. Chatt, J. R. Dilworth, R. L. Richards, Chem. Rev., 78,
   (1978) 589-625.
   (b) J. Chatt, G. L. M. da Camara Pina, and R. L. Richards (Ed.), New Trends
   in the Chemistry of Nitrogen Fixation, Academic Press., London, 1980.
4 For example; (a) M. E. Vol'pin and I. S. Kolomnikov, Pure Appl. Chem., 33
   (1973) 567-581.
   (b) R. P. A. Sneeden, Reactions of Carbon dioxide, in Ref. 1a, Vol. 8,
   pp. 225-283.
   (c) D. J. Darensbourg and R. A. Kudaroski, Ad. Organometal. Chem., 22
   (1983) 129-168.
   (d) D. A. Palmar and R. Van Eldik, Chem. Rev., 83 (1983) 651-731.
   (e) D. Walther, Coord. Chem. Rev., 79 (1987) 135-174.
5 (a) R. Eisenberg and C. D. Meyer, Acc. Chem. Res. 8 (1975) 26-33.
   (b) J. A. McCleverty, Chem. Rev., 79, (1979) 53-76.
6 R. R. Ryan, G. J. Kubas, D. C. Moody, and P. G. Eller, Structure and
   Bonding (Berlin), 46 (1981) 47. References are therein.
7 See following papers for theoretical investigations of this issue.
   (a) A. R. Rossi and R. Hoffmann, Inorg. Chem., 14 (1975), 365-374.
   (b) M. Elian and R. Hoffmann, Inorg. Chem., 14 (1975) 1058-1076.
   (c) S. Komiya, T. A. Albright, R. Hoffmann, and J. K. Kochi, J. Am. Chem.
   Soc., 98 (1976) 7255-7265.
   (d) R. Hoffmann, B. F. Beier, E. L. Muetterties, and A. R. Rossi, Inorg.
   Chem., 16 (1977), 512-521.
   (e) J. Demuynck, A. Strich, and A. Veillard, Nouv. J. Chim., 1 (1977) 217-
   228.
   (f) P. J. Hay, J. Am. Chem. Soc., 100 (1978), 2411-2417.
8 See following papers for theoretical investigations of this issue.
   (a) T. A. Albright, P. Hoffman, and R. Hoffmann, J. Am. Chem. Soc., 99
   (1977) 7546-7556.
   (b) T. A. Albright, R. Hoffmann, J. C. Thibeault, and D. L. Thorn, J. Am.
   Chem. Soc., 101 (1979) 3801-3812.
   (c) T. A. Albright, R. Hoffmann, Y. -C. Tse, and T. D'Ottavio, J. Am. Chem.
   Soc., 101, (1979) 3812-3821.
   (d) C. Bachmann, J. Demuynck, and A. Veillard, J. Am. Chem. Soc., 100
   (1978) 2366-2369.
   (e) S. Sakaki, K. Hori, and A. Ohyoshi, Inorg. Chem., 17 (1978) 3183-3188.
9 For example; (a) H. F. Schaefer (Ed.), Applications of Electronic Structure

Theory, Plenum, New York, 1977.

(b) Faraday Symp. Chem. Soc., Vol. 19 (1984). Many reviews are therein.

(c) A. Veillard (Ed.), Quantum Chemistry, The Challenge of Transition Metals and Coordination Chemistry, Nato ASI Series C, Vol. 176, Reidel, Dordrecht, 1985.

10 (a) A. Dedieu, M. -M. Rohmer, M. Benard, and A. Veillard, J. Am. Chem. Soc., 98 (1976) 3717-3718.

(b) A. Dedieu and M. -M. Rohmer, J. Am. Chem. Soc., 99 (1977) 8050-8051.

(c) M. -M. Rohmer, M. Barry, A. Dedieu, and A. Veillard, Int. J. Quant. Chem., Quant. Biol. Symp. 4 (1977) 337-342.

(d) A. Dedieu, M. -M. Rohmer, and A. Veillard, in; B. Pullman and N. Goldholm (Ed.), Metal-ligand Interactions in Organic Chemistry and Biochemistry, Reidel, Boston, 1977, pp. 101-129.

11 (a) S. Obara and H. Kashiwagi, J. Chem. Phys., 77 (1982) 3155-3165.

(b) M. Saito and H. Kashiwagi, J. Chem. Phys., 82 (1985) 848-855.

(c) M. Saito and H. Kashiwagi, J. Chem. Phys., 82 (1985) 3716-3721.

12 For example; (a) H. F. Schaefer (Ed.), Methods of Electronic Structure Theory, Plenum, New York, 1977.

(b) G. H. F. Diercksen and S. Wilson (Ed.), Methods in Computational Molecular Physics, NATO ASI Series Vol. 113, 1983. References are therein.

13 M. Wolfsberg and L. Hermholz, J. Chem. Phys., 20 (1952) 837-843.

14 R. Hoffmann, J. Chem. Phys., 39 (1963) 1397-1412. See refs. 7a-7d and 8a-8c for its applications to transition metal complexes

15 (a) R. Hoffmann, H. Fujimoto, J. R. Swenson, C. -C. Wan, J. Am. Chem. Soc., 95 (1973) 7644-7650.

(b) H. Fujimoto and R. Hoffmann, J. Phys. Chem., 78 (1974) 1874-1880.

16 (a) A. B. Anderson and R. Hoffmann, J. Chem. Phys., 60 (1974) 4271-4273.

(b) A. B. Anderson, J. Chem. Phys., 62 (1975) 1187-1188.

17 (a) A. B. Anderson and G. Fitzgerald, Inorg. Chem., 20 (1981) 3288-3291.

(b) A. B. Anderson, J. Am. Chem. Soc., 99 (1977) 696-707.

18 (a) R. F. Fenske, K. C. Caulton, D. D. Radtke, and C. C. Sweeney, Inorg. Chem., 5 (1966) 951-960.

(b) D. D. Radtke and R. F. Fenske, J. Am. Chem. Soc., 89 (1967) 2292-2296.

(c) R. F. Fenske and D. D. Radtke, Inorg. Chem., 7, (1968) 479-487.

(d) R. F. Fenske and R. L. DeKock, Inorg. Chem., 9 (1970) 1053-1060.

(e) R. F. Fenske and M. B. Hall, Inorg. Chem., 11 (1972) 768-775.

(f) M. B. Hall and R. F. Fenske, Inorg. Chem., 11 (1972) 1619-1624.

19 (a) J. A. Pople and D. L. Beveridge, Approximate Molecular Orbital Theory, McGraw-Hill, New York, 1970.

(b) J. A. Pople, D. P. Santry, and G. A. Segal, J. Chem. Phys., 43 (1965) s129-s135.

(c) J. A. Pople and G. A. Segal, J. Chem. Phys., 43 (1965) s136-s151.

(d) J. A. Pople and G. A. Segal, J. Chem. Phys., 44 (1966) 3289-3296.

(e) J. A. Pople, D. L. Beveridge, and P. A. Dobosh, J. Chem. Phys., 47 (1967) 2026-2033.

20 D. W. Clack, N. S. Hush, and J. R. Yandle, J. Chem. Phys., 57 (1972) 3503-3510.

(b) S. Sakaki, N. Hagiwara, N. Iwasaki, and A. Ohyoshi, Bull. Chem. Soc. Jpn., 50 (1977) 14-21.

21 (a) R. C. Bingham, M. J. S. Dewar, and D. H. Lo, J. Am. Chem. Soc., 97 (1975) 1285-1306.

(b) M. J. S. Dewar and W. Thiel, J. Am. Chem. Soc., 99 (1977) 4899-4907.

(c) W. Thiel, J. Am. Chem. Soc., 103 (1981) 1413-1420.

22 G. Blyholder and J. Springs, Inorg. Chem., 24 (1985) 224-227.

23 For example; J. C. Slater, The Calculation of Molecular Orbitals, John Wiley & Sons, New York, 1978; Ad. Quant. Chem., 6 (1972) 1.

24 For example; K. H. Johnson, Ad. Quant. Chem., 7 (1973) 143, J. Chem. Phys., 45 (1966) 3085-3095, Int. J. Quant. Chem., 1s (1967) 361-367.

25 For example; (a) E. J. Baerends and P. Ros, Chem. Phys. Lett., 23 (1973), 391-396.

172

(b) E. J. Baerends, D. E. Ellis, and P. Ros, Chem. Phys., 2 (1973) 41-51, 52-59.

(c) H. Sambe and R. H. Felton, J. Chem. Phys., 62 (1975) 1122-1126.

26 C. Satoko, Chem. Phys. Lett., 83 (1981) 111, Phys. Rev., B30 (1984) 1754-1764.

27 For example; T. H. Dunning and P. J. Hay, Gaussian Basis Sets for Molecular Calculations in ; H. F. Schaefer (Ed.), Methods of Electronic Structure Theory, Plenum, New York, 1977, pp. 1-27.

28 For example; (a) J. A. Pople, Apriori Geometry Predictions, in; H. F. Schaefer (Ed.), Applications of Electronic Structure Theory, Plenum, New York, 1977, pp. 1-27.

(b) P. W. Payne and L. C. Allene, Barriers to Rotation and Inversion, in; H. F. Schaefer (Ed.), Applications of Electronic Structure Theory, Plenum, New York, pp. 29-108.

29 (a) K. Faegri and H. J. Speis, J. Chem. Phys., 86 (1987) 7035-7040.

(b) H. P. Luth, J. Ammeter, J. Almlof, and K. Korsell, Chem. Phys. Lett., 69 (1980) 540-542.

(c) H. P. Luth, J. H. Ammeter, J. Almlof, and K. Faegri, J. Chem. Phys., 77 (1982) 2002-2009.

30 For example; (a) B. Roos, A. Veillard, and G. Vinot, Theoret. Chim. Acta, 20 (1971) 1-11.

(b) A. J. H. Wachters, J. Chem. Phys., 53 (1970) 1033-1036.

(c) P. J. Hay, J. Chem. Phys., 66 (1977) 4377- 4384.

(d) H. Tatewaki and S. Huzinaga, J. Chem. Phys., 71 (1979) 4339-4348.

(e) I. Hyla-Kryspin, J. Demuynck, and M. Benard, J. Chem. Phys., 75 (1981) 3954-3961.

(f) A. K. Rappe, T. A. Smedley, and W. A. Goddard, J. Chem. Phys., 85 (1981) 2607-2611.

(g) S. Huzinaga, J. Andzelm, M. Klobukowski, E. Radzio-Andzelm, Y. Sakai, and H. Tatewaki, Gaussian Basis Sets for Molecular Calculations, Elsevier, Amsterdam, 1984.

31 (a) S. Huzinaga, J. Chem. Phys., 42 (1965) 1293-1302.

(b) T. H. Dunning, J. Chem. Phys., 53 (1970) 2823-2833; 55 (1971) 716-723.

32 (a) H. Tatewaki and S. Huzinaga, J. Compt. Chem., 1 (1980) 205-228.

(b) Y. Sakai, H. Tatewaki, and S. Huzinaga, J. Compt. Chem., 2 (1981) 100-107.

33 (a) R. Ditchfield, W. J. Hehre, and J. A. Pople, J. Chem. Phys., 54 (1971) 724-728.

(b) J. S. Binkley, J. A. Pople, and W. J. Hehre, J. Am. Chem. Soc., 102 (1980) 939-947.

(c) W. J. Hehre, R. F. Stewart, and J. A. Pople, J. Chem. Phys., 51 (1969) 2657-2664.

34 (a) P. Pulay, Direct Use of the Gradient for Investigating Molecular Energy Surface, in; H. F. Schaefer (Ed.), Applications of Electronic Structure Theory, Plenum, New York, 1977, pp. 153-185.

(b) See following papers for its applications to transition metal complexes; M. E. Colvin and H. F. Schaefer, Faraday Symp. Chem. Soc., 19 (1984) 39-48. K. Morokuma, K. Ohta, N. Koga, S. Obara, and E. Davidson, Faraday Symp. Chem. Soc., 19 (1984) 49-61. S. Obara, K. Kitaura, and K. Morokuma, J. Am. Chem. Soc., 106 (1984) 7482-7492. N. Koga and K. Morokuma, J. Am. Chem. Soc., 108 (1986) 6136-6144. N. Koga, C. Daniel, J. Han, X. Y. Fu, and K. Morokuma, J. Am. Chem. Soc., 109 (1987) 3455-3456. D. Antolovic and E. R. Davidson, J. Am. Chem. Soc., 109 (1987), 977-985.

35 For example; (a) L. R. kahn, P. Baybutt, and D. G. Truhlar, J. Chem. Phys., 65 (1976) 3826-3853.

(b) P. J. Hay and W. R. Wadt, J. Chem. Phys., 82 (1985) 270-282.

36 See following reviews; (a) I. Shavitt, The Method of Configuration Interaction, in; H. F. Schaefer (Ed.), The Method of Electronic Structure Theory, Plenum, New York, 1977, pp. 189-275.

(b) B. O. Roos and P. E. M. Siegbhan, The Direct Configuration Interaction

Method from Molecular Integrals, in ; H. F. Schaefer (Ed.), Methods of Electronic Structure Theory, Plenum, New York, 1977 pp. 277-318.

37   (a) A. C. Wahl and G. Das, The Multiconfiguration Self-Consistent Field Method, in; H. F. Schaefer (Ed.), Methods of Electronic Structure Theory, Plenum, New York, 1977 pp. 51- 78.
    (b) B. O. Roos, The Multiconfigurational(MC) SCF Method, in; G. H. F. Diercksen and S. Wilson, Methods in Computational Molecular Physics, D. Reidel Publ. Co., Dordrecht, 1983, pp. 161-187.

38   (a) B. O. Roos, P. R. Taylor, and P. E. M. Siegbabn, Chem. Phys., 48 (1980) 157-173.
    (b) P. E. M. Siegbahn, J. Almlof, A. Heiberg, and B. O. Roos, J. Chem. Phys., 74 (1981) 2384-2396.

39   P. E. M. Siegbahn, Faraday Symp. Chem. Soc., 19 (1984) 97-107.

40   (a) J. A. Pople, J. S. Binkley, and R. Seeger, Int. J. Quant. Chem., Symp., 10 (1976) 1-19.
    (b) R. Krishnan and J. A. Pople, Int. J. Quant. Chem., 14 (1978) 91-100.

41   (a) J. S. Binkley, R. A. Whiteside, R. Krishnan, R. Seeger, D. J. DeFrees, H. B. Schregel, S. Topiol, L. R. Kahn, and J. A. Pople, Gaussian 80, Quantum Chemistry Program Exchange (Indiana University) 1980.
    (b) J. S. Binkley, M. Frisch, K. Raghavachari, D. DeFrees, H. B. Schlegel, R. Whiteside, E. Fluder, R. Seeger, and J. A. Pople, Gaussian 82, Carnegie-Mellon University Quantum Chemical Archive, 1982.

42   (a) S. R. Langhoff and E. R. Davidson, Int. J. Quant. Chem., 8 (1974) 61-72.
    (b) E. R. Davidson and D. W. Silver, Chem. Phys. Lett., 52 (1977) 403-406.

43   (a) K. Fukui and H. Fujimoto, Bull. Chem. Soc. Jpn., 41 (1968) 1989-1997; 42 (1969) 3399-3409.
    (b) H. Fujimoto, S. Yamabe, and K. Fukui, Bull. Chem. Soc. Jpn., 44 (1971) 2939-2941.
    (c) K. Fukui and S. Inagaki, J. Am. Chem. Soc., 97 (1975) 4445-4452.
    (d) S. Inagaki, H. Fujimoto, and K. Fukui, J. Am. Chem. Soc., 98 (1976) 4693-4061.

44   (a) H. Fujimoto, S. Kato, S. Yamabe, and K. Fukui, J. Chem. Phys., 60 (1974) 572-578.
    (b) H. Fujimoto and N. Kosugi, Bull. Chem. Soc. Jpn., 50 (1977) 2209-2214.

45   (a) K. Morokuma, J. Chem. Phys., 55 (1971) 1236-1243; Acc. Chem. Res., 10 (1977) 294-300.
    (b) K. Kitaura and K. Morokuma, Int. J. Quant. Chem., 10 (1976) 325-339.
    (c) K. Kitaura, S. Sakaki, and K. Morokuma, Inorg. Chem., 20 (1981) 2292-2297.

46   M. Dreyfus and A. Pullman, Theoret. Chim. Acta, 19 (1970) 20-37.

47   (a) P. S. Bagus, K. Hermann, and C. W. Bauschlicher, J. Chem. Phys., 80 (1984) 4378-4386; J. Chem. Phys., 81 (1984) 1966-1974.
    (b) C. W. Bauschlicher and P. S. Bagus, J. Chem. Phys., 81 (1984) 5889-5898.

48   J. A. Stone and R. W. Erskine, J. Am. Chem. Soc., 102 (1980) 7185-7192.

49   (a) T. Ziegler and A. Rauk, Theoret. Chim. Acta, 46, (1977) 1-10.
    (b) T. Ziegler and A. Rauk, Inorg. Chem., 18 (1979) 1558-1565.
    (c) T. Ziegler and A. Rauk, Inorg. Chem., 18 (1979) 1755-1759.

50   (a) D. Post and E. J. Baerends, J. Chem. Phys., 78 (1983) 5663-5681.
    (b) E. J. Baerends and A. Rozendaal, Analysis of -bonding, -(back)bonding and the synergic effect in $Cr(CO)_6$. Comparison of Hartree-Fock and X results for metal-CO bonding, in ref. 9c, pp. 159-177.

51   S. Nagase, T. Fueno, S. Yamabe, and K. Kitaura, Theoret. Chim. Acta, 49 (1978) 309-320.

52   (a) M. J. S. Dewar, Bull. Soc. Chim. Fr., 18 (1951) c79.
    (b) J. Chatt and L. A. Duncanson, J. Chem. Soc., (1953) 2339.

53   S. P. Walch and W. A. Goddard, J. Am. Chem. Soc., 98 (1976) 7908-7917.

54   K. Hermann and P. S. Bagus, Phys. Rev. B, 16 (1977) 4195-4208.

55   D. T. Clark, B. J. Cromarty, and A. Sgamelloti, Chem. Phys. Lett., 55

(1978) 482-487.
56 P. S. Bagus and B. O. Roos, J. Chem. Phys., 75 (1981) 5961-5962.
57 A. B. Rives and R. F. Fenske, J. Chem. Phys., 75 (1981) 1293-1302.
58 M. R. A. Blomberg, U. B. Brandemark, P. E. M. Siegbahn, K. B. Mathisen, and G. Karlstrom, J. Phys. Chem., 89 (1985) 2171-2180.
59 C. M. Rohlfing and P. J. Hay, J. Chem. Phys., 83 (1985) 4641-4649.
60 B. I. Dunlap, H. L. Yu, and P. R. Antoniewicz, Phys. Rev. A, 25 (1982) 7-13.
61 (a) A. E. Stevens, C. S. Feigerle, and W. C. Lineberger, J. Am. Chem. Soc., 104 (1982) 5026-5031.
(b) J. Ladell, B. Post, and I. Fankuchen, Acta Crystallogr., 5 (1952) 795-802.
(c) L. O. Brockway and P. C. Cross, J. Chem. Phys., 3 (1935) 828-833.
(d) L. Hedberg, T. Iijima, and K. Hedsberg, J. Chem. Phys., 70 (1979) 3224-3229.
(e) F. A. Cotton, A. K. Fischer, and G. Wilkinson, J. Am. Chem. Soc., 81 (1959) 800-803.
62 (a) D. Spangler, J. J. Wendoloski, M. Dupuis, M. M. L. Chen, and H. F. Schaefer, J. Am. Chem. Soc., 103 (1981) 3985-3990.
(b) H. F. Schaefer, J. Mol. Struct., 76 (1981) 117-135.
63 H. Jorg and N. Rosch, Chem. Phys. Lett., 120 (1985) 359-362.
64 P. Carsky and A. Dedieu, Chem. Phys., 103 (1986) 265-275.
65 (a) M. -H. Whangbo and R. Hoffmann, J. Chem. Phys., 68 (1978) 5498-5500.
(b) J. H. Ammeter, H. -B. Burgi, J. C. Thibeault, and R. Hoffmann, J. Am. Chem. Soc., 100 (1978) 3686-3692.
66 T. Ziegler, V. Tschinke, and C. Ursenbach, J. Am. Chem. Soc., 109 (1987) 4825-4837.
67 S. Sakaki, K. Kitaura, K. Morokuma, and K. Ohkubo, Inorg. Chem., 22 (1983) 104-108.
68 (a) R. J. Angelici, Organometal. Chem. Rev. A3 (1968) 173.
(b) W. D. Corey and T. L. Brown, Inorg. Chem., 12 (1973) 2820-2825.
(c) G. Centini, O. Gambino, Atti. Acad. Sci. Torino I, 97 (1963) 757,1197.
(d) H. Werner, Angew. Chem., Int. Ed. Engl., 7 (1968) 930-941.
(e) J. R. Graham and R. J. Angelici, Inorg. Chem., 6 (1967) 2082-2085.
(f) H. Werner and R. Prinz, Chem. Ber., 99 (1960) 3582-3592, J. Organometal. Chem., 5 (1966) 79.
69 M. Bernstein, J. D. Simon, and J. D. Peters, Chem. Phys. Lett., 100 (1983) 241-244.
70 K. E. Lewis, D. M. Golden, and G. P. Smith, J. Am. Chem. Soc., 106 (1984) 3906-3912.
71 T. Ziegler, Inorg. Chem., 25 (1986) 2721-2727.
72 C. W. Bauschlicher, Chem. Phys. Lett., 115 (1985) 387-391.
73 M. Aresta, C. F. Nobile, V. G. Albano, E. Forni, and M. Manassero, J. Chem. Soc., Chem. Comm., (1975) 636-637.
74 G. S. Bristow, P. B. Hitchcock, and M. F. Lappert, J. Chem. Soc., Chem. Comm., (1981) 1145-1146.
75 R. Alvarez, E. Carmona, J. M. Marin, M. L. Poveda, E. Gutierry-Puebla, and A. Mouge, J. Am. Chem. Soc., 108 (1986) 2286-2294.
76 (a) G. Fachinetti, C. Floriani, and P. F. Zanazzi, J. Am. Chem. Soc., 100 (1978) 7405-7406.
(b) S. Gambarotta, F. Arena, C. Floriani, and P. F. Zanazzi, J. Am. Chem. Soc., 104 (1982) 5082-5092.
77 J. C. Calabrese, T. Herskovitz, and J. B. Kinney, J. Am. Chem. Soc., 105 (1983) 5914-5915.
78 (a) J. W. Byrne, H. V. Blaser, and J. A. Osborn, J. Am. Chem. Soc., 97 (1975) 3871-3873.
(b) E. Carmona, J. M. Marin, M. L. Poveda, J. L. Atwood, and R. D. Rogers, J. Am. Chem. Soc., 105 (1983) 3014-3022.
79 C. R. Eady, J. J. Guy, B. F. G. Johnson, J. Lewis, M. C. Malatesta, and G. Sheldrick, J. Chem. Soc., Chem. Comm., (1976) 602-604.

80 W. Beck, K. Raab, U. Nagel, and M. Steimann, Angew. Chem., 94 (1982) 556–557; Angew. Chem., Int. Ed. Engl., 21 (1982) 526–527.
81 T. Herskovitz, J. Am. Chem. Soc., 99 (1977) 2391–2392.
82 T. Herskovitz and L. J. Guggenberger, J. Am. Chem. Soc., 98 (1976) 1615–1616.
83 V. D. Bianco, S. Doronzo, and N. Gallo, Inorg. Nucl. Chem. Lett., 15 (1979) 187–189.
84 M. Aresta and C. F. Nobile, J. Chem. Soc., Dalton Trans., (1977) 708–710.
85 M. Aresta and C. F. Nobile, Inorg. Chim. Acta, 24 (1977) L49–L50.
86 H. H. Karsch, Chem. Ber., 110 (1977) 2213–2221.
87 V. D. Bianco, S. Doronzo, and N. Gallo, J. Inorg. Nucl. Chem., 40 (1978) 1820–1821.
88 C. Bianchini and A. Meli, J. Am. Chem. Soc., 106 (1984) 2698–2699.
89 J. M. Maher, G. R. Lee, and N. J. Cooper, J. Am. Chem. Soc., 104 (1982) 6797–6799.
90 G. R. Lee and N. J. Cooper, Organometallics, 4 (1985) 794–798.
91 S. Sakaki, K. Kitaura, and K. Morokuma, Inorg. Chem., 21 (1982) 760–765.
92 S. Gambarotta, M. Pasquali, C. Floriani, A. Chiesi-Villa, and C. Guastini, Inorg. Chem., 20 (1981) 1173–1178.
93 S. Gambarotta and C. Floriani, J. Am. Chem. Soc., 104 (1982) 2019–2020.
94 (a) S. Sakaki and A. Dedieu, J. Organometal. Chem., 314 (1986) C63–C67.
    (b) S. Sakaki and A. Dedieu, Inorg. Chem., 26 (1987) 3278–3284.
95 M. Elian and R. Hoffmann, Inorg. Chem., 14 (1975) 1058–1076.
96 J. W. Lauher and R. Hoffmann, J. Am. Chem. Soc., 98 (1976) 1729–1742.
97 B. E. R. Schilling, R. Hoffmann, and J. W. Faller, J. Am. Chem. Soc., 101 (1979) 592–598.
98 S. Sakaki, to be published.
99 S. Sakaki, H. Sato, Y. Imai, K. Morokuma, and K. Ohkubo, Inorg. Chem., 24 (1985) 4538–4544.
100 M. Elian, M. M. L. Chen, D. M. P. Mingos, and R. Hoffmann, Inorg. Chem., 15 (1976) 1148–1155.
101 R, Hoffmann, M. M. L. Chen, M. Elian, A. R. Rossi, and M. P. Mingos, Inorg. Chem., 13 (1974) 2666–2675.
102 S. Sakaki, N. Kudou, and A. Ohyoshi, Inorg. Chem., 16 (1977) 202–205.
103 V. Branchadell and A. Dedieu, Inorg. Chem., 26 (1987) 3966–3968.
104 C. Mealli, R. Hoffmann, and A. Stockis, Inorg. Chem., 23 (1984) 56–65.
105 A. J. L. Pombeiro, Preparation, Structure, Bonding and Reactivity of Dinitrogen Complexes, in; J. Chatt, C. L. M. da Camara Pina, and R. L. Richards (Ed.), Nitrogen Fixation Through Stable Dinitrogen Complexes, Academic Press, London, 1980, pp.153–197.
106 D. L. Thorn, T. H. Tulip, and J. A. Ibers, J. Chem. Soc., Dalton Trans., (1979) 2022–2025.
107 T. Yoshida, T. Okano, D. L. Thorn, T. H. Tulip, S. Otsuka, and J. A. Ibers, J. Organomet. Chem., 181 (1979) 183–201.
108 H. W. Turner, J. D. Fellmann, S. M. Rocklage, R. R. Schrock, M. R. Churchill, and H. J. Wasserman, J. Am. Chem. Soc., 102 (1980) 7809–7811.
109 E. Carmona, J. M. Marin, M. L. Poveda, R. D. Rogers, J. L. Atwood, J. Organomet. Chem., 238 (1982) C63–C66.
110 (a) E. Carmona, J. M. Marin, M. L. Poveda, J. L. Atwood, R. D. Rogers, and G. Wilkinson, Angew. Chem. Suppl. (1982) 1116–1120.
    (b) E. Carmona, J. M. Marin, M. L. Poveda, J. L. Atwood, and R. D. Rogers, J. Am. Chem. Soc., 105 (1983) 3014–3022.
111 E. Carmona, J. M. Marin, M. L. Poveda, J. L. Atwood, and R. D. Rogers, Polyhedron, 2 (1983) 185–193.
112 R. H. Morris, J. M. Ressner, J. F. Sawyer, and M. Shiralian, J. Am. Chem. Soc., 106 (1984) 3683–3684.
113 S. N. Anderson, R. L. Richards, and D. L. Hughes, J. Chem. Soc., Dalton Trans., (1986) 245–252.
114 G. P. Pez, P. Apgar, R. K. Crissey, J. Am. Chem. Soc., 104 (1982) 482–490.
115 L. E. Sutton (Ed.), Tables of Interatomic Distances and Configuration in

176

Molecules and Ions, Special Publication No. 11 and Supplement Special Publication No. 18, The Chemical Society, London, 1958 and 1965.
116 K. Nakamoto, Infrared and Raman Spectra of Inorganic and Coordination Compounds, John-Wiley & Sons, New York, 1978, p. 110.
117 H. W. Schroetter, Naturwissenschaften, 54 (1967) 513.
118 J. R. Durig, S. F. Bush, and E. E. Mercer, J. Chem. Phys., 44 (1966) 4238 -4247.
119 M. Sato, T. Tastumi, T. Kodama, M. Hidai, T. Uchida, and Y. Uchida, J. Am. Chem. Soc., 100 (1978) 4447-4452.
120 B. R. Davis and J. A. Ibers, Inorg. Chem., 10 (1971) 578-585.
121 P. R. Hoffman, T. Yoshida, T. Okano, S. Otsuka, and J. A. Ibers, Inorg. Chem., 15 (1976) 2462-2466.
122 P. W. Jolly, K. Jonas, C. Kruger, and Y. -H.Tsay, J. Organomet. Chem., 33 (1971) 109-122.
123 (a) K. Jonas, Angew. Chem., 85 (1973) 1050; Angew. Chem., Int. Ed. Engl., 12 (1973) 997-998.
    (b) C. Kruger and Y. -H. Tsay, Angew. Chem., 85 (1973) 1051-1052.; Angew. Chem., Int. Ed., Engl., 12 (1973) 998-999.
    (c) K. Jonas, D. J. Brauer, C. Kruger, P. J. Roberts, and Y. -H. Tsay, J. Am. Chem. Soc., 98 (1976) 74-81.
124 J. M. Manriquez, D. R. McAlister, E. Rosenberg, A. M. Shiller, K. L. Williamson, S. I. Chan, and J. E. Bercaw, J. Am. Chem. Soc., 100 (1978) 3078-3083.
125 J. E. Bercaw, E. Rosenberg, and J. D. Roberts, J. Am. Chem. Soc., 96 (1974) 612-614.
126 J. A. Armor and H. Taube, J. Am. Chem. Soc., 92, (1970) 2560-2562.
127 (a) J. Jeffrey, M. F. Lappert, P. I. Riley, J. Organometal. Chem., 181 (1979) 25.
    (b) M. J. S. Gynane, J. Jeffrey, and M. F. Lappert, J. Chem. Soc., Chem. Comm., (1978) 34-36.
128 (a) D. Sellmann, Angew. Chem., Int. Ed. Engl., 13 (1974) 639-649.
    (b) J. Chatt and G. J. Leigh, Angew. Chem., Int. Ed. Engl., 17 (1978) 400-407.
129 D. J. Darensbourg, Inorg. Chem., 10 (1971) 2399-2403.
130 P. Finn and W. L. Jolly, Inorg. Chem., 11 (1972) 1434-1435.
131 H. Binder and D. Sellmann, Angew. Chem., Int. Ed. Engl., 12 (1973) 1017-1019.
132 J. Chatt, C. M. Elson, N. E. Hooper, and G. J. Leigh, J. Chem. Soc., Dalton Trans., (1975) 2392-2401.
133 K. G. Caulton, R. L. DeKock, and R. F. Fenske, Inorg. Chem., 12 (1970) 515-518.
134 M. J. Ondrechen, M. A. Ratner, and D. E. Ellis, J. Am. Chem. Soc., 103 (1981) 1656-1659.
135 S. Sakaki, K. Morokuma, and K. Ohkubo, J. Am. Chem. Soc., 107 (1985) 2686-2693.
136 D. L. DuBois and R. Hoffmann, Nouv. J. Chim., 1 (1977) 479-492.
137 H. Veillard, Nouv. J. Chim., 2 (1978) 215-224.
138 J. N. Murrell, A. Al-Derzi, G. J. Leigh, and M. F. Guest, J. Chem. Soc., Dalton Trans., (1980) 1425-1433.
139 T. Yamabe, K. Hori, and K. Fukui, Inorg. Chem., 21 (1982) 2046-2050.
140 R. Hoffmann, M. M. -L. Chen, and D. L. Thorn, Inorg. Chem., 16 (1977) 503-511.
141 S. Inagaki, H. Fujimoto, and K. Fukui, J. Am. Chem. Soc., 98 (1976) 4054-4061.
142 K. I. Goldberg, D. M. Hoffman, and R. Hoffmann, Inorg. Chem., 21 (1982) 3864-3868.
143 T.Yoshida, T.Adachi, M. Kaminaka, T.Veda, T. Higuchi, J. Am. CHem. Soc., 110 (1988) 4872-4873.
144 M. R. A. Blomberg, U. B. Brandemark, P. E. M. Siegbahrn, J. Wennerberg, C. W. Bauschlicher, J. Am. Chem. Soc., 110(1988) 6650-6655.

145 M. Rosi, A. Sgamellotti, F. Tarantelli, C. Floriani, Inorg. Chem.,
    26(1987), 3805-3811.
146 S. Sakaki, T. Aizawa, N. Koga, K. Morokuma, K. Ohkubo, Inorg. Chem.,
    28(1989) 108-109.
147 S. Sakaki, K. Ohkubo, Inorg. Chem., 27(1988), 2020-2021; 28(1989), in
    press.
148 C. Bo, A. Dedieu, Inorg. Chem., 28(1989) 304-309.
149 S. Sakaki and K. Ohkubo, J. Phys. Chem., in press.

ELECTROCHEMISTRY OF MONONUCLEAR COPPER COMPLEXES. STRUCTURAL REORGANIZATIONS ACCOMPANYING REDOX CHANGES

P. Zanello

ELECTROCHEMISTRY OF MONONUCLEAR COPPER COMPLEXES. STRUCTURAL REORGANIZATIONS ACCOMPANYING REDOX CHANGES

P.Zanello

1. INTRODUCTION

We have recently reviewed the electrochemistry of dinuclear copper complexes with compartmental ligands |1|. We intend now to give a panoramic survey of the redox behaviour of the wide field of mononuclear copper complexes, giving also evidence for the stereochemical reorganizations which accompany redox changes.

The central role of the Cu(II)/Cu(I) redox couple in enzymes involved in biological electron-transfer processes |2-4| has stimulated an impressive amount of work towards the characterization of low molecular weight copper complexes with coordination spheres able to model those now known for the cuproproteins. In this connection, Table 1 shows the donor atoms sets in single copper sites of different proteins, together with the redox potentials for the relevant Cu(II)/Cu(I) redox couples. If one considers that the aquocopper(II/I) redox couple exhibits an $E^{\circ\prime}= -0.08$ V (vs. S.C.E.), it can be deduced that in these protein sites access to the copper(I) form is markedly facilitated.

Two electrochemical parameters are useful in determining structural-redox relationships: (i) the thermodynamic standard electrode potential at which a redox change occurs; (ii) the kinetic rate of the heterogeneous electron transfer between the electrode and a copper complex.

Let us examine in detail each one of these parameters:

(i). In a classical thermodynamic sense, the value of the redox potential of a simple redox change, not complicated by

TABLE 1

Donor set and redox potentials (in Volt vs. S.C.E.) for monocopper active sites in different proteins

protein	ascertained or more reliable donor set	pH	$E°'$ Cu^{II}/Cu^{I}	reference
Polyporus laccase	N_2S_2	5.5	+0.54	5
Rusticyanin	N_2S_2	2.0	+0.44	5
Ceruloplasmin	N_2S_2	5.5	+0.34	5
		5.5	+0.25	
Superoxide dismutase	N_4 or N_4O	7.0	+0.18	6
Rhus laccase	N_2S_2	7.5	+0.15	5
	N_3O^2	7.5	+0.12	
Plastocyanin	N_2S_2	7.0	+0.13	5
Galactose oxidase	N_2O_2	7.0	+0.06	7
Azurin	N_2S_2 or N_2S_2O	7.0	+0.06	5
Mavicyanin	N_2S_2	7.0	+0.04	5
Stellacyanin	N_2S_2	7.1	-0.06	5
Cytochrome c oxidase	N_2S_2	7.0	-0.05	8

subsequent chemical reactions, such as:

$$Cu^{II}L \underset{-e}{\overset{+e}{\rightleftarrows}} Cu^{I}L$$

is expressed by the following relationship [9]:

$$E°'(Cu^{II}L/Cu^{I}L) = E°'(Cu^{2+}/Cu^{+}) + 0.059 \log \frac{K_{Cu^{I}L}}{K_{Cu^{II}L}}$$

This equation suggests that the more stable the copper(I) complex is with respect to the corresponding copper(II) complex, the more positive is the shift of the formal electrode potential of the complexed redox couple with respect to that of the free redox couple.

We can extend this relationship, by including in the concept of relative stability of two oxidation states of a given copper complex, structural features in the ligand caused by the presence of moieties able to exert electronic effects towards the metal center. In this connection, it is worth recalling that centres of electron donating ability (from peripheral groups as well as from the donor atoms themselves) stabilize copper(II) forms, while centres of electron withdrawing ability stabilize copper(I) forms.

(ii). As far as the rate of the electron transfer is concerned, it is summarily conditioned by two main factors: (a) an intrinsic (and unforeseeable) aspect of the interaction between the electrode material and the specific depolarizer (solvent and solvated electroactive species); (b) the occurrence of significant inner-sphere structural reorganizations within the redox active complex, which on account of the Franck-Condon activation barrier to the charge transfer, slow down the rate of the redox change |10|. As a matter of fact it is worth recalling that, in the absence of steric constraints, copper(II) ion commonly adopts six-coordinate octa-hedral or rhombic, five-coordinate square-pyramidal or trigonal-bipyramidal, or four-coordinate square-planar geometries, whereas copper(I) ion prefers the four-coordinate tetrahedral assembly. Since in organometallic systems the presence of interfacial pheno-mena can be often neglected, we will assume: the occurrence of a fast electron transfer (reversible in electrochemical terms|11|) as indicative of no significant structural reorganization accompany-ing the redox step; the occurrence of a relatively slow electron transfer (quasireversible in electrochemical terms |11|) as indica-tive of important stereochemical reorganizations; the occurrence of a notably slow electron transfer (irreversible in electrochemical terms |11|) as indicative of structural modifications so deep as to destroy the starting molecular framework.

Finally, however, H.B.Gray has authoritatively proved that

structural reorganizations accompanying redox changes (which in principle should affect the kinetic electrode aspects) may affect also the location of the thermodynamic electrode potential through ligand field stabilization energies |12|. In particular, he has pointed out that one-electron reduction of tetrahedrally distorted copper(II) complexes to pseudotetrahedral copper(I) forms must occur at quite positive redox potentials.

This last contribution is extremely important, since it will allow us to discuss the stereodynamic consequences of the redox changes taking into account both the thermodynamic and the kinetic aspects of the electrochemical steps.

Unless otherwise specified, throughout the review potential values will be referred to the saturated calomel electrode (S.C.E.). Copper complexes will be examined according to the nature of the donor atom sets formally present in the different ligands.

2. OXYGEN DONOR LIGANDS

2.1 O_4 donor set

Beyond any doubt the O_4 donor set is generally unapt to stabilize copper in oxidation states different from +2. In fact all CuO_4 complexes undergo cathodic processes accompanied by demetallation processes.

A wide series of bis-diketonato copper(II) complexes, schematized in Chart I, have been electrochemically examined in different solvents |13-19|.

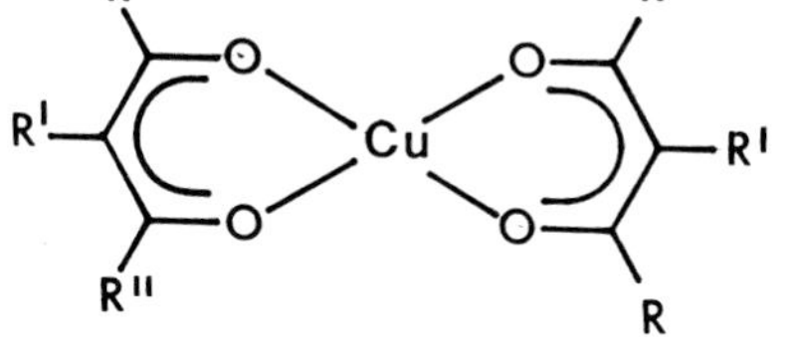

CHART I

All these complexes undergo an irreversible cathodic reduction through either a single two-electron step in relatively slow electrochemical techniques (polarography or voltammetry at rotating electrodes), or two distinct one-electron steps in faster electrochemical techniques (cyclic voltammetry).

The following scheme includes all possible electron transfer pathways for the reduction of a copper(II) complex:

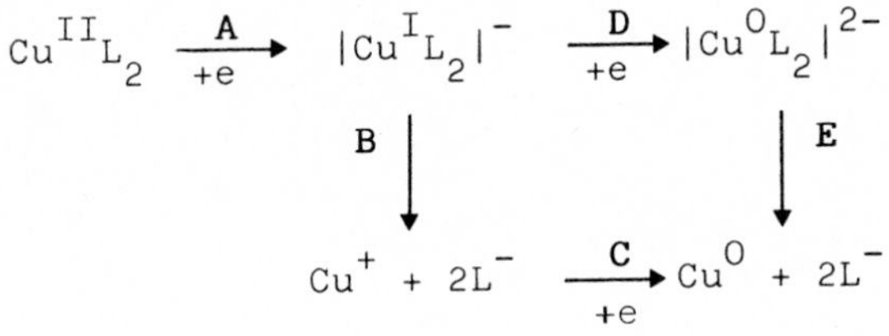

$$Cu^{II}L_2 \xrightarrow[+e]{A} |Cu^IL_2|^- \xrightarrow[+e]{D} |Cu^OL_2|^{2-}$$

$$\downarrow B \qquad\qquad \downarrow E$$

$$Cu^+ + 2L^- \xrightarrow[+e]{C} Cu^O + 2L^-$$

Schematically the occurrence of a single two-electron process in polarography or related techniques means that the instantaneously electrogenerated copper(I) complex $|Cu^IL_2|^-$ is rather unstable and decomposes following the pathway B,C. The merging of the two electron transfers A and C in a single process is due to the fact that the second charge transfer is thermodynamically favoured with respect to the first one (in aqueous solution $E^\circ_{(Cu^+/Cu^O)} = + 0.28$ V vs. S.C.E.).

The appearance of two distinct one-electron steps in cyclic voltammetry means that the copper(I) complex $|Cu^IL_2|^-$, even if labile in longer timeframes, is able to maintain intact, at least partially, its molecular framework in the short times of the technique; thus it may follow the further reduction pathway D,E. The irreversibility of the steps A and/or D is due to the occurrence of chemical complications B and/or E.

Table 2 summarizes the potential values for the cathodic reduction of such diketonato complexes. The irreversibility of the process makes these potential values of little significance. However good correlations between half-wave potentials and electronic effects of the substituents in the chelate rings (as expressed by Hammett σ values) have been found |17|; thus suggesting that the same electrode mechanism is operative in the reduction of such derivatives.

TABLE 2

Redox potentials (in Volt) for the irreversible cathodic reduction of the copper(II) complexes schematized in Chart I

R	R'	R"	$E_{1/2}$ [a] or E_p [b]	solvent	reference
CH_3	H	CH_3	− 0.50	aqueous-75% dioxane	14,15,17
CH_3	H	CH_3	− 0.13	aqueous-50% Py	13
CH_3	H	CH_3	− 0.85	DMF	18
CH_3	H	CH_3	− 0.57	MeCN	16
CH_3	H	CH_3	− 0.97	MeCN	19
CH_3	H	C_6H_5	− 0.38	aqueous-75% dioxane	14,15,17
CH_3	H	C_6H_5	− 0.11	aqueous-50% Py	13
CH_3	H	C_6H_5	− 0.58	DMF	18
C_6H_5	H	C_6H_5	− 0.38	aqueous-75% dioxane	14,15,17
C_6H_5	H	C_6H_5	− 0.06	aqueous-50% Py	13
C_6H_5	H	C_6H_5	− 0.67	DMF	18
CH_3	H	CF_3	− 0.17	aqueous-75% dioxane	15,17
CF_3	H	CF_3	+ 0.04	aqueous-75% dioxane	15,17
CH_3	H	C_2H_5	− 0.51	aqueous-75% dioxane	17
CH_3	H	$\underline{i}\text{-}C_3H_7$	− 0.55	aqueous-75% dioxane	17
CH_3	H	$\underline{i}\text{-}C_4H_9$	− 0.52	aqueous-75% dioxane	17
CH_3	H	$\underline{t}\text{-}C_4H_9$	− 0.60	aqueous-75% dioxane	17
CH_3	H	$\underline{n}\text{-}C_5H_{11}$	− 0.45	aqueous-75% dioxane	17
C_2H_5	H	C_2H_5	− 0.52	aqueous-75% dioxane	17
CF_3	H	$\underline{t}\text{-}C_4H_9$	− 0.34	aqueous-75% dioxane	17
$\underline{t}\text{-}C_4H_9$	H	$\underline{t}\text{-}C_4H_9$	− 0.69	aqueous-75% dioxane	17
CH_3	H	C_2H_5O	− 0.09	aqueous-50% Py	13
CH_3	H	$2\text{-}C_4H_3O$	− 0.34	aqueous-75% dioxane	17
CH_3	H	$2\text{-}C_4H_3S$	− 0.32	aqueous-75% dioxane	17
CF_3	H	$2\text{-}C_4H_3O$	− 0.13	aqueous-75% dioxane	17
CF_3	H	$2\text{-}C_4H_3S$	− 0.14	aqueous-75% dioxane	17
CF_3	H	C_6H_5	− 0.15	aqueous-75% dioxane	17
C_6H_5	H	$2\text{-}C_4H_3O$	− 0.32	aqueous-75% dioxane	17
C_6H_5	H	$2\text{-}C_4H_3S$	− 0.34	aqueous-75% dioxane	17

TABLE 2 (continued)

R	R'	R"	$E_{1/2}$ [a] or E_p [b]	solvent	reference
$2-C_4H_3O$	H	$2-C_4H_3O$	-0.29	aqueous-75% dioxane	17
$2-C_4H_3S$	H	$2-C_4H_3O$	-0.29	aqueous-75% dioxane	17
$2-C_4H_3S$	H	$2-C_4H_3S$	-0.31	aqueous-75% dioxane	17
CH_3	CH_3	CH_3	-0.56	aqueous-75% dioxane	14,15
CH_3	C_2H_5	CH_3	-0.56	aqueous-75% dioxane	14,15
CH_3	$\underline{n}-C_3H_7$	CH_3	-0.56	aqueous-75% dioxane	14,15
CH_3	C_6H_5	CH_3	-0.50	aqueous-75% dioxane	15
CH_3	$C_6H_5CH_2$	CH_3	-0.55	aqueous-75% dioxane	15

[a] Half-wave potential in polarography; [b] peak potential in cyclic voltammetry.

It is worth noticing that, in the presence of coordinating pyridine, the reduction process proceeds through two distinct one-electron steps; the second charge transfer is in fact shifted towards negative potential values, because of the stabilization of the released copper(I) ion according to |13|:

$$Cu^I(py)_m + e \longrightarrow Cu^0 + m\ py$$

Tentatively, an explanation for the irreversibility of the above reduction processes can be advanced on the basis of structural arguments.

In these complexes the CuO_4 moiety is square planar |20-22|, as shown in Figure 1, for the complexes with R = R"= CH_3, R'= C_6H_5, and R = CF_3, R'= H, R"= $2-C_4H_3S$, respectively.

It is likely that the six-membered chelate rings C_3O_2Cu are too rigid to allow the instantaneously electrogenerated copper(I) complexes to assume their stabilizing tetrahedral geometry.

188

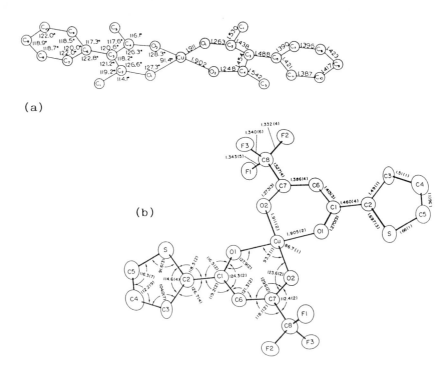

(a)

(b)

Fig.1. Molecular structure of: (a) bis(3-phenyl-2,4-pentanedionato) copper(II) (from Ref. 21); (b) (2-thenoyltrifluoroacetonate) copper(II) (from Ref. 22).

Another series of compounds exhibiting the planar CuO_4 moiety are derived from the bis-chelates of salicylaldehyde, schematized in Chart II.

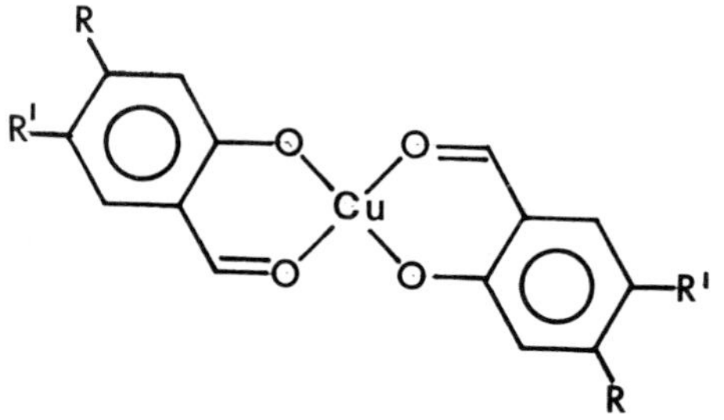

CHART II

As in the case of the preceding ß-diketonato complexes, these salicylato derivatives undergo irreversible cathodic reduction. Since they have been studied in aqueous–50% pyridine solutions, the Cu(II)/Cu(0) process occurs in two distinct one-electron steps, in accordance with previous discussion.

Table 3 reports the relevant redox potentials. As it can be seen, they are generally reduced at potentials more positive than the diketonato complexes. This can be attributed simply to the poor electron donating ability of the salicylic aromatic ring.

TABLE 3

Redox potentials (in Volt) for the irreversible reduction of the copper(II) complexes schematized in Chart II in aqueous-50% pyridine [13]

R	R'	$E_{1/2}$ [a]
H	H	+ 0.01
H	CH_3	0.00
OH	H	+ 0.01

[a] Half-wave potential in polarography.

A further group of planar CuO_4 coordination compounds is present in the catecholato complexes schematized in Chart III.

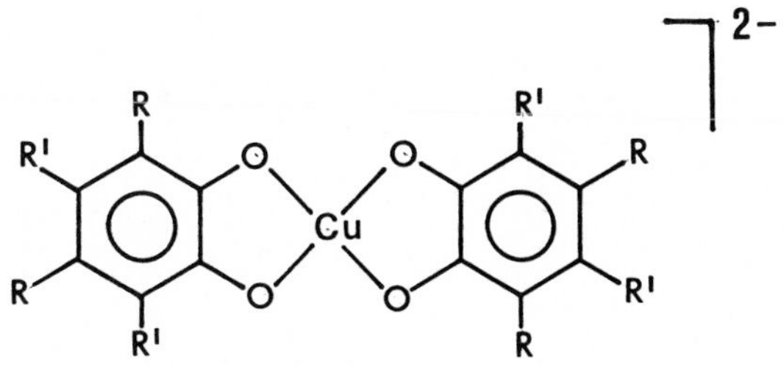

CHART III

The most important redox change is the one-electron oxidation 2-/1-, the potentials of which are reported in Table 4 for the complexes studied. This oxidation is attributed to the ligand-centred process [23]:

$$|Cu^{II}(catecholato)_2|^{2-} \underset{+e}{\overset{-e}{\rightleftharpoons}} |Cu^{II}(catecholato)(semiquinonato)|^{-}$$

The chemical reversibility of this electron transfer in the different complexes is likely associated to the stability of the relevant catecholato-semiquinonato-copper(II) product.

TABLE 4

Redox potentials (in Volt) for the one-electron oxidation of the bis(catecholato)-copper(II) complexes schematized in Chart III

R	R'	$E^{o\,\prime}$ 2-/1-	solvent	reference
H	H	$-\,0.09^{a}$	CH_2Cl_2	24
Cl	Cl	$+\,0.44^{a}$	CH_2Cl_2	24
H	$\underline{t}-C_4H_9$	$-\,0.35^{b}$	MeCN	23
H	$\underline{t}-C_4H_9$	$-\,0.31^{b}$	DMSO	23

[a]Irreversible process; [b]quasireversible process.

At variance with such ligand centred oxidation, an authentic copper(III)-O_4 coordination has been prepared by anodic oxidation of tetrakis(pyridine-N-oxide)copper(II) complex in liquid SO_2 |25|. Figure 2 shows the relevant cyclic voltammetric response.

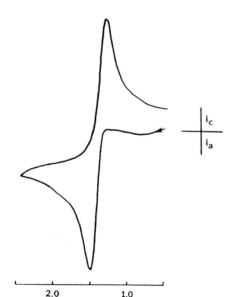

Fig. 2. Cyclic voltammogram recorded at −40°C at a platinum disk electrode in liquid SO_2 solution of $|Cu(C_5H_5NO)_4|(PF_6)_2$. Scan rate 0.1 Vs^{-1} (from Ref.25).

2.0   1.0

An uncomplicated, quasireversible, one-electron oxidation is obtained at the notably high formal electrode potential of +1.6 V. The electrogenerable $|Cu^{III}(C_5H_5NO)_4|(PF_6)_3$ has been chemically characterized. However there are no crystallographic data available to allow an evaluation of the extent of stereochemical reorganization accompanying this redox change. It must be pointed out that access to this copper(III) complex has been in good part due to the unique uncoordinating properties of the SO_2 solvent.

Finally, the last series of mononuclear CuO_4 compounds consists of derivatives of binucleating compartmental ligands |26|. The two types schematized in Chart IV have been studied by electrochemical methods.

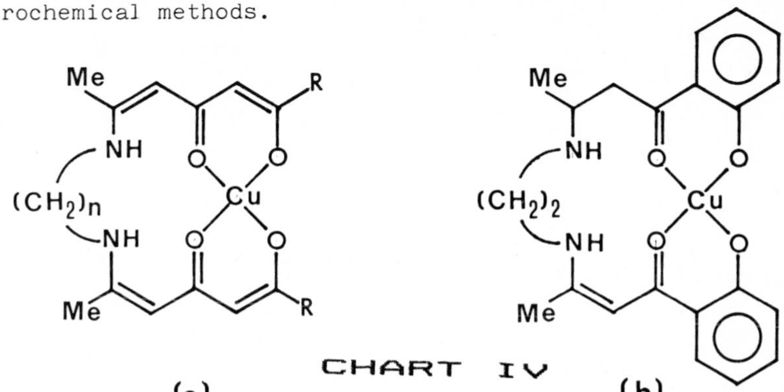

CHART IV

(a) **(b)**

X-ray determinations on type _a_ molecules (R = C_6H_5, n = 2) |27| show that the CuO_4 moiety is also essentially planar (Figure 3), the copper atom being displaced by 0.06 Å out of the O_4 plane by an weak axial interaction with the oxygen atom of an adjacent molecule in the crystal packing.

Fig. 3. Molecular structure of the type _a_ derivatives in Chart IV (R = C_6H_5, n = 2). Cu-O mean distance, 1.90 Å (from Ref. 27).

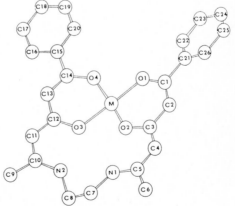

Consequently, as in previous cases, the Cu(II)/Cu(I) cathodic reduction proceeds through an irreversible cathodic process. Table 5 summarizes the relevant redox potentials.

TABLE 5

Redox potentials (in Volt) for the copper(II)/copper(I) reduction of the complexes schematized in Chart IV

complex	n	R	E_p [a]	solvent	reference
type _a_	2	C_6H_5	- 1.2	DMF	19
	3	C_6H_5	- 1.1	DMF	19
	3	$\underline{t}$-C_4H_9	- 1.2	DMF	19
type _b_	-	-	- 1.06	DMSO	28

[a] Peak potential value for irreversible process.

3. SULFUR DONOR LIGANDS

Two main groups of ligands act as sulfur donor chelating agents in copper coordination: dithiolates and thioethers.

3.1 S_3 donor set

It has been briefly reported that the 1:1 complex of Cu(II) with the |12|-ane-S_3 ligand (1,5,9-trithiacyclododecane), reported in Chart V, undergoes a near reversible, one-electron cathodic reduction |29, 30|.

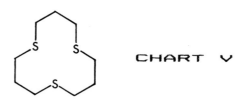

CHART V

In aqueous-80% methanol the copper(II)/copper(I) redox couple for this complex is located at +0.55 V (_vs_. S.C.E.) |29|, whereas in aqueous solution, it assumes the value of +0.49 V |30|.

We must note, however, that in the absence of synthetic data, the ligand to metal ratio of 1 in this complex seems somewhat surprising; specially, if one considers that nickel(II) forms a bis-ligand complex |32|. In addition, the complex $Cu(|12|-ane-S_3)_2$ Cl_2 has been recently characterized by X-ray diffraction; in it, only one sulfur atom of each ligand coordinates to the copper(II) ion |31|. Further, the strictly correlated ligand $|9|-ane-S_3$ (1,4,7-trithiacyclononane) (Chart VI) forms bis-ligand complexes with Cu(II), Ni(II), Co(II) |33|, Fe(II) |34|, Pd(II), Rh(III) |35|, Pt(II) |36| and Ru(II) |37|. Nevertheless, $|9|-ane-S_3$ forms a 1:1 complex with Cu(I), even if it is thought to be constituted of oligomers, in which a CuS_4 coordination occurs through bridging thioethers |38|.

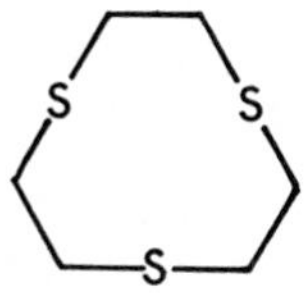

CHART VI

3.2 $\underline{S}_4$ donor set

In copper-sulfur complexes, the CuS_4 coordination is undoubtedly the most widely represented.

In chronological order, 1,2-dihiolene complexes (forming five-membered chelate rings) are those first examined from the redox viewpoint and were, in part, already reviewed |39|.

Ethene-1,2-dithiolates of the type represented in Chart VII undergo reversibly the one-electron oxidation 2-/1-.

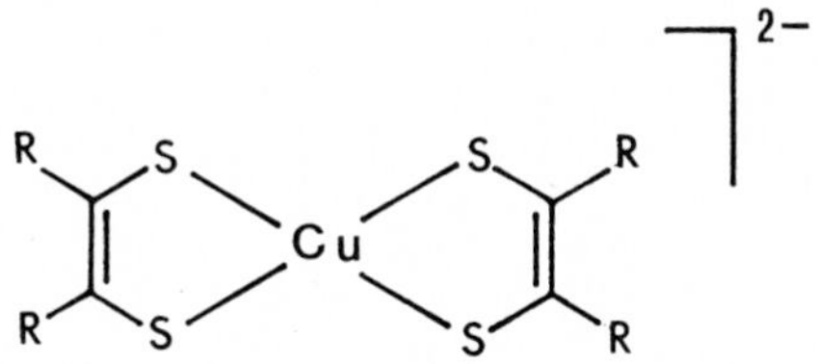

CHART VII

The redox potentials for this oxidation process in different copper(II) derivatives are summarized in Table 6.

TABLE 6

Redox potentials (in Volt) for the one-electron oxidation of the bis-1,2-dithiolato-copper(II) complexes schematized in Chart VII

R	$E^{0\prime}_{2-/1-}$	solvent	reference
H	− 0.74	DMSO	39
CN	+ 0.33	DMF	39
CN	+ 0.28	DMSO	40
CN	+ 0.33	MeCN	39
CN	+ 0.20	MeCN	41
CF$_3$	− 0.01	MeCN	39

The same redox behaviour holds for the substituted benzene-1,2-dithiolates schematized in Chart VIII.

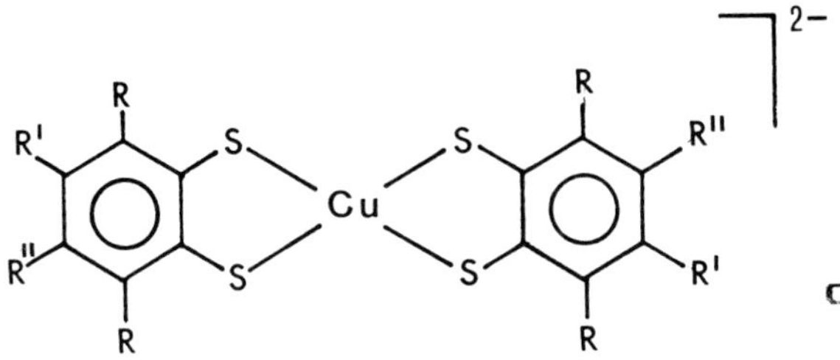

CHART VIII

Table 7 reports the redox potentials for the reversible, one-electron, oxidation of these copper(II) complexes. However, the redox activity of 1,2-dithiolato complexes of copper(II) is more extensive. Figure 4 shows the cyclic voltammograms recorded in MeCN solutions of $|Cu(maleonitrile-dithiolato)_2|^{2-}$ (a) and $|Cu(toluene-3,4-dithiolato)_2|^{-}$ (b).

TABLE 7

Redox potentials (in Volt) for the one-electron oxidation of the bis(arene-1,2-dithiolato)copper(II) complexes schematized in Chart VIII

R	R'	R"	$E^{\circ\,'}_{2-/1-}$	solvent	reference
H	H	H	− 0.61	DMF	39
Cl	Cl	Cl	− 0.22	DMF	39
CH_3	CH_3	CH_3	− 0.88	DMF	39
H	CH_3	CH_3	− 0.68	DMF	39
H	H	CH_3	− 0.62	DMF	39
H	H	CH_3	− 0.59	MeCN	42

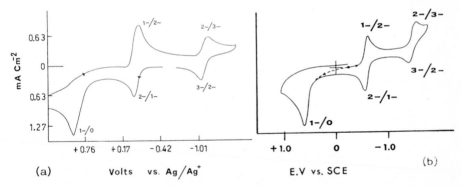

Fig. 4. Cyclic voltammograms recorded at platinum electrodes in MeCN solutions of: $|Cu\{S_2C_2(CN)_2\}_2|(Et_4N)_2$ (from Ref. 41) (a); $|Cu\{S_2(C_6H_3Me)_2\}_2|(Bu_4N)$ (from Ref. 42) (b).

In both cases, three sequential redox changes occur in the order shown below:

$$|Cu(ligand)_2|^- \underset{-e}{\overset{+e}{\rightleftharpoons}} |Cu(ligand)_2|^{2-} \underset{-e}{\overset{+e}{\rightleftharpoons}} |Cu(ligand)_2|^{3-}$$

$$\downarrow {-e}$$

$$Cu(ligand)_2 \xrightarrow{\text{fast}} \text{demetallation}$$

The relevant potentials are listed in Table 8.

TABLE 8

Redox potentials (in Volt) for the four-membered electron-transfer sequence of 1,2-dithiolato-copper(II) complexes in acetonitrile solution

complex	redox changes			reference
	0/1-	1-/2-	2-/3-	
$\|Cu\{S_2C_2(CN)_2\}_2\|^{2-}$	+ 1.19[a]	+ 0.20[b]	- 0.84[c]	41
$\|Cu\{S_2(C_6H_3Me)_2\}_2\|^{-}$	+ 0.62[a]	- 0.59[b]	- 1.48[c]	42

[a]Irreversible process; [b]reversible process; [c]quasireversible process.

The self-explanatory nature of the decomplexing step 1-/0 deserves no further comments; but, the electrochemical reversibility of the one-electron redox change $\|Cu(ligand)_2\|^{-}/\|Cu(ligand)_2\|^{2-}$ suggests that no gross structural change accompanies this electron transfer. This is, in fact, the case for the maleonitrile-dithiolato species, for which X-ray crystal structures of both, dianion and monoanion, derivatives have been determined |43, 44|. Both species have an essentially planar architecture, of the type reported in Figure 5 for the monoanion (stacking interations in these types of complexes have been recently reviewed |45|). However, a significantly tetrahedrally distorted CuS_4 coordination has been demonstrated for $\|Cu\{S_2C_2(CN)_2\}_2\|\|MB^+\|_2 \cdot (Me_2CO)$, ($MB^+$ = 3,9-bis-(dimethylamino) phenazothionium = methylene blue cation) |46|, for which the nonplanarity of the dianion has to be mainly ascribed to the presence of the large methylene blue cations.

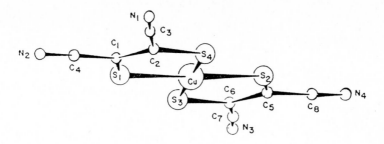

Fig. 5. Molecular structure of $|Cu\,S_2C_2(CN)_2\,_2|(Bu_4N)$ (from Ref.44).

Table 9 compares those mean distances and angles in the CuS_4 moiety of the two congeners which undergo significant variations as consequence of the redox change. As these data show, the one-electron oxidation mainly causes a significant shortening of the Cu-S bond lenght (~ 0.1 Å), so suggesting that the electron is removed from a metal-sulfur antibonding orbital.

TABLE 9

Mean distances (Å) and angles (deg) in the CuS_4 moiety of the redox couple $|Cu\{S_2C_2(CN)_2\}_2|(Bu_4N)_2 / |Cu\{S_2C_2(CN)_2\}_2|(Bu_4N)$

complex	Cu-S	S$\overset{\frown}{-}$Cu-S (five-membered rings)	reference
dianion	2.28	90.7	43
monoanion	2.17	92.3	44

Analogously, the same geometry is maintained as far as the dianion and monoanion couple of toluene-3,4-dithiolato are concerned. In fact, the spectroscopic behaviour of the dianion |47| and the X-ray crystal structure of the monoanion |42| (Figure 6) attest for the square-planarity of the CuS_4 moiety in both species.

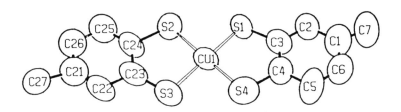

Fig. 6. X-ray molecular structure of $|Cu(toluene-3,4-dithiolate)_2|$ (Bu_4N). Cu-S mean distance, 2,16 Å; S-$\widehat{Cu}$-S (five-membered ring) 91.6° (from Ref. 42).

At this stage, a question arose: taken for granted that in both the dianionic complexes, $|Cu(maleonitrile-dithiolato)_2|^{2-}$ and $|Cu(toluene-3,4-dithiolato)_2|^{2-}$, the copper has +2 oxidation state, what is the nature of the site of the one-electron removal in the 2-/1- redox change in these complexes? As for the maleonitrile-dithiolato-copper(II) complex, it is accepted that its HOMO ($d_{x^2-y^2}$) has mainly metal character; thus, the one-electron removal formally leads to $|Cu^{III}(maleonitrile-dithiolato)_2|^-$ |44, 45, 48, 49|. On the contrary, in the case of toluene-3,4-dithiolato-copper(II) complex it is thought that the electron is removed from a ligand-based orbital |42, 47|, so that the 2-/1- redox change is formally represented by |42|:

$$|Cu^{II}(toluene-3,4-dithiolato)_2|^{2-} \underset{+e}{\overset{-e}{\rightleftarrows}} |Cu^{II}(toluene-3,4-dithiolato)(toluene-3,4-dithiolato)^{\cdot}|^{-}$$

$(toluene-3,4-dithiolato)^{\cdot}$ being a semiquinonato-like radical arising from a redox process similar to that illustrated in the previous section, 2.1, for the oxidation of $|Cu(catecholato)_2|^{2-}$.

Finally, it is evident that in both compounds the second reduction process $|Cu(ligand)_2|^{2-}/|Cu(ligand)_2|^{3-}$, even if theoretically chemically reversible, mimicks the features of a quasireversible electron transfer. This observation implies that, in the redox change, an important molecular reorganization takes place. This is not unexpected if we realize that this cathodic step, leading to a $|Cu^I(ligand)_2|^{3-}$ species, must involve a structural change around the copper centre; e.g., from square-planar to the tetrahedral (or pseudotetrahedral) coordination typical of copper(I) ions.

Interestingly, the copper(II) complex of the geometrical isomer of maleonitrile-dithiolato, schematized in Chart IX, undergoes irreversible anodic oxidation |50,51|.

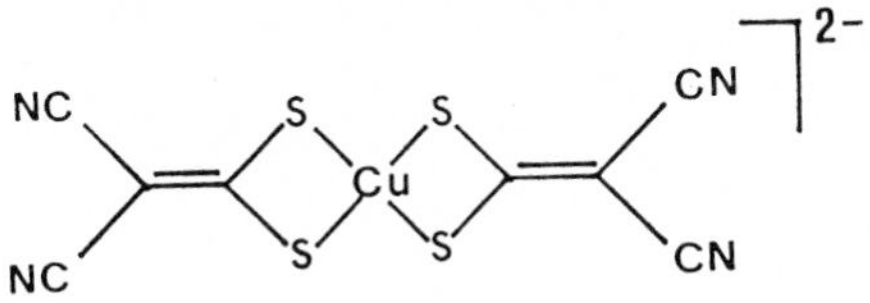

CHART IX

This result suggests that the stabilization of |Cu(maleonitrile-dithiolato)$_2$|$^-$ arises from the π-electron delocalization in the CuS_2C_2 five-membered rings |50|.

Strictly correlated to 1,2-dithiolato-copper complexes, is the copper(II) complex of quinoxaline-2,3-dithiol ligand (Chart X).

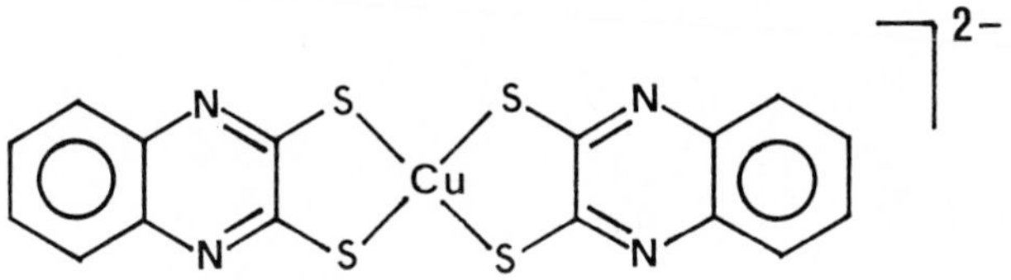

CHART X

This complex exhibits a redox behaviour quite similar to the one just now illustrated for the complexes $|Cu(1,2-dithiolato)_2|^{2-}$. In fact, as shown in Figure 7, it undergoes, in dimethylformamide solution, both a reversible one-electron oxidation ($E^{\circ\prime} = -0.18$ V), and a quasireversible one-electron reduction ($E^{\circ\prime} = -1.28$ V) according to the scheme |52|:

$$|Cu(ligand)_2|^- \underset{-e}{\overset{+e}{\rightleftharpoons}} |Cu(ligand)_2|^{2-} \underset{-e}{\overset{+e}{\rightleftharpoons}} |Cu(ligand)_2|^{3-}$$

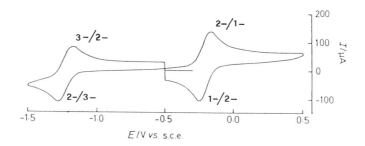

Fig. 7. Cyclic voltammogram recorded at a glassy carbon electrode in a DMF solution of $|Cu(quinoxaline-2,3-dithiolato)_2|^{2-}$ (from Ref. 52).

Once again, the electrochemical reversibility of the 2-/1- redox change nicely agrees with the X-ray diffraction results which show a retention of the square planarity in the CuS_4 moiety of the two congeners $|Cu^{II}(quinoxaline-2,3-dithiolato)_2|^{2-}$ and $|Cu^{III}(quinoxaline-2,3-dithiolato)_2|^{-}$ |52|, Figure 8.

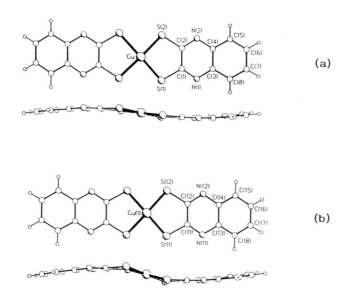

Fig. 8. Molecular structures of: (a) $|Cu(quinoxaline-2,3-dithiolato)_2|(PPh_4)_2$; (b) $|Cu(quinoxaline-2,3-dithiolato)_2|(PPh_4)$ (from Ref. 52).

Data in Table 10 offers evidence that the removal of one-electron (likely from the metal $d_{x^2-y^2}$ orbital, antibonding with respect to the Cu–S bond) causes a shortening of the metal-sulfur bonds, decidedly minor, compared with that observed for the corresponding maleonitrile-dithiolato species, probably because of the large extent of aromatic character in the quinoxaline which spreads the change over the entire ligand. It can be also noted from the bottom views, Figure 8a and 8b, that the slight molecular contraction accompanying the one-electron oxidation causes an increase of the angle between the CuS_4 and $C_8N_2S_2$ planes (from 10° to about 23°) [52].

TABLE 10

Mean structural parameters in the CuS_4 moiety of the redox couple $|Cu(S_2C_2N_2C_6H_4)_2|(PPh_4)_2 / |Cu(S_2C_2N_2C_6H_4)_2|(PPh_4)$ [52]

complex	Cu–S (Å)	S–Cu–S (deg) (five-membered rings)
dianion	2.19	91.5
monoanion	2.18	92.2

Also in this case, the reduction of the Cu(II) complex to the corresponding Cu(I) complex (2–/3–), likely involving a significant planar-to-tetrahedral molecular reorganization, proceeds through a quasireversible electron transfer.

An analog of 1,2-dithiolato-copper complexes is the bis(dithiooxalato)-copper(II) complex schematized in Chart XI.

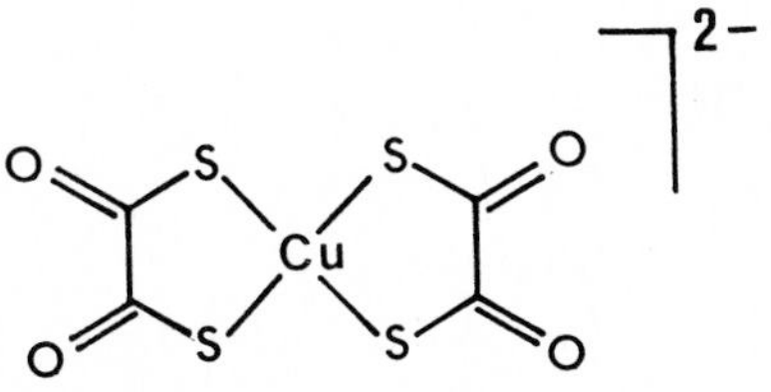

CHART XI

This copper(II) complex undergoes a one-electron, reversible, anodic oxidation in dichloromethane according to the step |53,54|:

$$|Cu^{II}(dithiooxalato\text{-}S,S')_2|^{2-} \underset{+e}{\overset{-e}{\rightleftharpoons}} |Cu^{III}(dithiooxalato\text{-}S,S')_2|^{-}$$

$$E°' = +0.13 \text{ V}$$

Slight departures from pure electrochemical reversibility are attributed to uncompensated solution resistance, quite commonly arising in dichloromethane solutions. As usual, this electrochemical reversibility preludes a retention of geometry during the redox change. In fact, X-ray investigations allowed the planar geometry of both reduced and oxidized anions to be ascertained |55|; and, as shown in Figure 9, the main consequence of the one-electron oxidation results in a marked shortening of the Cu-S bond lengths (from 2.26 Å to 2.17 Å), in agreement with the removal of the electron residing in the oft-cited antibonding orbital, having essentially metal character.

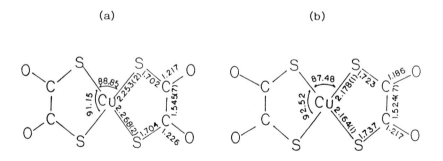

(a)   (b)

Fig. 9. Structural parameters in the planar molecules of: (a) |Cu(dithiooxalate)$_2$|$^{2-}$; (b) |Cu(dithiooxalate)$_2$|$^{-}$ (from Ref. 54).

A significant exception to the planarity of the copper(II) complexes of 1,2-dithiolates was observed in bis|4,5-dimercapto-1, 3-dithiole-2-thionato(2-)|cuprate(II), schematized in Chart XII.

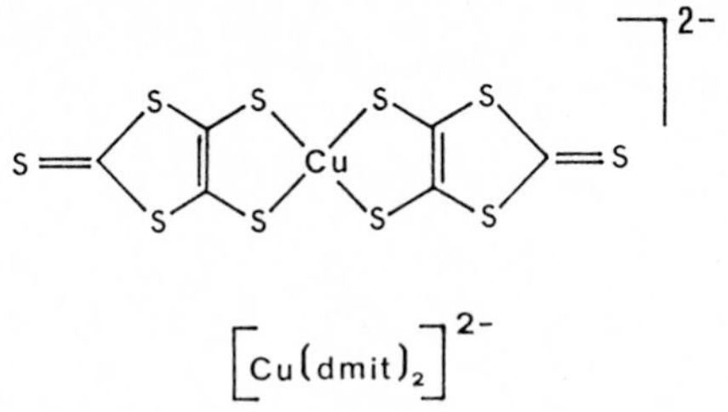

CHART XII

$$\left[Cu(dmit)_2 \right]^{2-}$$

As shown in Figure 10, this complex displays a CuS_4 coordination higly distorted from square-planarity |56|.

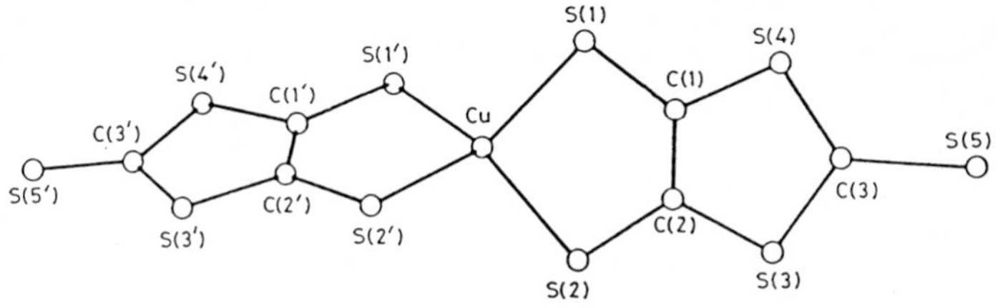

Fig. 10. Perspective view of |Cu(dmit)$_2$|$^{2-}$. N-Ethylpyridinium counteranion. Cu-S, 2.27 Å (from Ref. 56).

In this substance, the simplest index of the extent of tetra-hedral geometry is the dihedral angle between the mean planes of the two dmit ligands. Its value of 57.3°, compared with the value of 90° for a regular tetrahedral geometry, shows the non-trivial extent of the distortion.

As shown in Figure 11, |Cu(dmit)$_2$|$^{2-}$ undergoes easy oxidation processes ($Ep_{2-/-}$ = + 0.02 V; $Ep_{-/0}$ = + 0.08 V) |56|. However the presence of electrode adsorption effects makes difficult the analysis of these electrochemical steps; nevertheless, in view of the closeness of the two subsequent one-electron removals (one of which at least ligand centred), it seems unlikely that the corresponding copper(III) complex (likely planar) may be stable.

Fig. 11. Cyclic voltammogram recorded at a platinum electrode on a MeCN solution of $|Cu(dmit)_2|^{2-}$. Scan rate 0.1 Vs^{-1} (from Ref. 56).

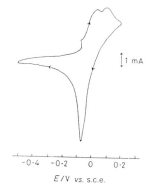

E/V vs. s.c.e.

Another class of dithiolates is illustrated by 1,1-dithiolates (forming four-membered chelate rings) |57|.

One of these derivatives, whose redox behaviour has been examined, is the 1,1-dicarboethoxy-2,2-ethylenedithiolate-copper complex represented in Chart XIII.

EtO$_2$C ... S ... S ... CO$_2$Et   CHART XIII

EtO$_2$C ... S ... Cu ... S ... CO$_2$Et

This copper(III) complex possesses the usual square-planar CuS$_4$ array, as illustrated in Figure 12 |58,59|.

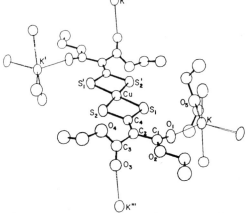

Fig. 12. Molecular structure of $K|Cu(S_2C=C\{COOEt\}_2)_2|\cdot OEt_2$. Cu-S mean distance, 2.19 Å; S-Cu-S (four-membered ring) 78.1° (from Ref. 59).

In dichloromethane solution, it undergoes a reversible, one-electron reduction ($E°' = -0.64$ V) |58,59|, leading to the corresponding dianion, air sensitive, the expected planarity of which is proved by its isomorphism to the dianion $|Ni(S_2C=C\{COOEt\}_2)_2|^{2-}$ |58|.

The reversibility of the 2-/1- redox change in this complex raises some doubts on the accuracy of the electrochemistry of the related complex in Chart IX. This view is re-enforced by the fact that dithiocarbamate-copper(II) complexes (Chart XIV), member of 1,1-dithiolates, also, are redox active.

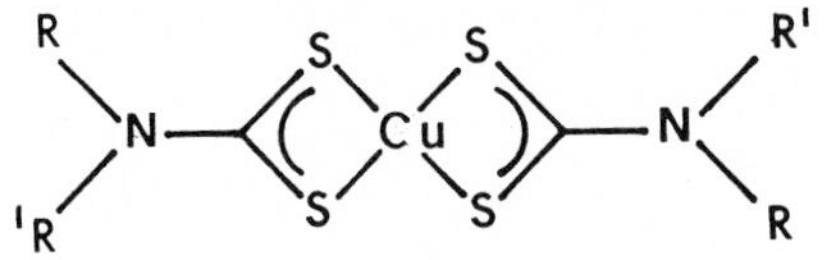

CHART XIV

$Cu(RR'dtc)_2$

They commonly undergo a reversible, both chemically and electrochemically, one-electron oxidation as well as a chemically reversible, but electrochemically quasireversible, one-electron reduction, generating the corresponding copper(III) and copper(I) complexes, respectively |60,61|. Table 11 summarizes the relevant redox potentials for a wide series of such complexes.

Figure 13 shows the cyclic voltammogram recorded in an acetone solution of $|Cu^I((\underline{i}-Pr)_2dtc)_2|^-$ (obtained by exhaustive one-electron reduction of the neutral $Cu^{II}((\underline{i}-Pr)_2dtc)_2$.

The planarity of the CuS_4 coordination in $Cu((\underline{i}-Pr)_2dtc)_2$ is shown in Figure 14. The Cu-S mean distance is of 2.29 Å, fully comparable with the distance of 2.31 Å in $Cu((Et)_2dtc)_2$ |64|.

TABLE 11

Formal electrode potentials (in Volt) for the redox changes of the copper(II)-dithiocarbamate schematized in Chart XIV

R	R'	$E^{o\prime}_{Cu^{II}/Cu^{I}}$	$E^{o\prime}_{Cu^{II}/Cu^{III}}$	solvent	reference
CH_3	CH_3	-0.38	$+0.65$	Me_2CO	61
C_2H_5	C_2H_5	-0.41	$+0.66$	Me_2CO	61
C_2H_5	C_2H_5	-0.52	$+0.42$	MeCN	60
$\underline{i}\text{-}C_3H_7$	$\underline{i}\text{-}C_3H_7$	-0.52	$+0.56$	Me_2CO	61
$\underline{n}\text{-}C_4H_9$	$\underline{n}\text{-}C_4H_9$	-0.44	$+0.64$	Me_2CO	61
$\underline{n}\text{-}C_4H_9$	$\underline{n}\text{-}C_4H_9$	$-$	$+0.47$	CH_2Cl_2	62
$\underline{i}\text{-}C_4H_9$	$\underline{i}\text{-}C_4H_9$	-0.45	$+0.62$	Me_2CO	61
C_6H_{11}	C_6H_{11}	-0.54	$+0.53$	Me_2CO	61
$C_6H_5CH_2$	$C_6H_5CH_2$	-0.27	$+0.73$	Me_2CO	61
C_6H_5	C_6H_5	-0.31	$+0.67$	Me_2CO	61
CH_3	$\underline{n}\text{-}C_4H_9$	-0.41	$+0.67$	Me_2CO	61
C_6H_5	CH_3	-0.33	$+0.67$	Me_2CO	61
C_6H_5	C_2H_5	-0.35	$+0.66$	Me_2CO	61

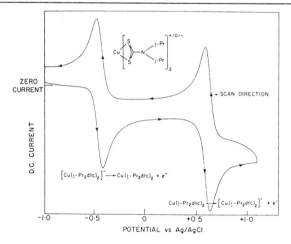

Fig. 13. Cyclic voltammogram recorded at a platinum electrode in an acetone solution of $|Cu((\underline{i}\text{-}Pr)_2dtc)_2|^-$. Scan rate 0.2 Vs^{-1} (from Ref. 61).

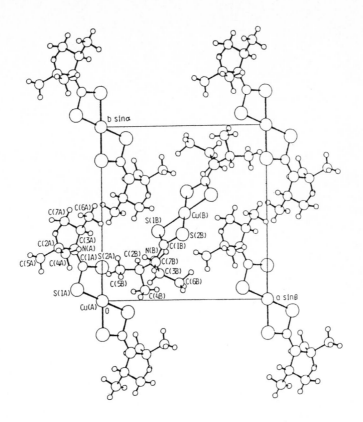

Fig. 14. Crystal structure of Cu((i-Pr)₂dtc)₂ (from Ref.63).

In view of the planarity of the bis(dithiocarbamate) copper(III) species |65,66|, the electrochemical reversibility of the copper(II)/copper(III) oxidation is well congruent (in $|Cu^{III}((n-Bu)_2dtc)_2|^+$ the Cu-S distance shortens to 2.22 Å |66|). Also, the quasireversibility of the copper(II)/copper(I) reduction agrees well with the likely planar-to-tetrahedral reorganization.

Thiaethers, the second main class of donor sulfur ligands, give rise to a wide series of tetrathia-copper(II) complexes.

Chart XV depicts the cyclic tetrathiaether ligands forming 1:1 copper(II) complexes studied also from the electrochemical viewpoint (stereochemistry of these complexes has been widely discussed in Vol.2 of this series by J.C.A.Boeyens and S.M.Dobson and by K.E.Matthes and D.Parker).

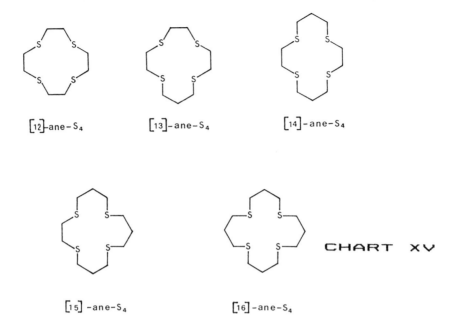

[12]-ane-S₄ · [13]-ane-S₄ · [14]-ane-S₄

[15]-ane-S₄ · [16]-ane-S₄

CHART XV

Likewise, Chart XVI schematizes some open-chain ligands able to form similar tetrathia-copper(II) complexes.

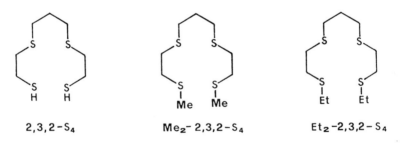

2,3,2-S₄ · Me₂-2,3,2-S₄ · Et₂-2,3,2-S₄

CHART XVI

All complexes of both classes undergo a Cu(II)/Cu(I) reduction step, quasireversible in character |29,30,67|. In some cases this step appears complicated by coupled processes possibly due to the presence of isomeric complexes |30|. A further Cu(I)/Cu(0) step (irreversible because of the decomplexing process of the metal) is likely present for all complexes, but it has been described only for the |14|-ane-S₄ complex |67|. The redox potentials of these processes are summarized in Table 12.

TABLE 12

Redox potentials (in Volt) for the cathodic reductions of the copper(II) complexes of the ligands schematized in Charts XV and XVI

ligand	$E^{\circ\prime}_{Cu^{II}/Cu^{I}}$	$E_{p Cu^{I}/Cu^{0}}$	solvent	reference
$\lvert 12 \rvert$-ane-S_4	+ 0.48	–	aqueous–80% MeOH	30
$\lvert 12 \rvert$-ane-S_4	+ 0.40	–	H_2O	30
$\lvert 13 \rvert$-ane-S_4	+ 0.43	–	aqueous–80% MeOH	30
$\lvert 13 \rvert$-ane-S_4	+ 0.35	–	H_2O	30
$\lvert 14 \rvert$-ane-S_4	+ 0.45	–	aqueous–80% MeOH	30
$\lvert 14 \rvert$-ane-S_4	+ 0.36	–	H_2O	30
$\lvert 14 \rvert$-ane-S_4	+ 0.61	– 0.69	MeCN	67
$\lvert 14 \rvert$-ane-S_4	+ 0.55	–	$MeNO_2$	67
$\lvert 15 \rvert$-ane-S_4	+ 0.54	–	aqueous–80% MeOH	30
$\lvert 15 \rvert$-ane-S_4	+ 0.39	–	H_2O	72
$\lvert 16 \rvert$-ane-S_4	+ 0.56	–	aqueous–80% MeOH	30
$\lvert 16 \rvert$-ane-S_4	+ 0.44	–	H_2O	72
$2,3,2$-S_4	+ 0.60	–	aqueous–80% MeOH	30
Me_2-$2,3,2$-S_4	+ 0.65	–	aqueous–80% MeOH	30
Me_2-$2,3,2$-S_4	+ 0.58	–	H_2O	30
Et_2-$2,3,2$-S_4	+ 0.65	–	aqueous–80% MeOH	30

All authors have described the Cu(II)/Cu(I) redox change as quasireversible. This suggests that significant structural changes take place as consequence of the electron transfer. The X-ray structures of $\lvert Cu^{II}(\lvert n \rvert$-ane-$S_4)\rvert(ClO_4)_2$ (n = 12, 13, 14, 15, 16) and $\lvert Cu^{II}(Et_2$-$2,3,2$-$S_4)\rvert(ClO_4)_2$ have been solved $\lvert 68$-$70 \rvert$, and these are here illustrated in Figure 15.

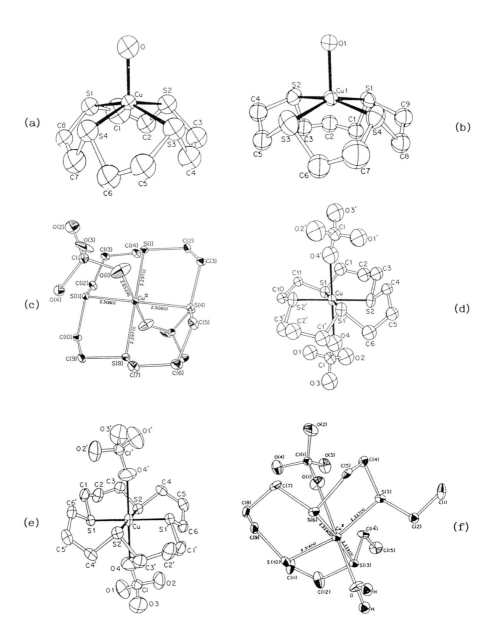

Fig. 15. Molecular structure of: (a) $|Cu(|12|$-ane-$S_4)|(ClO_4)_2$; (b) $|Cu(|13|$-ane-$S_4)|(ClO_4)_2 \cdot H_2O$; (c) $|Cu(|14|$-ane-$S_4)|(ClO_4)_2$; (d) $|Cu(|15|$-ane-$S_4)|(ClO_4)_2$; (e) $|Cu(|16|$-ane-$S_4)|(ClO_4)_2$; (f) $|Cu(Et_2$-2,3,2-$S_4)|(ClO_4)_2 \cdot H_2O$ (from Refs. 69,70).

As far as the macrocyclic complexes are concerned, two main features have been pointed out |69|: (i) when the ring contains 14 or more atoms, the CuS$_4$ group forms a square-planar arrangement, with a perchlorate oxygen axially coordinated, to form a tetragonal octahedral geometry; (ii) for rings containing 12 or 13 atoms, the cavity size is too small to accomodate the copper (II) ion, which is forced out of the S$_4$ plane, coordinating with a water oxygen in a square-pyramidal geometry. Concerning the open-chain complex of the ligand Et$_2$-2,3,2-S$_4$, the copper atom and three sulfur atoms are coplanar, while the fourth sulfur atom is out of the CuS$_3$ plane; a water oxygen and a perchlorate oxygen complete the tetragonal assembly.

All these data are quantitatively summarized in Table 13.

TABLE 13

Structural parameters of some copper(II)-tetrathiaether complexes |69|

ligand	donor atoms set	averaged bond length (Å)		geometrical features of the equatorial coordination
|12|-ane-S$_4$	4 S 1 O (H$_2$O)	Cu-S Cu-O	2.33 2.11	S coplanar; Cu +0.53 Å
|13|-ane-S$_4$	4 S 1 O (H$_2$O)	Cu-S Cu O	2.32 2.15	S nearly coplanar; Cu +0.38 Å
|14|-ane-S$_4$	4 S 2 O (ClO$_4$)	Cu-S Cu-O	2.30 2.65	S coplanar;Cu coplanar
|15|-ane-S$_4$	4 S 2 O (ClO$_4$)	Cu-S Cu-O	2.32 2.53	S coplanar;Cu coplanar
|16|-ane-S$_4$	4 S 2 O (ClO$_4$)	Cu-S Cu-O	2.36 2.48	S coplanar;Cu coplanar
Et$_2$-2,3,2-S$_4$	4 S 1 O (H$_2$O) 1 O (ClO$_4$)	Cu-S Cu-O Cu-O	2.33 2.30 2.81	3 S coplanar; 1 S +0.78 Å; Cu coplanar

To date, only two X-ray investigations dealt with the molecular structure of copper(I)-tetrathiaether complexes, namely $|Cu(|14|-ane-S_4)|(ClO_4)$ |71| and $|Cu(Et_2-2,3,2-S_4)|(ClO_4)$ |70|. In both cases, the need for copper(I) ion to assume a pseudotetra-hedral geometry causes deep reorganizations between copper ion and the nearest ligands, affording, in the crystalline state, linear polymeric chains. $|Cu^I(|14|-ane-S_4)|^+$ forms a 3:1 coordination polymer, in which the Cu(I) atom is coordinated to the three sulfur atoms from a single ligand, and to a fourth sulfur atom from an adjacent ligand, Figure 16a. $|Cu^I(Et_2-2,3,2-S_4)|^+$ forms a 2:2 coordination polymer, containing a Cu(I) atom coordinated to two sulfur atoms, from a single ligand, and to two sulfur atoms from an adjacent ligand, Figure 16b. In both cases the perchlorate anions are no longer coordinating.

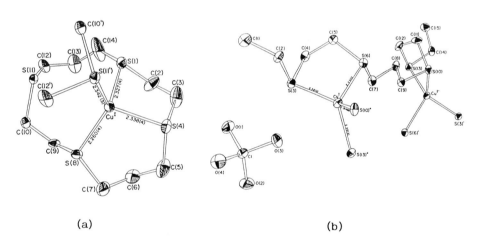

(a)   (b)

Fig. 16. Molecular structures of: (a) $|Cu(|14|-ane-S_4)|_x(ClO_4)_x$; (b) $|Cu(Et_2-2,3,2-S_4)|_x(ClO_4)_x$ (from Ref. 70).

Comparing the Cu-S distances for these copper(I) complexes, reported in Table 14, with those reported in Table 13, one concludes that one-electron additions cause only small variations.

Although the crystal packing affords polymeric Cu(I) products, there is evidence that, in dilute aqueous solution, the above copper(I) complexes dissociate to monomers |70|.

TABLE 14

Mean bond distances (Å) in some copper(I)-tetrathiaether complexes
|70|

ligand	Cu-S
|14|-ane-S$_4$	2.32
Et$_2$-2,3,2-S$_4$	2.31

Extrapolations of crystallographic structural data to solution
conditions has allowed to summarize the structural changes
accompanying the one-electron tranfer Cu(II)/Cu(I) in tetrahia
complexes according to the picture schematized in Figure 17 |72|.

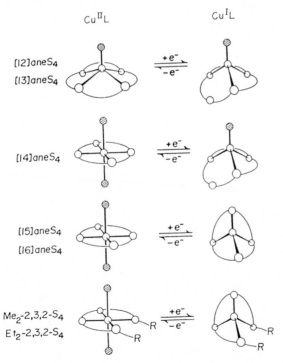

Fig. 17. Schematization of the structural changes accompanying the
Cu(II)/Cu(I) redox step in tetrahia complexes. Shaded atoms represent oxygen atoms from solvent or anions. Unshaded atoms represent
the central copper atoms and the sulfur atoms (from Ref. 72).

The fact that, despite the deep molecular reorganizations taking place in the Cu(II)/Cu(I) redox change, the formal electrode potentials remain quite positive served the authors to dispute that in blue copper proteins the positive potentials arise as consequence of slight molecular rearrangements occurring during the electron transfer.

Finally, a formal S_4 donation can be achieved through the bis-chelate-copper complex formation of thiather ligands |73|. The ligands represented in Chart XVII have been shown to be suitable for the formation of redox active copper(II) complexes.

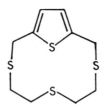

2,5-dithiahexane   **2,5,8-trithia[9](2,5) thiophenophane**

CHART XVII

In acetonitrile solution $|Cu(2,5-dithiahexane)_2|^{2+}$ undergoes a quasireversible one-electron reduction (Cu(II)/Cu(I)) at $E°' = +0.75$ V and a further irreversible step (Cu(I)/Cu(0)) at $E_p = -1.4$ V |67|.

The quasireversibility of the Cu(II)/Cu(I) redox change agrees well with the relevant conformational change occurring in the CuS_4 group. As shown in Figure 18, the one electron reduction causes the usual tetragonal-copper(II) to tetrahedral-copper(I) rearrangement |74,75|.

Some selected bond lengths in these two congeners are reported in Table 15. As can be seen, the structural change is accompanied by a significant shortening of the Cu-S bond length.

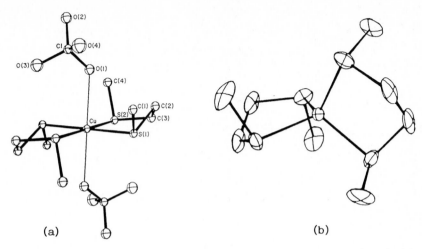

Fig. 18. Crystal structure of: (a) |Cu(2,5-dithiahexane)$_2$|(ClO$_4$)$_2$; (b) |Cu(2,5-dithiahexane)$_2$|(ClO$_4$) (from Ref. 75).

TABLE 15

Selected bond lengths (Å) in |Cu(2,5-dithiahexane)$_2$|$^{2+}$ and |Cu(2,5-dithiahexane)$_2$|$^+$

complex	mean distances		reference
|Cu(2,5-dithiahexane)$_2$|(ClO$_4$)$_2$	Cu–S	2.34	75
	Cu–O (ClO$_4$)	2.55	
|Cu(2,5-dithiahexane)$_2$|(BF$_4$)$_2$	Cu–S	2.32	74
	Cu–F (BF$_4$)	2.58	
|Cu(2,5-dithiahexane)$_2$|(ClO$_4$)	Cu–S	2.26	75

Interestingly, a mixed-valent species, of stoichiometry |Cu$_2$^{I}Cu^{II}(2,5-dithiahexane)$_6$|(ClO$_4$)$_4$, has been also crystallographycally characterized; its Cu(I) fragment is closely related, structurally, to that shown in Fig. 18b |75,76|.

2,5,8-trithia-thiophenophane, even if apparently closely related to the 12-membered macrocycle |12|-ane-S$_4$, seems not able to form 1:1 complexes with copper(II) |77| or copper(I) |78|.

216

Figure 19 shows the crystal structure of the bis(2,5,8-trithia-thiophenophane)-copper(I) complex.

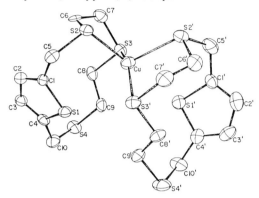

Fig. 19. Perspective view of $|Cu(2,5,8$-trithia-thiophenophane$)_2|^+$. $Cu-S_{(averaged)}$, 2.34 Å (from Ref. 78).

The inability of this ligand to encircle the copper(I) ion does not arise from its cavity size, but from the presence of hydrogens on C9,C7,C8,C6 which shield the access to the cavity |78|.

As shown in Figure 20, the copper(I) complex undergoes, in dimethylsulfoxide, a quasireversible copper(I)/copper(II) oxidation.

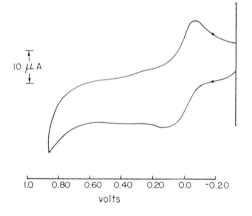

Fig. 20. Cyclic voltammogram of $|Cu(2,5,8$-trithia-thiophenophane$)_2|^+$ recorded at a glassy carbon electrode on a DMSO solution (from Ref. 78).

The apparent chemical reversibility of the oxidation step (i_{pc}/i_{pa} = 1) seems to suggest that the 1:2 metal/ligand ratio is maintained (in contrast to a previously reported 3:2 ratio |77|), even if some significant structural reorganization must occur (likely towards a tetragonal geometry) in view of the electrochemical quasireversibility.

3.3 $\underline{S_5}$ donor set

The only ligand able to form an electrochemically characte-
rized CuS$_5$ assembly is the polycyclic thiather |15|-ane-S$_5$, shown
in Chart XVIII.

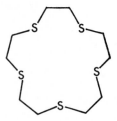

CHART XVIII

$[15]$ - ane - S$_5$

In aqueous-80% methanol solution the Cu(II)/Cu(I) redox change
occurs through a quasireversible electron transfer at E°'= +0.61 V,
while in aqueous solution at E°'= +0.51 V |29,30|.

The structural changes occurring in the one-electron addition
are easily deducible from Figure 21.

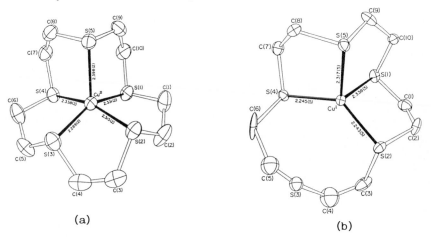

(a)   (b)

Fig. 21. Molecular structure of: (a) |Cu(|15|-ane-S$_5$)|(ClO$_4$)$_2$;
(b) |Cu(|15|-ane-S$_5$)|(ClO$_4$) (from Ref. 79).

The square-pyramidal copper(II) complex (Cu-S mean distance,
2.33 Å) rearranges to the corresponding tetrahedral copper(I)
complex (Cu-S mean distance, 2.28 Å), by simply leaving one sulfur
atom uncoordinated (Cu----S, 3.5 Å) |79|.

The authors pointed out that this redox change involves the lowest reorganization energy among all the studied copper-polythia ether complexes, likely because of the presence of internal strains in both the oxidized and reduced species.

3.4 S_6 donor set

A series of polycyclic thiathers, here schematized in Chart XIX, give redox active copper complexes.

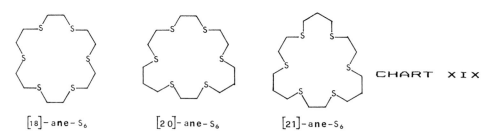

[18]-ane-S_6 [20]-ane-S_6 [21]-ane-S_6 CHART XIX

All the relevant copper(II) complexes undergo a Cu(II)/Cu(I) electron transfer quasireversible in character.

The cyclic voltammogram illustrated in Figure 22 shows the Cu(I)/Cu(II) one-electron oxidation of $|Cu^I(|18|-ane-S_6)|^+$.

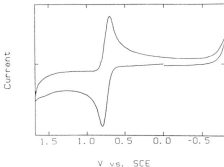

V vs. SCE

Fig. 22. Cyclic voltammogram recorded at a glassy carbon electrode on a MeNO$_2$ solution of $|Cu(|18|-ane-S_6)|^+$. Scan rate 0.02 Vs^{-1} (from Ref. 80).

Even if the authors describe this redox change as reversible |80|, the dependence of ΔEp with scan rate testifies to its quasireversibility |81|.

In Table 16, the redox potentials of hexathia complexes are reported.

TABLE 16

Redox potentials (in Volt) for the one-electron reduction of the copper(II) complexes of the ligands schematized in Chart XIX

ligand	$E^{o'}_{Cu^{II}/Cu^{I}}$	solvent	reference
|18|-ane-S_6	+ 0.72	$MeNO_2$	80
|20|-ane-S_6	+ 0.56	aqueous-80% MeOH	29, 30
|20|-ane-S_6	+ 0.48	H_2O	30
|21|-ane-S_6	+ 0.61	aqueous-80% MeOH	29, 30

The structural changes occurring in the one-electron reduction process in hexathia complexes are typically exemplified by Figure 23, which depicts the \|18\|-ane-S_6 complex \|80\|.

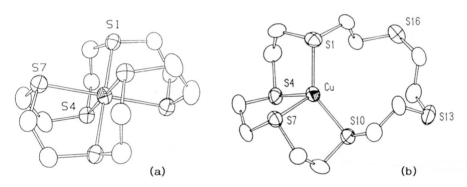

(a)   (b)

Fig. 23. ORTEP drawing of the cationic species: (a) \|Cu(\|18\| -ane-S_6)\|(picrate)$_2$; (b) \|Cu(\|18\|-ane-S_6)\|(BF_4) (from Ref. 80).

The Cu(II) complex has a distorted octahedral geometry (Cu-S mean bond length, 2.43 Å), whereas the Cu(I) complex, leaving two sulfur atoms unbonded, assumes a distorted tetrahedral geometry (Cu-S mean bond length, 2.30 Å).

Finally a CuS_6 coordination was observed in the bis-ligand-copper(II) complex of the trithia ether ligand in Chart VI.

The X-ray structure of \|Cu(trithiacyclononane)$_2$\|$^{2+}$ is shown in Figure 24 \|33\|.

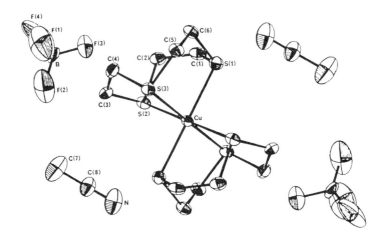

Fig. 24. Perspective view of |Cu(trithiacyclononane)$_2$|(BF$_4$)$_2$. Cu-S mean distance, 2.43 Å (from Ref.33).

In nitromethane solution, the octahedral |Cu(trithiacyclo-nonane)$_2$|$^{2+}$ undergoes a quasireversible Cu(II)/Cu(I) step at E°'= +0.61 V, thus suggesting that the corresponding copper(I) complex could be prepared and allowing the foreseeable tetrahedral reorganization to be forecasted in detail.

4.   NITROGEN DONOR LIGANDS

A wide range of ligand molecules are available for coordination to copper ions through nitrogen donation.

4.1 N$_3$ donor set

Copper complexes with open-chain and/or cyclic triamines in 1:1 ratio are rare enough |82,83| and only in few cases have electrochemical investigations been performed.

In an electrochemical investigation of the carbonyl-copper(I) complexes with open-chain triamines illustrated in Chart XX, we have pointed out that, in dimethylsulfoxide solutions, the relevant anodic oxidation Cu(I)/Cu(II) involves a fast decarbonylation step leading to the corresponding copper(II)-amino complexes |84|.

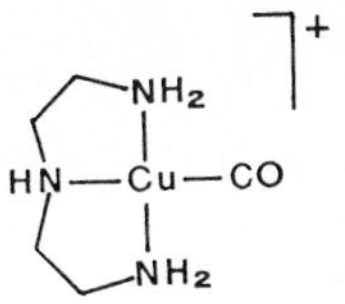

CHART XX

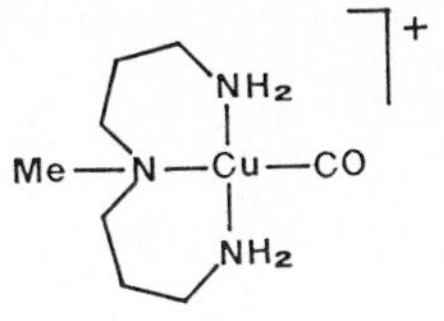

$|Cu(dien)CO|^+$
dien=diethylenetriamine

$|Cu(Medpt)CO|^+$
Medpt=N,N-bis(3-aminopropyl)methylamine

These copper(II) complexes have been identified as the solvates $|Cu(dien)(DMSO)_3|^{2+}$ and $|Cu(Medpt)(DMSO)_3|^{2+}$. These solvates undergo an uncomplicated, quasireversible, one-electron reduction according to:

$$|Cu(dien)|^{2+} \underset{-e}{\overset{+e}{\rightleftharpoons}} |Cu(dien)|^+$$

$$E°' = -0.52 \text{ V}$$

$$|Cu(Medpt)|^{2+} \underset{-e}{\overset{+e}{\rightleftharpoons}} |Cu(Medpt)|^+$$

$$E°' = -0.20 \text{ V}$$

It is interesting to note that, despite the presence of an electron donating methyl group, $|Cu(Medpt)|^{2+}$ is more easily reducible than $|Cu(dien)|^{2+}$. Even if structural informations are lacking, it is likely that the greater size, and hence flexibility, of the Medpt cavity allows the copper(II) complex to adopt a geometry which favours the tetrahedral rearrangement following the copper(I) generation to a greater extent than possible in the dien cavity.

An ill-defined, irreversible reduction, at Ep = -0.07 V has been reported for $|Cu(dien)|^{2+}$ in MeCN solvent [85]. We experienced, and these authors confirm, a difficult reproducibility for cyclic voltammograms on $|Cu(dien)|^{2+}$; this datum, and the differences in solvation, could account for the discrepancy in the electrode potential of the redox couple Cu(II)/Cu(I) in the dien complex.

222

The copper(II) complexes of the cyclic triamines, illustrated in Chart XXI, have been studied by electrochemical methods.

CHART XXI

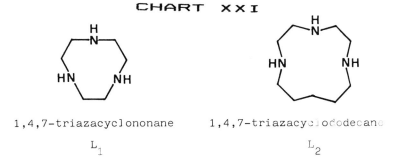

1,4,7-triazacyclononane   1,4,7-triazacyclododecane

L_1 L_2

In acetate buffer, pH 4.0, the complex $|CuL_1|^{2+}$ displays a polarographic wave at $E_{1/2}$= -0.11 V, attributed to the two-electron decomplexing step |86|:

$$|CuL_1|^{2+} + 2e \longrightarrow Cu + L_1$$

It has been also briefly reported that in acetonitrile $|CuL_1|^{2+}$ can be oxidized to the corresponding copper(III) complex; which, however, is likely a transient species |83|.

Figure 25 shows the cyclic voltammetric response exhibited by $|CuL_2|^{2+}$ in acetonitrile solution |85|. A quasireversible one-electron reduction is displayed.

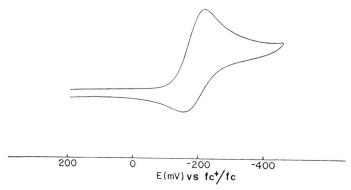

200   0   -200   -400
E(mV) vs fc⁺/fc

Fig. 25. Cyclic voltammogram recorded at a platinum electrode on a MeCN solution of $|CuL_2|^{2+}$. Scan rate 0.1 Vs^{-1} (from Ref.85).

The redox couple Cu(II)/Cu(I) is located at $E^{o'}$= +0.19 V. This relatively high redox potential suggests that the coordination sphere of copper(II) easily rearranges to the tetrahedral one. The crystal structure of $|CuL_2(NO_3)|^+$, shown in Figure 26, indicates the copper(II) moiety has a trigonal bipyramidal assembly, the atoms N(4), N(4)' and O(1) lying in the equatorial plane, and the O(2) atom of the axial position being displaced toward the O(1) atom |85|.

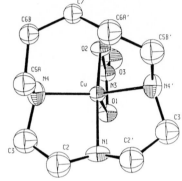

Fig. 26. Crystal structure of $|CuL_2(NO_3)|(NO_3)$. Mean distances: Cu–N, 1.99 Å; Cu–O, 2.16 Å (from Ref.85).

The i_{pa}/i_{pc} ratio of about 0.5 for the cathodic step in Figure 26, if not due to electrode poisoning phenomena, suggests however that the corresponding copper(I) complex is not long-lived.

Finally, a formal CuN_3 coordination is present in the copper(II) complexes of the ligands schematized in Chart XXII, derived from condensation of diacetylpyridine with phenylalanine methyl ester or tyrosine ethyl ester.

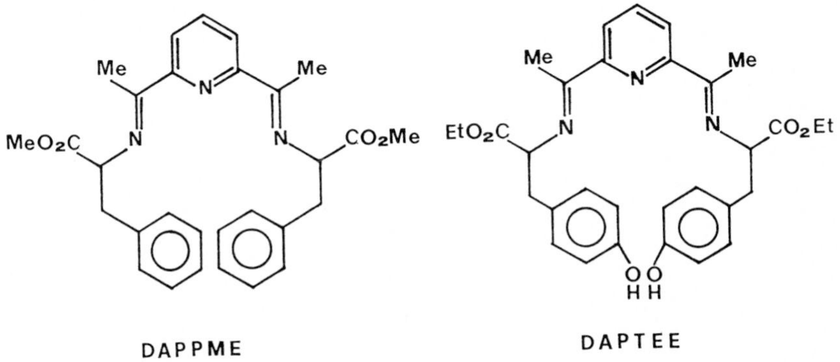

DAPPME   DAPTEE

CHART XXII

It has been briefly reported that such complexes undergo, in dimethylsufoxide, a copper(II)/copper(I) reduction at rather positive potential values ($|Cu(DAPPME)|^{2+}$, $E°'$ = +0.22 V; $|Cu(DAPTEE)|^{2+}$, $E°'$ = +0.14 V). It is however likely that the coordination sphere of the copper centre also includes side-arm oxygen donor atoms |87|.

4.2 $\underline{N_4}$ donor set

As in the case of the CuS_4 coordination within the sulfur donor ligands, the CuN_4 coordination is the commonest one with nitrogen donor ligands.

4.2.1 Ammonia. The simplest CuN_4 coordination is present in the tetraammino complex $|Cu(NH_3)_4|^{2+}$. It has been noted polarographically that such a species undergoes, in aqueous solution containing an excess of free NH_3, a reversible one-electron reduction followed by ligand dissociation according to |88|:

$$|Cu(NH_3)_4|^{2+} + e \longrightarrow |Cu(NH_3)_2|^+ + 2NH_3$$

$$E_{1/2} = - 0.23 \text{ V}$$

4.2.2 Pyridine. The redox couple Cu(II)/Cu(I) for the tetra-pyridinate complex $|Cu(py)_4|(ClO_4)_2$ in aqueous-50% DMF is located at + 0.10 V |89|. No information about the extent of reversibility of this electron-transfer is available. However in view of this relatively high value, the regular tetrahedral geometry of $|Cu^I(py)_4|ClO_4$ |90,91| (see Figure 27), and the real CuN_4 coordination in $|Cu^{II}(py)_4|^{2+}$|89|, it does not seem ventured to think the geometry of copper(II)-tetrapyridinate as simply tetrahedrally distorted, even if the relevant triflate complex $Cu^{II}(py)_4(CF_3SO_3)_2$ shows a pseudooctahedral geometry, as a result of trans-axially coordinated trifluoromethanesulfonate molecules (Cu-N, 2.03 Å; Cu-O, 2.43 Å)|92|.

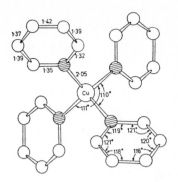

Fig. 27. Molecular structure of |Cu(py)$_4$|ClO$_4$; Cu···Cl(O$_4$), 6.40 Å (from Ref. 90).

4.2.3 <u>Diamines</u>. Copper(II) bis-diamine complexes are generally unstable towards redox changes.

Bis(ethylenediamine)- and bis(propylenediamine)-copper(II) complexes in aqueous solutions undergo the single two-electron reduction |93,94|:

$$|Cu(diamine)_2|^{2+} + 2e \longrightarrow Cu + 2 \text{ diamine}$$

Because of the chemical irreversibility of this electron transfer, we do not consider it further.

By contrast, the copper(II) complex of the deprotonated bis(N-acetylethylenediamine) (Chart XXIII) undergoes in aqueous solution both a one-electron oxidation and a one-electron reduction, according to |95|:

$$|Cu(H_{-1}Aen)_2|^{+} \underset{-e}{\overset{+e}{\rightleftharpoons}} Cu(H_{-1}Aen)_2 \underset{-e}{\overset{+e}{\rightleftharpoons}} |Cu(H_{-1}Aen)_2|^{-}$$

$$E°' = +0.91 \text{ V} \qquad\qquad E°' = -0.17 \text{ V}$$

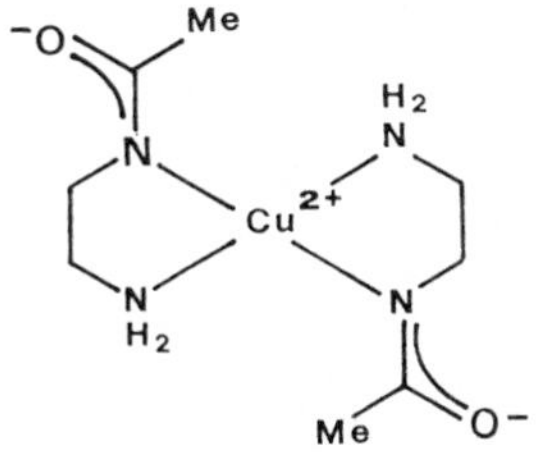

CHART XXIII

Cu(H$_{-1}$ Aen)$_2$

Even if both, the one-electron removal and the one-electron addition, are relatively difficult to perform, the occurrence of both the steps indicates that the starting copper(II) complex is sufficiently flexible to assume either the planar geometry of the copper(III) species or the tetrahedral one of the copper(I) species.

4.2.4 <u>Tetramines</u>. The redox behaviour of the copper(II) complexes of the linear tetramines, illustrated in Chart XXIV, has been investigated.

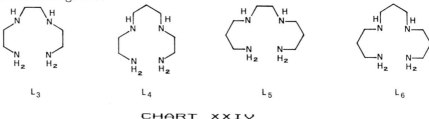

L_3 L_4 L_5 L_6

CHART XXIV

Figure 28 shows the cyclic voltammetric behaviour of $|Cu(L_3)|^{2+}$ in acetonitrile solution $|96|$.

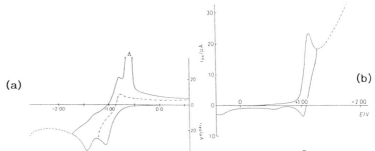

Fig. 28. Cyclic voltammetric responses of $|Cu(L_3)|^{2+}$ in MeCN solution at a platinum electrode. a) Cathodic scan; b) anodic scan (from Ref. 96).

It can be seen that both the cathodic Cu(II)/Cu(I) and the anodic Cu(II)/Cu(III) steps display a directly correletable response in the reverse scan, indicating a relative stability of $|Cu^I(L_3)|^+$ and $|Cu^{III}(L_3)|^{3+}$ species. For longer times of electrolysis, however, only the Cu(I) species is stable. The Cu(I)/Cu(0) reduction is completely irreversible.

While the instability of the copper(III) complexes is common to all L_3-L_6 complexes, the cathodic behaviour for the other complexes is slightly different; in fact L_4 and L_5 complexes give rise to a single two-electron reduction leading to demetallation, whereas the L_6 complex behaves as the L_3 complex. We attempted to explain this behaviour on the basis of the difficult reorganization of the planar copper(II) geometry to the tetrahedral copper(I) geometry. In fact spectroscopic evidence shows that the medium-sized L_4 and L_5 ligands can accomodate the copper(II) ion, through strong in-plane interactions, in a square-planar arrangement; on the contrary, L_3 and L_6 ligands are too small and too large, respectively, to dispose the copper(II) ion coplanar with the four nitrogen atoms, so favouring the access to the tetrahedral copper(I) assembly.

Table 17 summarizes the redox potentials for the copper(II) complexes of the ligands L_3-L_6. Also included are the redox potentials for the copper(II)/copper(III) redox change in the copper(II) complexes of both the doubly deprotonated dioxatetramines |95, 97|, analogues of L_3 and L_4, and the triply deprotonated trioxatetramine, somewhat analogous to L_3 |98| (Chart XXV).

TABLE 17

Formal electrode potentials (in Volt) for the redox changes of the copper(II) complexes of the ligands schematized in Charts XXIV and XXV

ligand	$E°'$ Cu^{II}/Cu^{I}	$E°'$ Cu^{II}/Cu^{III}	solvent	reference
L_3	- 0.50	+ 1.45[a]	MeCN	96
L_3	- 0.52	-	aqueous-80% MeOH	29
L_4	- 0.6[b]	+ 1.47[a]	MeCN	96
L_5	- 0.6[b]	+ 1.43[a]	MeCN	96
L_6	- 0.41	+ 1.48[a]	MeCN	96
diox-L_3	-	+ 0.43	H_2O	95
diox-L_4	-	+ 0.70	H_2O (pH 9.3)	97
triox-L_3	-	+ 0.38	H_2O (pH 10.0)	98

[a]Copper(III) species are transient; [b]coupled to the Cu^{I}/Cu^{0} step.

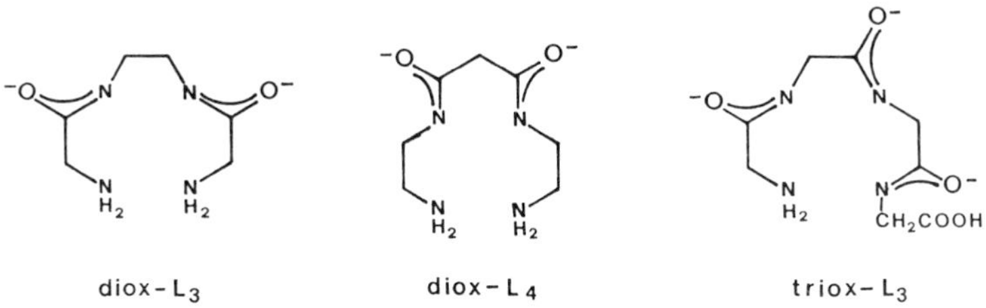

228

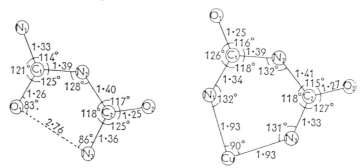

diox-L₃   diox-L₄   triox-L₃

CHART XXV

The remarkable shift of the redox potentials towards an easier one-electron removal with the increase of deprotonated amido groups and the stability of the relevant copper(III) species has to be emphasized, and it will be discussed in more detail in the Section 4.2.18, devoted to similar closed-ring macrocycles.

4.2.5 <u>Biureto</u>. Bis-biureto cuprate(II) complexes possess a square planar CuN_4 coordination.

Figure 29 shows the dimensions of $|Cu(NHCONHCONH)_2|^{2+}$, as deduced from X-ray investigations |99|.

Fig.29. Structural data for the planar complex $K_2|Cu(NHCONHCONH)_2| \cdot 4H_2O$ (from Ref.99).

Biuretato-copper(II) complexes can be oxidized, both chemically |100| or electrochemically |101|, to the corresponding copper(III) species. In aqueous alkaline solutions, the redox potentials for the copper(II)-copper(III) biuretato couple occur in the range 0.6-0.8 V. Thermodynamic potential values have been measured, polarographycally, in nonaqueous solutions |101,102| for the copper(III) complexes of 3-propyl-biureto and <u>o</u>-phenylenebis-(biureto) schematized in Chart XXVI.

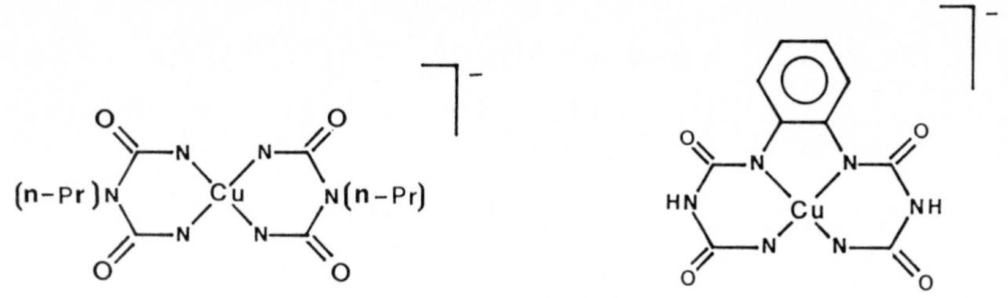

CHART XXVI

In dimethylsulfoxide solutions, these copper(III) complexes undergo an almost reversible one-electron reduction at $E^{o'}$ = -0.36 V for the 3-propyl derivative, and at $E^{o'}$ = +0.04 V for the phenylenebis-derivative |101, 102|. The substantial electrochemical reversibility of this redox change suggested that these copper(III) complexes should maintain the planar geometry typical for the corresponding copper(II) complexes. In fact X-ray investigations, Figure 30, proved that in o-phenylenebis(biuretate) cuprate(III) the copper ion has a planar four-coordinate geometry.

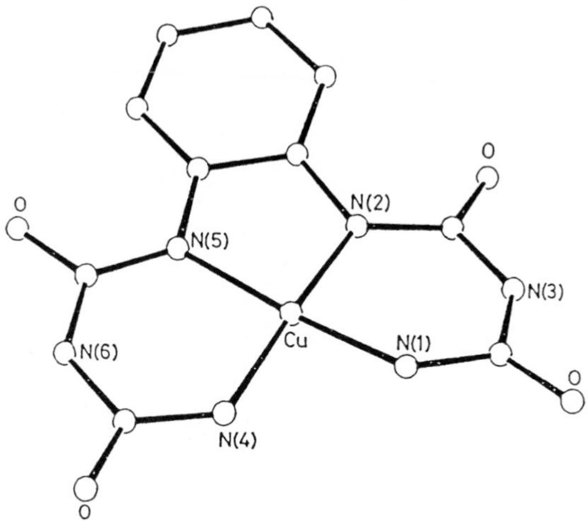

Fig. 30. Structure of $|(n-C_4H_9)_4N||Cu\{o-C_6H_4(HNCONHCONH)_2\}|$. Cu-N mean distance, 1.85 Å (from Ref. 103).

Comparison with the previously cited copper(II) complex suggests that the removal of one electron from the antibonding $d_{x^2-y^2}$ orbital causes a significant shortening (~ 0.08 Å) of the Cu-N bonds.

4.2.6 _Imidazoles_. Tetraimidazolo-copper(II) complexes, which also display a CuN_4 coordination, crystallize in an elongated octahedral geometry, in which the equatorial plane is constituted by the cation $|Cu(C_3H_4N_2)_4|^{2+}$, while the two apical positions are occupied by atoms of the relevant anions |104, 105|. In aqueous solutions, containing an excess of free imidazole, this copper(II) complex undergoes two subsequent one-electron reductions, coupled to ligand losses |106|, according to:

$$|Cu(imidazole)_4|^{2+} + e \xrightarrow{\quad E_{1/2} = -0.20 \text{ V} \quad} |Cu(imidazole)_2|^{+} + 2 \text{ imidazole}$$

$$|Cu(imidazole)_2|^{+} + e \xrightarrow{\quad E_{1/2} = -0.58 \text{ V} \quad} Cu + 2 \text{ imidazole}$$

The instability towards decomposition of the Cu(II)/Cu(I) redox change is not unexpected in view of the planarity of the starting copper(II) species.

It has been briefly reported that tetrakis(N-methylimidazole)-copper(II) complex undergoes, in nitromethane solution, a reduction process at + 0.15 V |67|.

Interestingly, the X-ray structures of both copper(II) and copper(I) complexes of the ligand 2,2'-bis(2-imidazolyl)biphenyl (Chart XXVII) have been recently reported |107|, Figure 31.

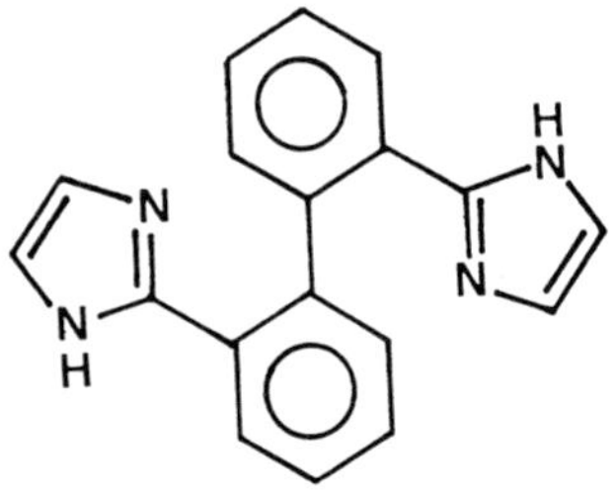

CHART XXVII

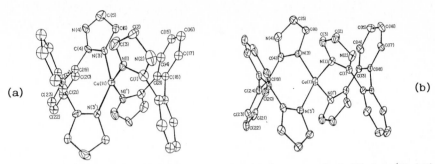

(a)

(b)

Fig. 31. Crystal structures of the complexes of 2,2'-bis(2-imida-zolyl)biphenyl. (a) Copper(II) complex; Cu-N mean distance, 1.96 Å.
(b) Copper(I) complex; Cu-N mean distance, 2.03 Å (from Ref. 107).

In both complexes the CuN_4 coordination is nearly tetrahedral; in fact, the dihedral angle between the CuN_2 planes of the two ligands are: 86.4° for the copper(II) complex; 87.9° for the copper(I) complex.

In these species the copper(II)/copper(I) redox change is essentially reversible and in acetonitrile occurs at +0.08 V |108|. In view of the tetrahedral geometry of both oxidation states, we suggest there are unfavourable electronic effect which keep the electrode potential to a limitedly positive value.

4.2.7   Biimidazole. The bis(2,2'-biimidazole)copper(II) com-plex, schematized in Chart XXVIII, undergoes in acetonitrile solution the copper(II)/copper(I) reduction at 0.00 V |67|.

No further details on this electron transfer are available. A planar CuN_4 coordination is assigned to the starting copper(II) complex on the basis of EPR spectra |67|, as was analogously ascertained for the nickel(II) complex |109|.

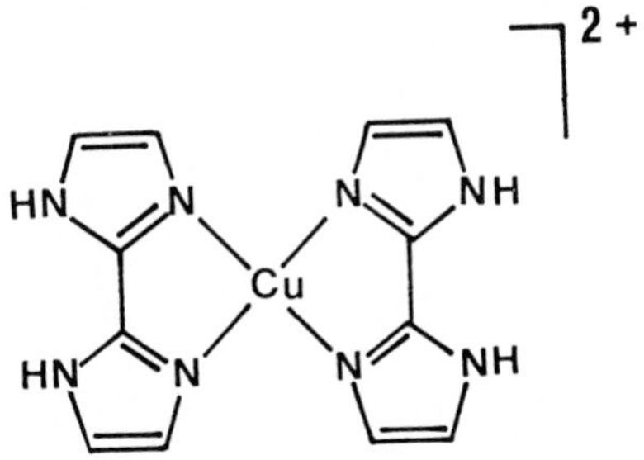

CHART XXVIII

4.2.8   Bipyridines. An interesting trend in copper(II)/ copper(I) redox potentials is exhibited by the 2,2'-bipyridyl copper(II) complex schematized in Chart XXIX.

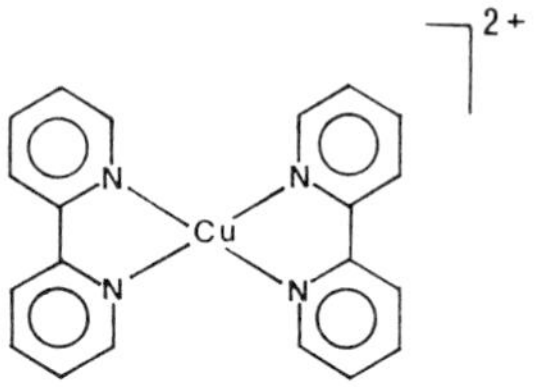

$$\left[Cu(bpy)_2\right]^{2+}$$

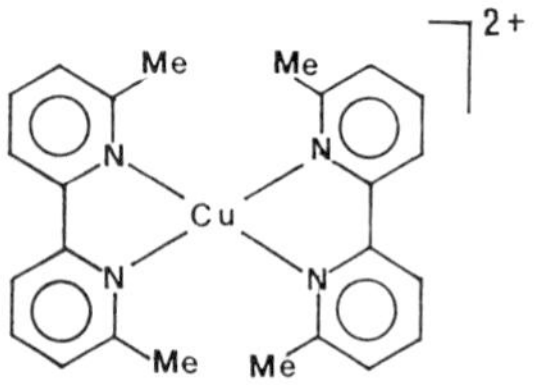

$$\left[Cu(dmbp)_2\right]^{2+}$$

CHART XXIX

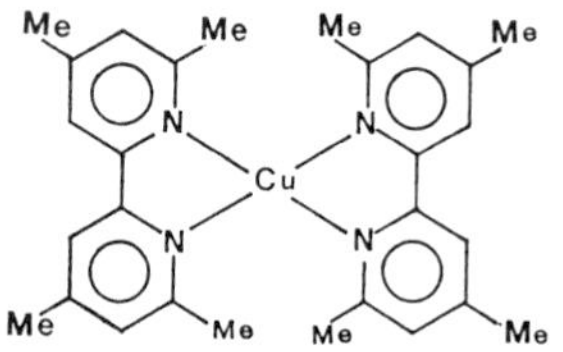

$$\left[Cu(tmbp)_2\right]^{2+}$$

Even if electrochemical details on the quasireversible copper(II)/copper(I) reduction are scanty, the potentials reported in Table 18 are at a first glance rather surprising, in that the presence of an increasing number of methyl groups in the pyridine rings of dmbp and tmbp ligands would have been expected to disfavour access to the copper(I) derivative, as compared with the unsubstituted bpy ligand. We feel this apparent contradiction can be accounted for by stereochemical arguments.

X-ray investigations on unsubstituted 2,2'-bipyridil-copper(II) complexes revealed generally the copper(II) ion to be engaged in five- or six-coordinations, because of the presence of copper-anions interactions in the solid state |112|. Nevertheless, the CuN_4 chromophores are always tetrahedrally distorted from the planar geometry.

TABLE 18

Formal electrode potentials (in Volt) for the one-electron reduction of the copper(II) complexes schematized in Chart XXIX

complex	$E°'_{Cu^{II}/Cu^{I}}$	solvent	reference
$|Cu(bpy)_2|^{2+}$	+ 0.01	Me_2CO	110
$|Cu(dmbp)_2|^{2+}$	+ 0.33	Me_2CO	110
$|Cu(dmbp)_2|^{2+}$	+ 0.35	aqueous-20% MeOH	111
$|Cu(tmbp)_2|^{2+}$	+ 0.36	aqueous-20% MeOH	111

A further confirmation of the non-planar solution geometry of $|Cu(bpy)_2|^{2+}$ comes from the crystal structure of $|Cu(bpy)_2|(PF_6)_2$, which is four-coordinate |113|, Figure 32a. It exhibits a flattened tetrahedral CuN_4 assembly; in fact, the dihedral angle between the mean planes of the two bipyridyl ligands, is 44.6°. As a consequence of the one-electron reduction, $|Cu(bpy)_2|^{+}$ assumes the pseudotetrahedral geometry illustrated in Figure 32b, with a short elongation of the Cu-N distance |110|; correspondingly the interligand dihedral angle increases to 75.2°.

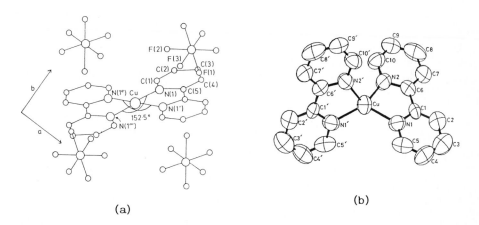

(a)   (b)

Fig. 32. Crystal structure of: (a) $|Cu(bpy)_2|(PF_6)_2$, Cu-N 1.98 Å (from Ref. 113); (b) $|Cu(bpy)_2|ClO_4$, Cu-N 2.02 Å (from Ref. 110).

234

Concerning the structural change involved in the $|Cu(dmbp)_2|^{2+}/|Cu(dmbp)_2|^+$ reduction, unfortunately crystal data for the divalent complex are unavailable. We only know the molecular structure of the copper(I) complex, Figure 33, |114|.

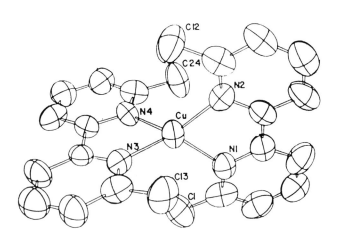

Fig. 33. Crystal structure of $|Cu(dmbp)_2|^+$. Cu-N mean distance, 2.03 Å (from Ref. 114).

In it, the interligand dihedral angle is 80.9°, *i.e.*, not very different from that of $|Cu(bpy)_2|^+$. If one considers that in $|Cu(dmbp)_2|^{2+}$ the extent of tetrahedral geometry must be notably greater than in $|Cu(bpy)_2|^{2+}$, because of the steric interactions of the methyl groups at the 6 and 6' positions, it can be deduced that the structural reorganization accompanying the $|Cu(dmbp)_2|^{2+}/|Cu(dmbp)_2|^+$ reduction must be less energy demanding than that of $|Cu(bpy)_2|^{2+}/|Cu(bpy)_2|^+$. Hence, the redox potential of the former couple is expected, as it happens, to be more positive than that of the latter couple.

The same arguments are likely valid for the $|Cu(tmbp)_2|^{2+}/|Cu(tmbp)_2|^+$ couple. In $|Cu(tmbp)_2|^+$, Figure 34, the distortion from the tetrahedral geometry is even greater than in the two preceding copper(I) complexes, since the dihedral angle between the symmetry related ligands is equal to 68° |115|.

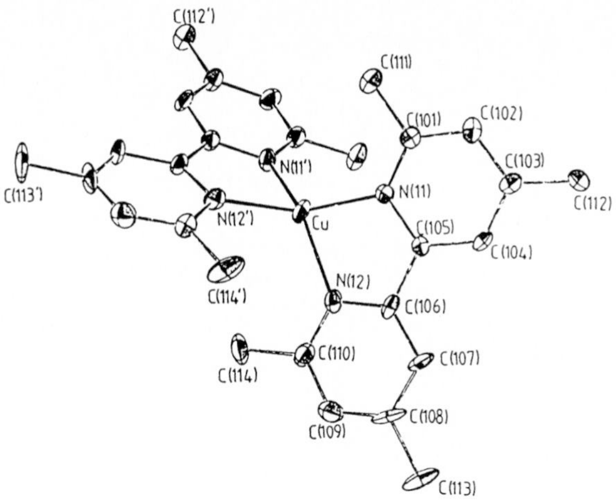

Fig. 34. Crystal structure of $|Cu(tmbp)_2|^+$. Cu-N mean distance, 2.06 Å (from Ref. 115).

Speculatively, the smaller angle in this last copper(I) complex should facilitate the reorganization starting from $|Cu(tmbp)_2|^{2+}$, whose likely pseudotetrahedral solution geometry |116| should be roughly similar to that of $|Cu(dmbp)_2|^{2+}$. Unfortunately, $|Cu(tmbp)_2| (ClO_4)_2$ in the solid state crystallizes in a distorted trigonal bipyramid, with the fifth position occupied by a coordinating perchlorate oxygen atom |115|, and it does not reflect the structure in solution.

A copper complex, somewhat correletable to the dipyridine derivatives, has been prepared from the 2,2'-bis(6{2,2'-bipyridyl})-biphenyl ligand (TET) in Chart XXX.

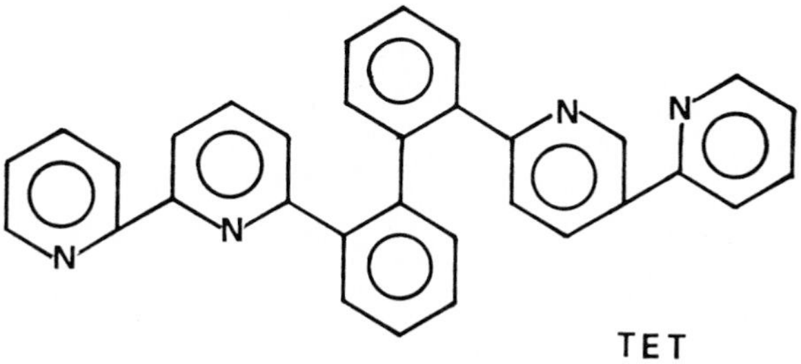

TET

CHART XXX

The redox pair $|Cu^{II}(TET)|^{2+}$, $|Cu^{I}(TET)|^{+}$ has been examined |117|. The copper(II)/copper(I) redox change occurs through a quasireversible step at $E^{o\prime}$ = + 0.48 V in MeCN, and $E^{o\prime}$ = + 0.70 V in CH_2Cl_2. This high redox potential suggests that both the members of the redox couple have a great aptitude to assume a tetrahedral geometry. As shown in Figure 35, the copper(I) complex has a pseudotetrahedral CuN_4 coordination, the dihedral angle between the two dipyridyl planes being 74.9°.

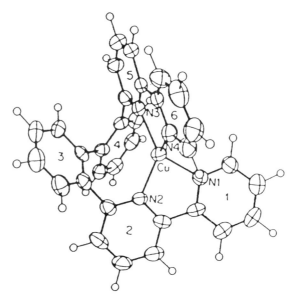

Fig. 35. Stereoview of the copper(I) complex $|Cu^{I}(TET)|ClO_4$. Cu-N mean distance, 2.03 Å (from Ref. 117).

As in the case of the similar complex of 2,2'-bis-(2-imidazolyl)biphenyl previously described (Section 4.2.6), the corresponding copper(II) complex is expected to possess a not very different geometry (at present, it has been crystallographically characterized only as the pentacoordinate copper(II) species $|Cu(TET)Cl|^{+}$ |117|).

4.2.9 <u>Phenanthrolines</u>. Strictly correlated to dipyridil-copper(II) complexes are the bis(phenanthroline)copper(II) complexes schematized in Chart XXXI.

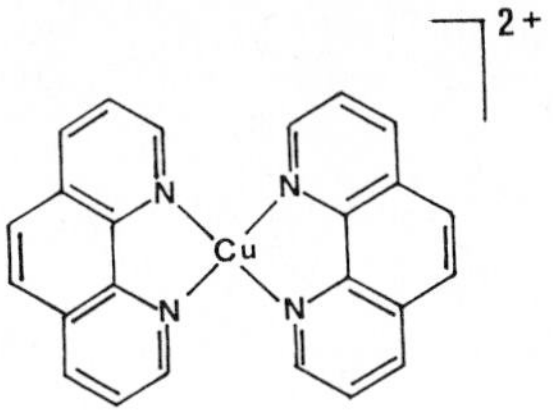

$[Cu(phen)_2]^{2+}$

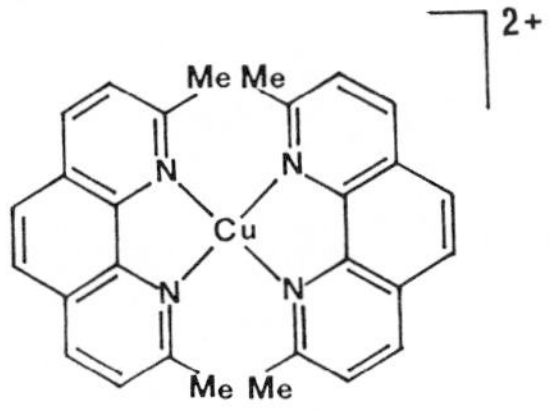

$[Cu(dmp)_2]^{2+}$

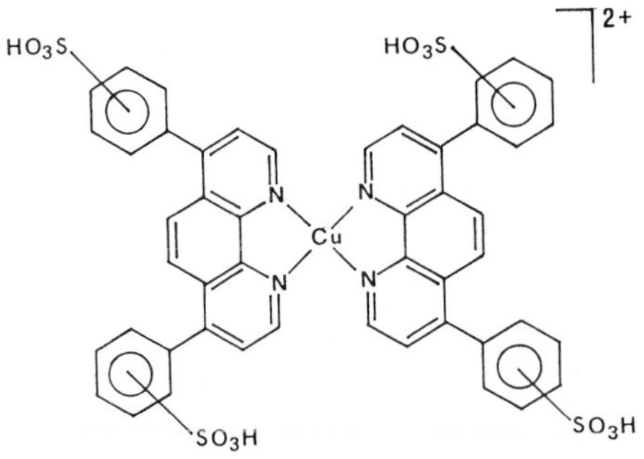

$[Cu(dpmp)_2]^{2+}$

CHART XXXI

Figure 36 shows the cyclic voltammetric response of $|Cu(phen)_2|^{2+}$ in dimethylformamide solvent |118|.

5 μA
+0.5
0
-0.5
-0.9
E vs. Ag/AgCl, volt

Fig. 36. Cyclic voltammograms recorded at a platinum electrode at different scan rates (10, 20, 50,100 mV s^{-1}) in a DMF solution of $|Cu(phen)_2|^{2+}$ (from Ref. 118).

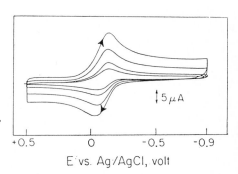

A simple quasireversible, one-electron reduction, attributable to the $|Cu(phen)_2|^{2+}/|Cu(phen)_2|^+$ couple, is displayed.

The same qualitative behaviour holds as far as the $|Cu(dmp)_2|^{2+/+}$ and $|Cu(dpmp)_2|^{2+/+}$ couples are concerned. The relevant redox potentials are reported in Table 19.

TABLE 19

Redox potentials (in Volt) for the copper(II)/copper(I) couple of the complexes schematized in Chart XXXI

complex	$E^{\circ\prime}$	solvent	reference
$|Cu(phen)_2|^{2+}$	− 0.04	H_2O (pH 5.6)	118
$|Cu(phen)_2|^{2+}$	− 0.13	DMF	118
$|Cu(dmp)_2|^{2+}$	+ 0.37	H_2O (pH 4 − 7)	119
$|Cu(dmp)_2|^{2+}$	+ 0.64	DMF	120, 121
$|Cu(dmp)_2|^{2+}$	+ 0.68	MeCN	121
$|Cu(dmp)_2|^{2+}$	+ 0.67	$MeNO_2$	121
$|Cu(dmp)_2|^{2+}$	+ 0.76	Me_2CO	121
$|Cu(dmp)_2|^{2+}$	+ 0.66	MeOH	121
$|Cu(dmp)_2|^{2+}$	+ 0.72	EtOH	121
$|Cu(dmp)_2|^{2+}$	+ 0.71	ACAN	121
$|Cu(dmp)_2|^{2+}$	+ 0.73	ACAC	121
$|Cu(dpmp)_2|^{2+}$	+ 0.38	H_2O (pH 4 − 8)	119, 122

Even if the amount of crystallographic data is less unequivocal than in the dipyridyl complexes, we can explain the positive shift of the redox potentials for the methylated ligands on the basis of easier stereochemical reorganization.

The structure of $|Cu(phen)_2|^{2+}$ can be deduced from that of $|Cu(H_2O)(phen)_2|(BF_4)_2$ |123|. Even if the presence of a coordinated water molecule induces a distorted trigonal bipyramidal geometry around the copper(II) centre, the $Cu(phen)_2$ moiety, Figure 37, can be viewed as a first approximation as tetrahedrally distorted, with an interligand dihedral angle of 47°.

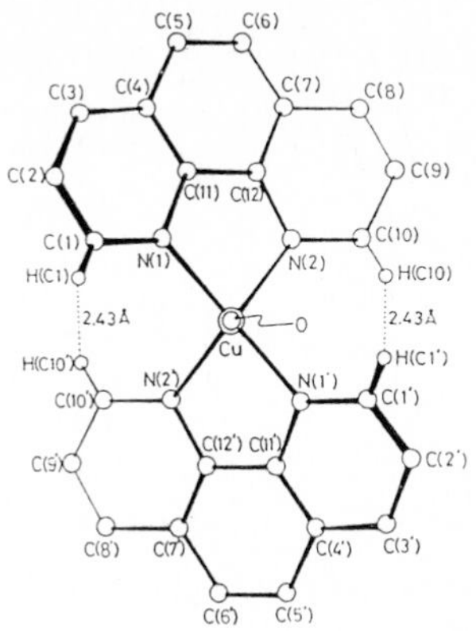

Fig. 37. Perspective view of the Cu(phen)$_2$ moiety in $|Cu(H_2O)$(phen)$_2|^{2+}$. Cu-N mean distance, 2.01 Å (from Ref. 123).

A more pronounced tetrahedral distortion should have to be present in the four-coordinate copper(I) complex $|Cu(phen)_2|ClO_4|124|$. Unexpectedly, the dihedral angle in this complex is unusually low (49.9°) to agree with the rather high reorganizational barrier suggested by the low redox potential for the $|Cu(phen)_2|^{2+/+}$ couple. In this connection, however, it seems more realistic to think the real solution structure of $|Cu(phen)_2|^+$ as that reported for $|Cu(phen)_2||CuBr_2|$ $|124|$, which exhibits a dihedral angle of 76.8°.

Let us now examine the case of the couple $|Cu(dmp)_2|^{2+/+}$. In the copper(I) form, the geometry, as depicted from different investigations $|125-127|$, is tetrahedrally distorted, see for instance Figure 38.

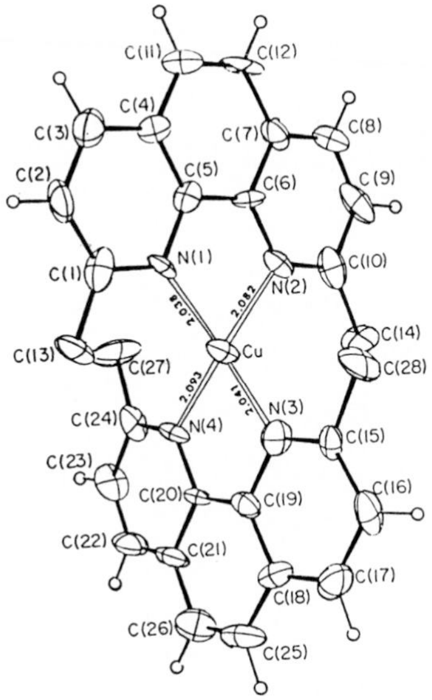

Fig. 38. Perspective view of $|Cu(dmp)_2|^+$. Cu-N mean distance, 2.06 Å (from Ref. 125).

The interligand dihedral angle in the $|Cu(dmp)_2|^+$ cation of the dimer $|Cu(dmp)_2|_2|TCNQ|_2$ ($|TCNQ|^- =$ anion of tetracyanoquinodimethane) is 82.8° |125|, however in $|Cu(dmp)_2|NO_3$, which more likely reflects the solution structure, is 67.6° |126|.

Now, if we consider that the dihedral angle in $|Cu(dmp)_2|^{2+}$ is certainly greater than 47° (the dihedral angle in $|Cu(phen)_2|^{2+}$), because of the higher steric repulsions exerted between methyl groups in position 2,9 compared to those exerted by hydrogen atoms, we must deduce that the reorganizational work operative during the one-electron reduction is lower for the methyl-substituted complex, thus increasing its reduction potential.

The same arguments should hold for the $|Cu(dpmp)_2|^{2+/+}$ couple, the phenyl-sulfonic substituents being outside the interaction sphere directly affecting the stereochemistry of copper ions.

The copper(I) complex depicted in Chart XXXII merits special mention.

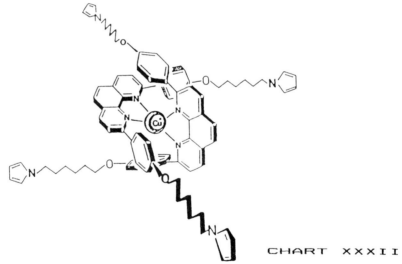

CHART XXXII

Its tetrahedral assembly leads to a copper(II)/copper(I) redox couple with a quite positive redox potential of + 0.74 V in acetonitrile solution |128|.

This complex can be considered as a precursor of a catenate copper(I) complex we shall examine in Section 4.2.18.

4.2.10 <u>Pyrroles</u>. A wide series of pyrrole derivatives give rise to a CuN_4 coordination. They are reported in Chart XXXIII.

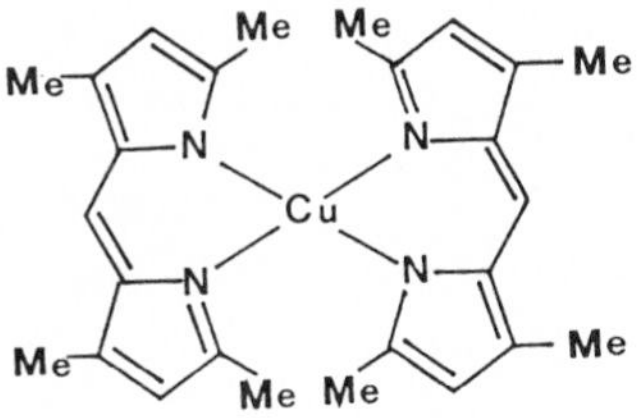

$Cu(Me_4-dipyrm)_2$

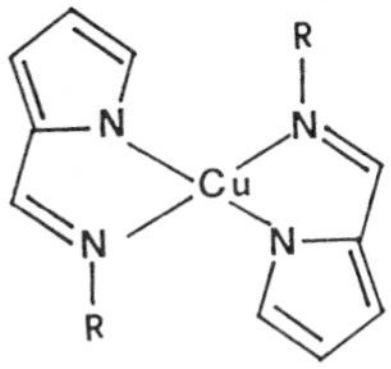

$Cu(R-pa)_2$

CHART XXXIII

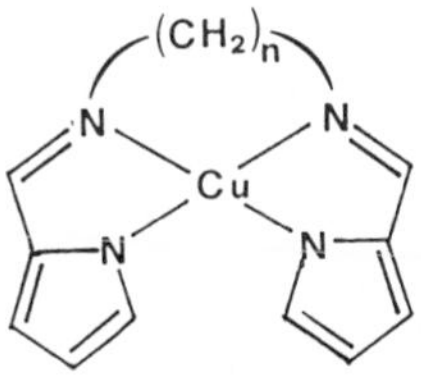

$Cu[(CH_2)_n(pa)_2]$

$Cu(Me_4-dipyrm)_2$ is assumed to have a pseudotetrahedral geometry by analogy with the similar complex 3,3',5,5'-tetramethyl-dipyrromethene-4,4'-dicarboxylate, Chart XXXIV, in which a dihedral angle of 68° exists between the two CuN_2 fragments of the chelate rings |133|.

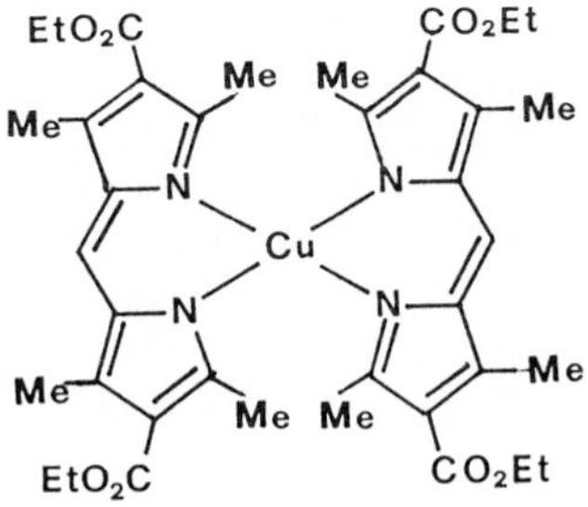

CHART XXXIV

242

Although this is a distorted tetrahedral geometry, the relevant Cu(II)/Cu(I) redox couple is located at a rather negative potential (-0.70 V, see Table 20). It is not unreasonable to attribute this relative difficulty of access to copper(I) to the presence of the eight electron donating methyl groups.

As far as the Cu(R-pa)$_2$ derivatives are concerned, the precursor bis-(H-pyrrole-2-aldimine) compound has a square planar geometry, Figure 39.

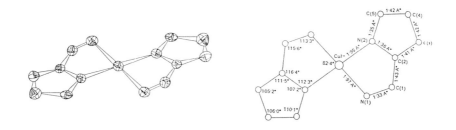

Fig. 39. X-ray molecular structure of Cu(R-pa)$_2$, R = H (from Ref. 134).

As the bulkyness of the R substituent increases, distortions towards tetrahedral geometry are expected, as evidenced by spectroscopic data |129|. As a matter of fact, for R = t-C$_4$H$_9$ a pseudotetrahedral geometry is reached, with a dihedral angle of 61.3° between the two CuN$_2$ wings (triclinic form) Figure 40, as well as an angle of 60° in the tetragonal form |135|.

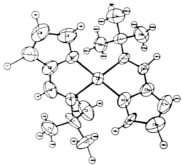

Fig. 40. X-ray structure of the triclinic form of Cu(R-pa)$_2$, R = t-Bu (from Ref. 135).

Correspondingly, as documented in Table 20, the redox potential for the copper(II)/copper(I) couple shifts towards less negative values of about 0.2 V.

TABLE 20

Redox potentials (in Volt) for the copper(II)/copper(I) reduction of the complexes schematized in Chart XXXIII

complex	$E^{0'}_{Cu^{II}/Cu^{I}}$	solvent	reference
$Cu(Me_4\text{-dipyrm})_2$	- 0.70	DMF	120

$Cu(R\text{-pa})_2$

R			
H	- 0.69	DMF	120
H	- 0.73	MeCN	129
CH_3	- 0.61	DMF	120
CH_3	- 0.69	MeCN	129
C_2H_5	- 0.67	MeCN	129
CF_3CH_2	- 0.41	MeCN	129
$\underline{n}\text{-}C_3H_7$	- 0.69	MeCN	129
$\underline{i}\text{-}C_3H_7$	- 0.60	MeCN	129
$\underline{i}\text{-}C_3H_7$	- 0.53	DMF	131
$\underline{n}\text{-}C_4H_9$	- 0.68	MeCN	129
$\underline{i}\text{-}C_4H_9$	- 0.67	MeCN	129
$\underline{s}\text{-}C_4H_9$	- 0.58	MeCN	129
$\underline{t}\text{-}C_4H_9$	- 0.55	MeCN	129
$\underline{t}\text{-}C_4H_9$	- 0.46	DMF	120
C_6H_{11}	- 0.60	MeCN	129
C_7H_{11} (norb)	- 0.54	MeCN	129
$C_{10}H_{15}$ (adam)	- 0.56	MeCN	129
$(C_6H_5)_2CH$	- 0.44	MeCN	129
$(CH_3)_2C_6H_3$	- 0.57	MeCN	129
$C_6H_5CH_2C(CH_3)_2$	- 0.50	MeCN	129

TABLE 20 (continued)

complex	$E°'$ Cu^{II}/Cu^{I}	solvent	reference
C_6H_5	- 0.46	MeCN	132
$C_6H_4-NO_2-p$	- 0.32	MeCN	132
$C_6H_4-CH_3CO-p$	- 0.40	MeCN	132
$C_6H_4-C_2H_5CO_2-p$	- 0.39	MeCN	132
$C_6H_4-CF_3-p$	- 0.39	MeCN	132
C_6H_4-Cl-p	- 0.42	MeCN	132
C_6H_4-Br-p	- 0.43	MeCN	132
C_6H_4-I-p	- 0.43	MeCN	132
C_6H_4-F-p	- 0.45	MeCN	132
$C_6H_4-C_6H_5-p$	- 0.45	MeCN	132
$C_6H_4-CH_3-p$	- 0.48	MeCN	132
$C_6H_4-OCH_3-p$	- 0.50	MeCN	132
$C_6H_4-N(CH_3)_2-p$	- 0.54	MeCN	132

$$Cu\{(CH_2)_n(pa)_2\}$$

n			
2	- 0.82	MeCN	129
2	- 0.74	DMF	120
3	- 1.01	MeCN	129
3	- 0.86	DMF	130
4	- 0.85	MeCN	129

We can hence conclude that the progressive increase of the redox potentials is indicative of increasing tetrahedral distortions, associated with the increasing bulkyness of the R substituents. In addition, as put in evidence by $Cu(p-X-C_6H_4-pa)_2$ derivatives, the presence of electron withdrawing units promote further the reduction to copper(I).

Finally, the complexes $Cu\{(CH_2)_n(pa)_2\}$ exhibit quite negative reduction potentials. It is clearly evident that the methylenic bridge not only imposes planarity to the copper(II) complexes, but also tends to prevent the copper(I) species from assuming a tetrahedral assembly. Indeed, it would be expected the increasing of the carbon chain from n =2 to n =4 to favour progressively tetrahedral distortion. This does not seem the case.

4.2.11 ß-Iminoaminates. The ß-iminoaminate copper(II) complexes schematized in Chart XXXV, although likely to exhibit a distorted tetrahedral geometry |136|, are cathodically reduced, in dimethylformamide solution, to the corresponding copper(I) complexes in a near reversible reduction at the rather negative potential values $E°'_{Cu^{II}/Cu^I}(R=Me) = -1.52$ V; $E°'_{Cu^{II}/Cu^I}(R=Ph) = -0.88$ V |120|.

R= Me, Ph   CHART XXXV

The higher electron donating ability of methyl substituents with respect to phenyls is well reflected by the relevant redox potentials. On the other hand, since steric arguments cannot be invoked, it is conceivable that the unexpected negative location of both redox potentials may be ascribed to a deep conjugation of the outer electron donating methyl groups.

4.2.12 Mixed pyridine-aza functions. CuN_4 coordination can be also reached through ligands obtained by condensation of pyridyl compounds with different aza functions. We include here the copper(II) complexes schematized in Chart XXXVI.

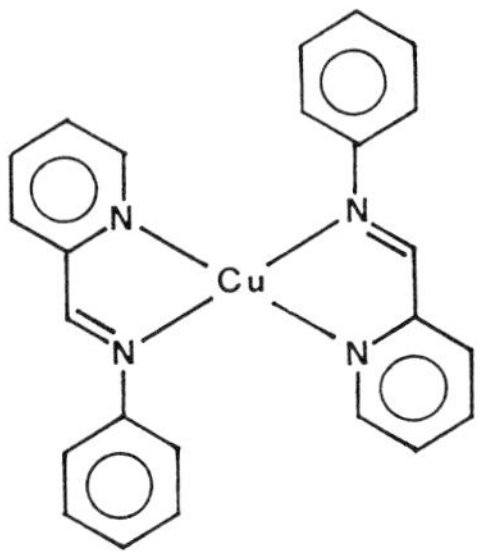

246

$$\left[Cu(L_7)\right]^{2+}$$

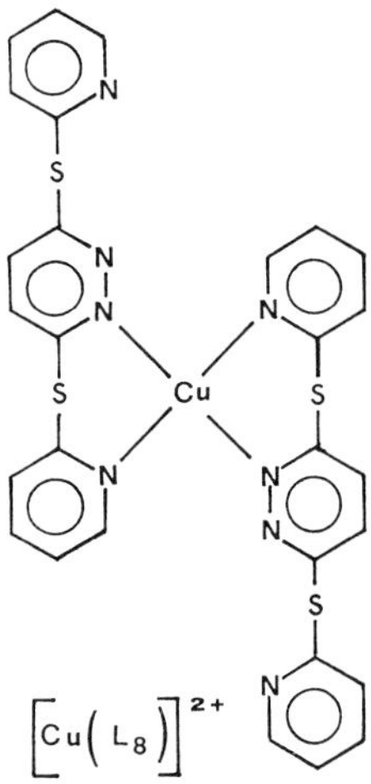

$$\left[Cu(L_8)\right]^{2+}$$

CHART XXXVI

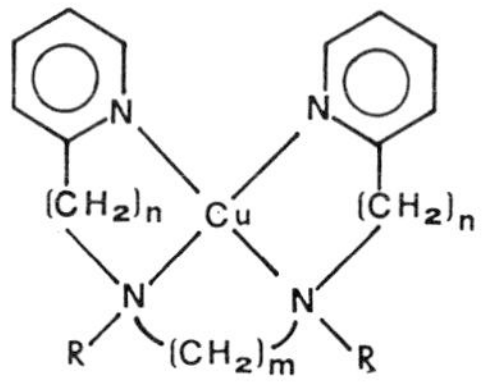

$$\left[Cu(L_9 - L_{12})\right]^{2+}$$

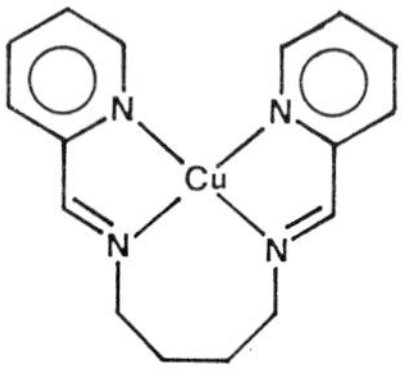

$$\left[Cu(L_{13})\right]^{2+}$$

Concerning the complex with L_7, both the bis(2-phenyl-azopyridine)copper(II) and the bis(2-phenylazopyridine) copper(I) species are stable |137|. On the basis of spectroscopic evidence, an essentially planar geometry is assigned to the copper(II) complex, and an essentially tetrahedral geometry is assigned to the copper(I) complex.

Figure 41 shows the cyclic voltammogram obtained by $|Cu(L_7)|^+$ in methanol |137|.

The copper(I)/copper(II) transition occurs at the unusually high redox potential of +0.63 V (+0.67 V in aqueous solution).

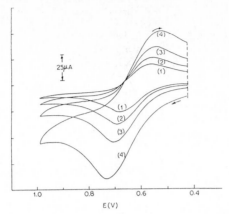

Fig. 41. Cyclic voltammograms obtained at a platinum electrode on a MeOH solution of $|Cu(L_7)|^+$. Scan rates (mVs^{-1}): (1) 50, (2) 100, (3) 200, (4) 500 (from Ref. 137).

The tetrahedral $\rightleftharpoons$ planar conversion is thought to be responsible for the quasireversibility of this one-electron process. The extended ligand unsaturation could be in part responsible for the difficulty of the relevant one-electron removal.

The copper(II) complex $|Cu(L_8)|^{2+}$ undergoes, in acetonitrile solution, a cathodic reduction at the very positive redox potential of +0.57 V, likely complicated by formation of a binuclear copper(I) species. In the solid state this copper(II) derivative crystallizes with a square-pyramidal geometry, in which the fifth position is occupied by an oxygen atom of a coordinated water molecule |138|. Also, in this case the high redox potential could be ascribed to the extended π-delocalization.

As far as the copper(II) complexes with the ligands L_9-L_{12} are concerned, they give rise to a quasireversible copper(II)/copper(I) reduction step |139|; the relevant redox potentials are summarized in Table 21. (From the electrochemical parameters reported, it is unlikely that the copper(I) complexes with the ligands L_9 and L_{10} are stable for times greater than some tens of seconds).

TABLE 21

Formal electrode potentials (in Volt) for the one-electron reduction of the copper(II) complexes of the ligands L_9 - L_{12} schematized in Chart XXXVI |139|

Ligand	n	m	R	$E^{\circ\prime}$ Cu^{II}/Cu^{I}	solvent
L_9	1	2	H	− 0.44	H_2O
L_9	1	2	H	− 0.19[a]	MeCN
L_{10}	1	3	H	− 0.35	H_2O
L_{10}	1	3	H	− 0.17[a]	MeCN
L_{11}	2	2	CH_3	− 0.14	H_2O
L_{11}	2	2	CH_3	+ 0.06[a]	MeCN
L_{12}	2	3	CH_3	+ 0.03	H_2O
L_{12}	2	3	CH_3	+ 0.29[a]	MeCN

[a] In Ref.139 the ferrocenium/ferrocene couple is assumed to be located at +0.40 V vs. N.H.E. in MeCN solution. In this author's laboratory, the ferrocenium/ferrocene couple in MeCN solution is located at +0.38 V vs. S.C.E..

As it can be seen, increasing of the ligand size of the chelate rings in $|5,5,5(L_9)-5,6,5(L_{10})-6,5,6(L_{11})-6,6,6(L_{12})|$ progressively favours the reduction process.

Spectroscopic evidence imply a substantial planarity of the CuN_4 inner-coordination of these tetragonal copper(II) complexes |139| such that the positive shift of the redox potentials, as consequence of the increased molecular flexibility, has to be ascribed to an easier reorganization of copper(I) to a pseudo-tetrahedral geometry, rather than to increasing tetrahedral distortions of the starting copper(II) units. Furthermore, these stereochemical arguments must predominate over electronic effects, since the methylated complexes with L_{11} and L_{12}, which would be expected to disfavour the reduction process, reduce at quite positive potential values.

The copper(II) complex with the pyridyl Schiff base ligand L_{13} has been electrochemically investigated by two groups |140,141|.

In aqueous solution, the recorded cyclic voltammogram is indicative of an electrochemical process very far from the claimed reversibility ($E°'_{Cu^{II}/Cu^I}$ = -0.05 V) |140|. The stability of the corresponding copper(I) complex seems unlikely.

In dimethylformamide solution, a quasireversible one-electron reduction is reported to occur at $E°'_{Cu^{II}/Cu^I}$ = -0.15 V |141|.

Since spectroscopy reveals an essential planar geometry of the copper(II) complex, with a possible minor tetrahedral distortion |142|, it is likely the quasireversibility of the copper(II)/ copper(I) couple arises from the molecular flexibility induced by the tetramethylene bridge, which favours the pseudotetrahedral reorganization of the copper(I) complex.

4.2.13 <u>Mixed ethylenediamine-aza functions</u>. The copper(II) compounds schematized in Chart XXXVII derive from condensation of ethylenediamines with different keto compounds.

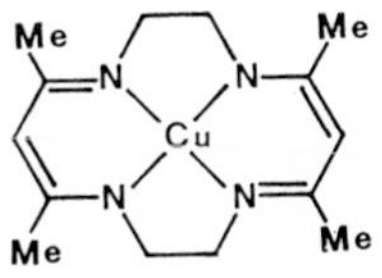

$Cu\left(MeHMe\right)_2\left(en\right)_2$

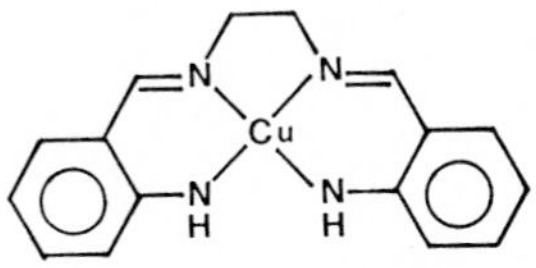

$Cu\left(oab\right)_2 en$

CHART XXXVII

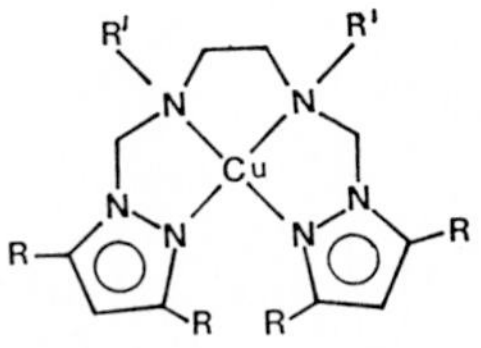

$R = R' = H$, $\left[Cu(edbp)\right]^{2+}$

$R = Me$; $R' = H$, $\left[Cu(edbd)\right]^{2+}$

$R = H$; $R' = Me$, $\left[Cu(debp)\right]^{2+}$

They undergo the copper(II)/copper(I) reduction in a quasireversible step |120, 143|. The copper(I) complexes of the ligands edbp, edbd, and debp do not seem completely stable |143|. Table 22 reports the redox potentials for the relevant copper(II)/copper(I) couples.

TABLE 22

Formal electrode potentials (in Volt) for the copper(II)/copper(I) couple of the complexes schematized in Chart XXXVII

complex	$E^{\circ\prime}_{Cu^{II}/Cu^{I}}$	solvent	reference
$Cu(MeHMe)_2(en)_2$	− 2.26	DMF	120
$Cu(oab)_2en$	− 1.54	DMF	120
$|Cu(edbp)|^{2+}$	+ 0.09	MeOH	143
$|Cu(edbd)|^{2+}$	+ 0.12	MeOH	143
$|Cu(debp)|^{2+}$	+ 0.17	MeOH	143

The lowest potential value belongs to the macrocyclic complex $Cu(MeHMe)_2(en)_2$. Such a parameter, because of the strong in-plane interactions between the copper atom and the four nitrogens, reflects the difficulty to reorganize the quite stable planar geometry of the starting copper(II) form to the final, more or less tetrahedrally distorted, copper(I) form. In addition, the copper(II)/copper(I) reduction is also disfavoured by the presence of the electron donating methyl groups. Further examination of this derivative will be performed in the next section, devoted to macrocyclic complexes.

As far as $Cu(oab)_2en$ is concerned, there are not significant structural differences with respect to $Cu\{(CH_2)_2(pa)_2\}$ previously discussed (Table 20), which may account for its more difficult reduction. In fact $Cu\{(CH_2)_2(pa)_2\}$ is described as planar, whereas

Cu(oab)$_2$en is thought to be very similar to the umbrella-shaped |1,2-bis(2-aminobenzylidene)amino|propanato copper(II) complex, in which the Cu atom is displaced by only 0.04 Å from the central N$_4$ plane |144|. Therefore, the markedly easy access to copper(I) for the pyrrole derivative is likely to be due to a better electron withdrawing ability of the directly coordinated pyrrole groups, as compared with the aminobenzylidene moiety.

Finally, the pyrazolyl copper(II) complexes |Cu(edbp)|$^{2+}$, |Cu(edbd)|$^{2+}$, |Cu(debp)|$^{2+}$ are likely to display a pseudo-tetrahedral geometry, the tetrahedral distortion increasing with the bulkyness of the R substituents. This suggestion agrees with the increased thermodynamic stability of their kinetically labile copper(I) complexes |143|.

4.2.14 <u>Tripodal ligands</u>. The beneficial effect of the pyrazolyl group in stabilizing copper(I) complexes is quite evident in the redox behaviour of the copper(I) derivatives schematized in Chart XXXVIII.

$$R = H \quad \left[Cu \left(L_{14} \right) \right]^+$$
$$R = Me \quad \left[Cu \left(L_{15} \right) \right]^+$$
$$R = \underline{t} - Bu \left[Cu \left(L_{16} \right) \right]^+$$

CHART XXXVIII

As shown in Figure 42, these copper(I) species undergo, in acetonitrile solution, a quasireversible copper(I)/copper(II) oxidation at potential values which are among the highest ones for CuN$_4$ complexes (Table 23) |145|.

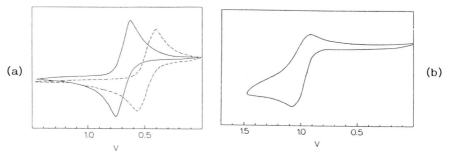

Fig. 42. Cyclic voltammograms recorded at a platinum electrode on MeCN solution of: (a)(------)$|Cu(L_{14})|^+$, (———)$|Cu(L_{15})|^+$; (b) $|Cu(L_{16})|^+$. Scan rate 0.2 Vs^{-1} (from Ref. 145).

TABLE 23

Redox potentials (in Volt) for the copper(II)/copper(I) couple of the complexes schematized in Charts XXXVIII and XXXIX

complex	$E^{\circ\prime}_{Cu^{II}/Cu^{I}}$	solvent	reference
$|Cu(L_{14})|^+$	+ 0.49	MeCN	145
$|Cu(L_{15})|^+$	+ 0.67	MeCN	145
$|Cu(L_{16})|^+$	+ 0.94	MeCN	145
$|Cu(L_{17})|^+$	+ 0.05	DMF	146
$|Cu(L_{18})|^{2+}$	+ 0.15	MeCN	147
$|Cu(L_{18})|^{2+}$	- 0.10	MeOH	148
$|Cu(L_{18})|^{2+}$	- 0.13[a]	DMF	147
$|Cu(L_{18})|^{2+}$	- 0.10	DMF	149
$|Cu(L_{19})|^{2+}$	- 0.13	MeOH	148
$|Cu(L_{20})|^{2+}$	+ 0.01	MeOH	148
$|Cu(L_{21})|^{2+}$	- 0.04	MeOH	148
$|Cu(L_{22})|^{2+}$	- 0.13	DMF	149
$|Cu(L_{23})|^{2+}$	- 0.13	DMF	149
$|Cu(L_{24})|^{2+}$	- 0.14	DMF	149
$|Cu(L_{25})|^{2+}$	- 0.07	DMF	149

[a]Peak potential value for an irreversible process.

In the case of the t-butyl-substituted derivative, the electrogenerated copper(II) complex is not long lived. However, the use of high sweep rates (>5 Vs^{-1}) allows one to overcome the associated complications, thus making possible a reliable evaluation of the copper(I)/copper(II) formal electrode potential |145|.

Figure 43 shows the molecular structure of $|Cu(L_{15})|^{+}$.

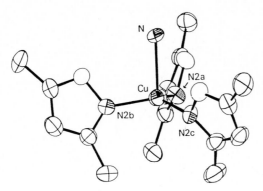

Fig. 43. X-ray structure of $|Cu(L_{15})|^{+}$. The ethylene bridging unities have been omitted for clarity. Cu-N$_{(amine)}$, 2.26 Å; Cu-N$_{(pyrazole)}$(mean), 2.00 Å (from Ref. 145).

Upon one-electron removal, this irregular tetrahedral geometry (the Cu atom is placed 0.30 Å over the N$_{3(pyrazole)}$ plane, compared with a 0.667 Å expected for a tetrahedral assembly) likely converts to a square pyramidal geometry (with the fifth position occupied by a solvent molecule in the basal plane) |145|. This geometrical reorganization accounts for the quasireversibility of the oxidation step.

Since the highest Cu(II)/Cu(I) potentials are in principle expected for the reversible tetrahedral copper(II)/tetrahedral copper(I) redox change, it is likely that both the nonplanar geometries of the copper(II)-copper(I) couples, and the electronic effects of the pyrazolyl groups, act in concert to produce the notably positive location of the redox potentials of these complexes.

Having discussed the behaviour of the complexes schematized in Chart XXXVIII, we now turn our attention to similar complexes of tripodal ligands, schematized in Chart XXXIX.

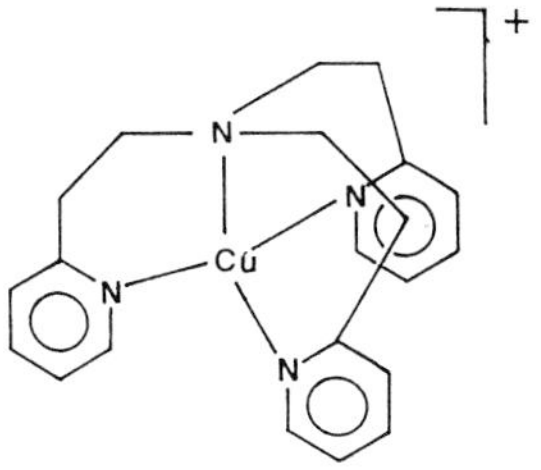

254

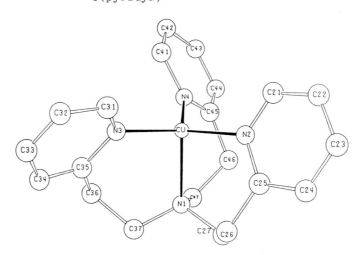

$$\left[Cu\left(L_{17}\right)\right]^+$$

CHART XXXIX

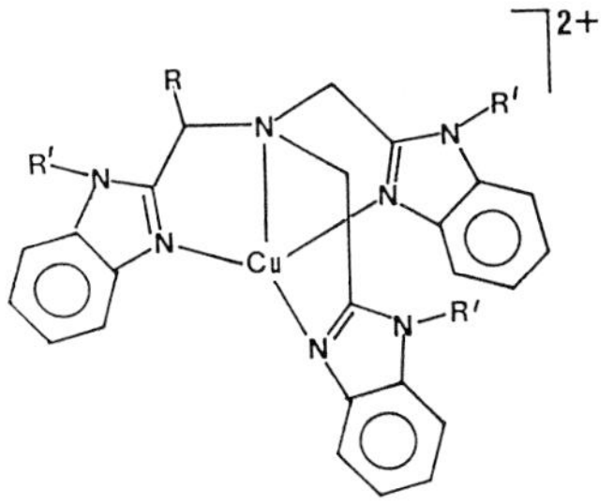

R=R'=H $\left[Cu(L_{18})\right]^{2+}$

R=Me ; R'=H $\left[Cu(L_{19})\right]^{2+}$

R=$\underline{i}$-Pr; R'=H $\left[Cu(L_{20})\right]^{2+}$

R=$\underline{i}$-Bu; R'=H $\left[Cu(L_{21})\right]^{2+}$

R=H ; R'=Me $\left[Cu(L_{22})\right]^{2+}$

R=H ; R'=Et $\left[Cu(L_{23})\right]^{2+}$

R=H ; R'=CH$_2$C$_6$H$_5$ $\left[Cu(L_{24})\right]^{2+}$

R=H ; R'=CH$_2$C$_6$H$_4$−$\underline{o}$−CH$_3$ $\left[Cu(L_{25})\right]^{2+}$

As shown in Figure 44, the copper(I) complex $|Cu(L_{17})|^+$ possesses a pseudo-tetrahedral geometry similar to that of $|Cu(L_{15})|^+$, distorted towards pyramidal. In fact the copper atom is 0.31 Å above the $N_{3(pyridyl)}$ plane.

Fig. 44. X-ray structure of $|Cu(L_{17})|^+$. Cu-N$_{(amine)}$, 2.19 Å; Cu-N$_{(pyridyl)}$ (mean), 2.02 Å (from Ref. 146).

This pyramidal distortion is expected to favour a fifth coordination upon one-electron removal. Accordingly, the anodic step results quasireversible in character |146|. The relevant redox potential (Table 23) is, however, significantly lower than those of the pyrazolyl derivatives.

The copper(II) complexes of ligands L_{18}-L_{25} are commonly trigonal bipyramidal because of the occurrence of a fifth coordination site occupied by a solvent molecule or anion, except for L_{20}, which is thought to be four-coordinate, tetrahedrally distorted |148|.

As an example, Figure 45 shows the cyclic voltammogram exhibited by $|Cu(L_{18})(H_2O)|^{2+}$ in MeCN solution.

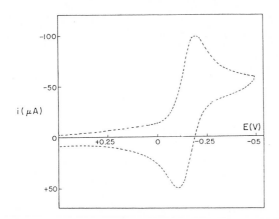

Fig. 45. Cyclic voltammogram recorded at a platinum electrode on a MeCN solution of $|Cu(L_{18})(H_2O)|^{2+}$. Scan rate 0.1 Vs^{-1} (from Ref. 147).

The electrochemical reversibility of this step is assumed to indicate that MeCN constitutes the fifth ligand (substituting the H_2O molecule) in both copper(II) and copper(I) complexes, which on the other hand maintain the same geometry |147|.

Table 23 summarizes the redox potentials for all copper complexes with tripodal ligands. References 147,149 report the redox potentials of a number of copper(II) complexes of L_{18}-L_{25} ligands with the fifth coordination site occupied by different anions too.

4.2.15 <u>Peptides</u>. A variety of peptide-deprotonated copper(II) complexes, which display a CuN$_4$ coordination, have been shown to be oxidized to the corresponding copper(III) derivatives |150|.

These copper(II) complexes are considered essentially planar on the basis of the crystal structure of the triply deprotonated tetraglicino cuprate(II) complex, shown in Figure 46.

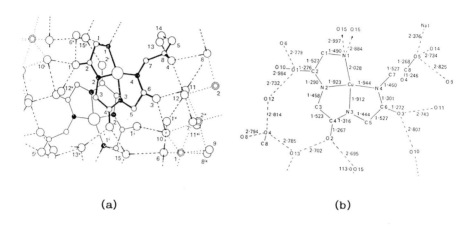

(a)   (b)

Fig. 46. Structure (a) and interatomic distances (b) of $|Cu^{II}(H_{-3}G_4)|^{2-}$ (from Ref. 151).

As shown in Figure 47, which refers to the deprotonated glycylglycylglycylamide copper(II) complex, they undergo a near reversible one-electron anodic step.

Fig. 47. Cyclic voltammogram of $|Cu^{II}(H_{-3}G_3a)|^{-}$ in aqueous solution (pH 9.5). Carbon paste electrode. Scan rate 0.1 Vs^{-1} (from Ref. 150).

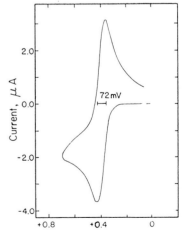

As expected on the basis of the electrochemical reversibility of this copper(II)/copper(III) redox change and inferred by solution properties, the copper(III)-peptide complexes retain the square-planar geometry. This is confirmed by the structure of the copper(III) complex of tri- α -aminoisobutyric acid |152|, which indeed exhibits a CuN_3O coordination.

Provided that comparison between the two cited X-ray structures may be generalized, it can be deduced that the one-electron oxidation, while maintaining the planar geometry, causes a significant shortening of the Cu-N bonds (the Cu-N mean distance shortens from 1.95 Å to 1.83 Å). This suggests that the electron is removed from an antibonding orbital, metal-ligand in character.

Redox potentials for the copper(II)/copper(III) oxidation in complexes with dipeptide amides, tripeptides, tripeptide amides and higher order peptides span from + 0.1 to + 1.0 V |150,153|. Further discussions on related complexes can be found in Sections 4.2.4 and 4.2.18.

4.2.16 <u>Porphyrins</u>. Porphyrins constitute an important class of biologically important ligands able to give CuN_4 coordination.

Porphyrins themselves are redox active and their electrochemistry has been reviewed. The commonest redox changes of the porphyrin ring are two distinct one-electron reductions as well as two distinct one-electron oxidations |154,156|.

The tetraphenylporphinato-copper(II) complexes schematized in Chart XL have been electrochemically studied.

X-ray structure of Cu(TPP) shows that the CuN_4 moiety is roughly planar (Cu-N, 1.98 Å), with the plane of phenyl groups rotated of about 90° with respect to the plane of the porphyrin ring |157|.

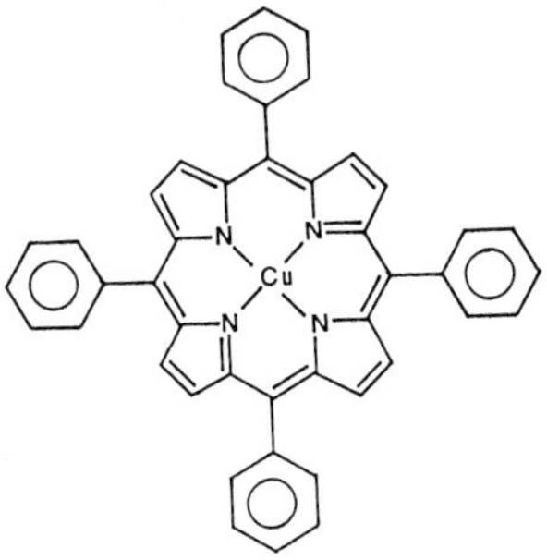

Cu(TPP)

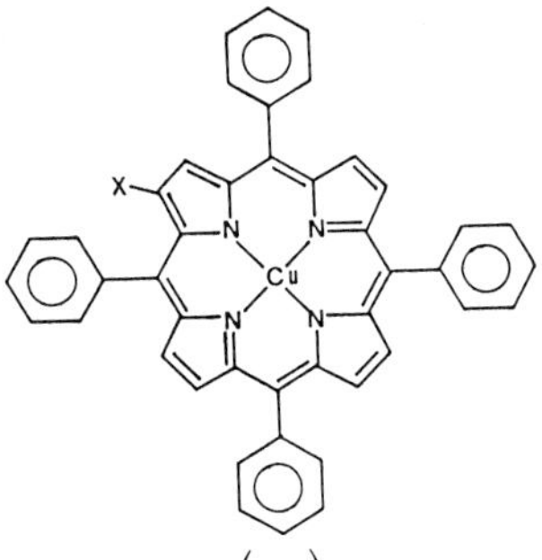

Cu(TPP)X
(X=CN, Br, NO₂, SCN)

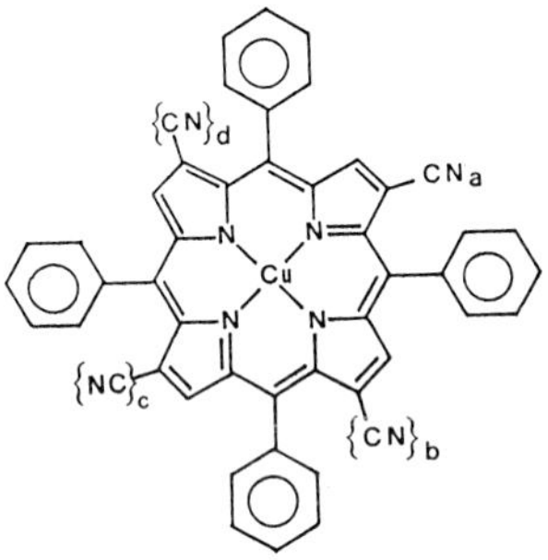

Cu(TPP)(CN)₁₋₄

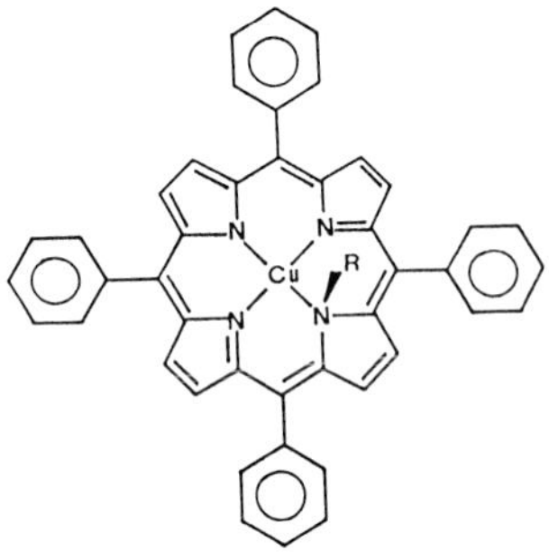

[Cu(TPP)R]⁺
(R=Me, Et, Ph)

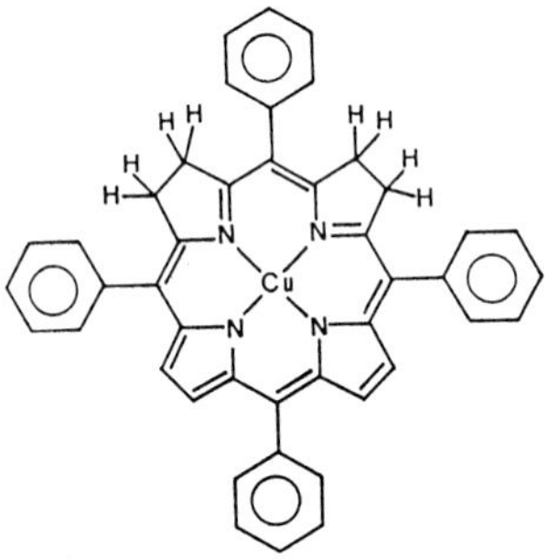

Cu(TPiBC)

CHART XL

As shown in Table 24, in comparison with the free tetraphenyl-porphyrine, the presence of the central copper(II) ion simply causes the sequential addition of the two electrons to become more difficult, whereas the sequential removal of the two electrons is less significantly affected.

TABLE 24

Redox potentials (in Volt) for the redox changes of tetraphenyl-porphyrine (H_2TPP) and some copper(II) complexes |Cu(TPP), Cu(TPP)X and Cu(TPP)(CN)$_{1-4}$ (see Chart XL)|

complex	$E^{o'}$ 2+/1+	$E^{o'}$ 1+/0	$E^{o'}$ 0/1-	$E^{o'}$ 1-/2-	solvent	reference
H_2TPP	+1.35	+1.08	-1.21	-	CH_2Cl_2	155
H_2TPP	-	+1.02	-1.20	-1.55	CH_2Cl_2	156
H_2TPP	-	-	-1.05	-1.47	DMSO	158
H_2TPP	-	-	-1.03	-	DMSO	156
H_2TPP	+1.20[a]	+1.00[a]	-	-	PhCN	159
H_2TPP	-	-	-1.13	-	PhCN	156
H_2TPP	-	-	-1.08	-1.46	DMF	160,161
H_2TPP	-	-	-1.08	-1.52	DMF	162
H_2TPP	-	-	-1.05	-1.43	DMF	156
H_2TPP	-	-	-1.08	-1.45	DMF	163
H_2TPP	-	+1.11	-	-	DMF	164
H_2TPP	-	+1.06	-	-	CH_2Br_2	165
Cu(TPP)	+1.33	+1.06	-1.35	-	CH_2Cl_2	155
Cu(TPP)	-	-	-1.20	-1.68	DMSO	158
Cu(TPP)	+1.33	+0.99	-	-	PhCN	159
Cu(TPP)	-	-	-1.20	-1.68	DMF	160,161
Cu(TPP)	-	+1.04	-	-	CH_2Br_2	165
Cu(TPP)Br	-	-	-1.12	-1.59	DMF	161
Cu(TPP)NO$_2$	-	-	-0.87	-1.31	DMF	161
Cu(TPP)SCN	-	-	-1.05	-1.33	DMF	161
Cu(TPP)CN	-	-	-0.94	-1.44	DMF	161
Cu(TPP)CN	+1.48	+1.23	-1.10	-	CH_2Cl_2	155
Cu(TPP)(CN)$_{2(a,b)}$	+1.61	+1.29	-0.86	-	CH_2Cl_2	155
Cu(TPP)(CN)$_{2(a,b)}$	-	-	-0.70	-1.19	DMF	160
Cu(TPP)(CN)$_{2(a,c)}$	-	-	-0.66	-1.15	DMF	160
Cu(TPP)(CN)$_3$	+1.56	+1.33	-0.61	-	CH_2Cl_2	155
Cu(TPP)(CN)$_3$	-	-	-0.50	-0.93	DMF	160
Cu(TPP)(CN)$_4$	+1.70	+1.41	-0.34	-	CH_2Cl_2	155
Cu(TPP)(CN)$_4$	-	-	-0.34	-0.76	DMF	166

[a] Peak potential for irreversible process.

260

It is commonly accepted that both, the addition and the removal of these electrons are ligand centred |158, 159, 167|, the site of the oxidation processes being the pyrrolic nitrogens, while the site of the reduction processes being the π-electron system of the porphyrin ring |155|; theoretical calculations however do not exclude the possibility of a metal centred reduction |170|.

The fact that the cathodic reductions of Cu(TPP) appear essentially reversible in character provides further support for the postulated ligand centred character of these steps; in fact, the eventual generation of a Cu(I) species would be expected to involve a deep geometrical reorganization (planar $\rightleftarrows$ tetrahedral), thus causing a significant departure from electrochemical reversibility.

The presence of electron-withdrawing ß-pyrrole substituents in |Cu(TPP)X| makes easier, as expected, the sequential reduction steps |161|. This effect is particularly evident in the series Cu(TPP)(CN)$_{1-4}$, in which the gradual increase of cyano groups linearly favours the reduction process. On the other hand, these substituents make more difficult the removal of electrons even if it is reported that the increased difficulty for the oxidation does not linearly correlate with the number of cyano substituents. This behaviour is ascribed to the cited difference in the nature of the oxidation site with respect to the reduction one |155|.

Interestingly, in Cu(TPP)(CN)$_4$ the ligand-centred addition of the first two electrons is followed by a further, likely metal-centred, one-electron addition leading to $|Cu^{I}(TPP)(CN)_4|^{3-}$ |166|. Figure 48 illustrates the cyclic voltammogram showing this unusual third step.

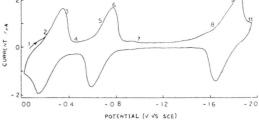

Fig. 48. Thin-layer cyclic voltammogram recorded at a gold minigrid electrode in a DMF solution of Cu(TPP)(CN)$_4$ containing |NBu$_4$|ClO$_4$ (0.1 mol dm^{-3}) as supporting electrolyte (from Ref. 166).

As it can be judged from the backscan response, the copper(I) trianion is relatively long-lived (however, at longer times of electrolysis, it decomposes). Thin-layer voltammetry, owing to the presence of uncompensated solution resistance, prevents a detailed analysis of the degree of reversibility of this Cu(II)/Cu(I) reduction step, which however is expected to be quasireversible. This third step ($E^{o\prime}$ = -1.95 V) precedes in its turn the usual two successive two-electron steps, irreversible because of the associated uptake of protons |163|.

Interestingly, the theoretically foreseen possibility of a metal centred reduction in copper porphines is supported by the redox behaviour of N-substituted Cu(TPP)R derivatives |168|. It has been reported that such complexes display a one-electron reduction step well before the usual first two ligand-centred one-electron cathodic steps (see Table 25). A somewhat puzzling question is the electrochemical reversibility (or nearly) of this assumed copper(II)/copper(I) reduction. A tentative explanation could be found in the fact that N-substituents cause significant distortions from the Cu(II)-N$_4$ planarity owing to the weakening of the metal-N alkylated bond |169|, thus demanding only a minor molecular reorganization upon reduction to the (pseudo)tetrahedral copper(I) assembly.

TABLE 25

Formal electrode potentials (in Volt) for the redox changes of N-substituted derivatives |Cu(TPP)R|(ClO$_4$)(see Chart XL) in acetonitrile solution |168|.

R	$E^{o\prime}$ 3+/2+	$E^{o\prime}$ 2+/1+	$E^{o\prime}$ 1+/0 (Cu^{II}/Cu^{I})	$E^{o\prime}$ 0/1-	$E^{o\prime}$ 1-/2-
CH$_3$	+ 1.48	+ 1.24	- 0.33	- 1.20	- 1.41
C$_2$H$_5$	+ 1.47	+ 1.27	- 0.29	- 1.19	- 1.45
C$_6$H$_5$	+ 1.42	+ 1.20	- 0.36	- 1.05	- 1.31

Finally, concerning the hydroporphyrine complex Cu(TPiBC) (tetraphenyl isobacteriochlorin copper(II)), its redox behaviour in dichloromethane indicates that, compared with the free ligand, the presence of the metal centre favours the two successive one-electron removals, as illustrated by |171, 172|:

$$E^{\circ\prime} = +0.57 \text{ V} \qquad\qquad E^{\circ\prime} = +0.95 \text{ V}$$

$$H_2TPiBC \underset{+e}{\overset{-e}{\rightleftharpoons}} |H_2TPiBC|^+ \underset{+e}{\overset{-e}{\rightleftharpoons}} |H_2TPiBC|^{2+}$$

$$E^{\circ\prime} = +0.40 \text{ V} \qquad\qquad E^{\circ\prime} = +0.75 \text{ V}$$

$$Cu(TPiBC) \underset{+e}{\overset{-e}{\rightleftharpoons}} |Cu(TPiBC)|^+ \underset{+e}{\overset{-e}{\rightleftharpoons}} |Cu(TPiBC)|^{2+}$$

Not only is the H_2TPiBC/Cu(TPiBC) system more easily oxidizable than the H_2TPP/Cu(TPP) system, but also the lowering of the oxidation potential caused by the presence of copper(II) ion seems appreciably greater.

Another series of copper-porphyrin complexes, the redox properties of which have been studied, is shown in Chart XLI.

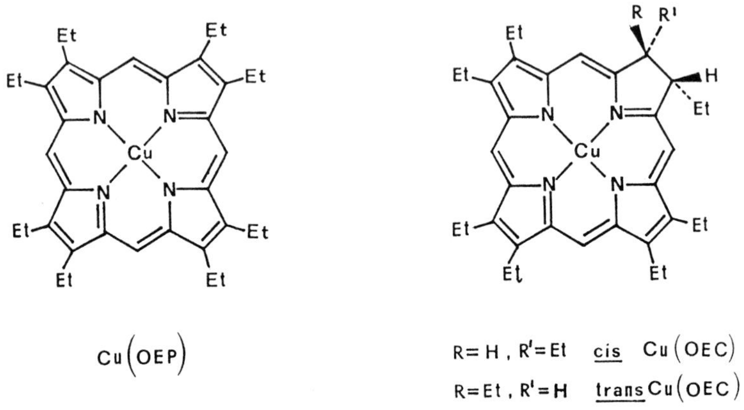

Cu(OEP)

R=H , R'=Et   cis Cu(OEC)
R=Et , R'=H   trans Cu(OEC)

CHART XLI

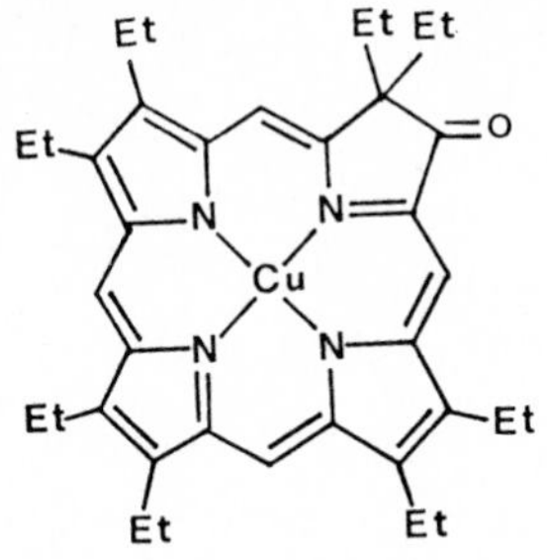

Cu(OEOx)

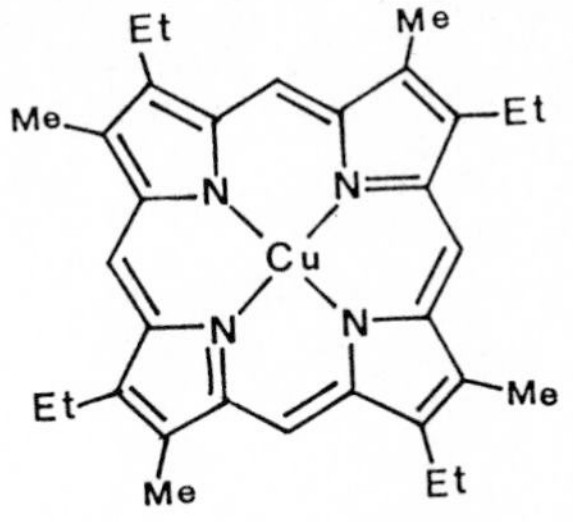

Cu(Etio I)

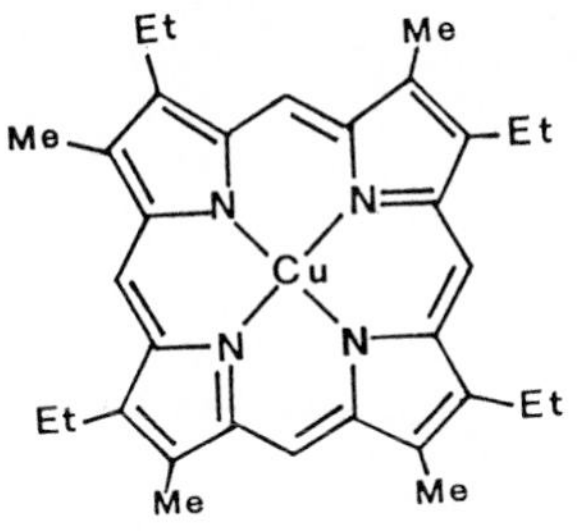

Cu(Etio IV)

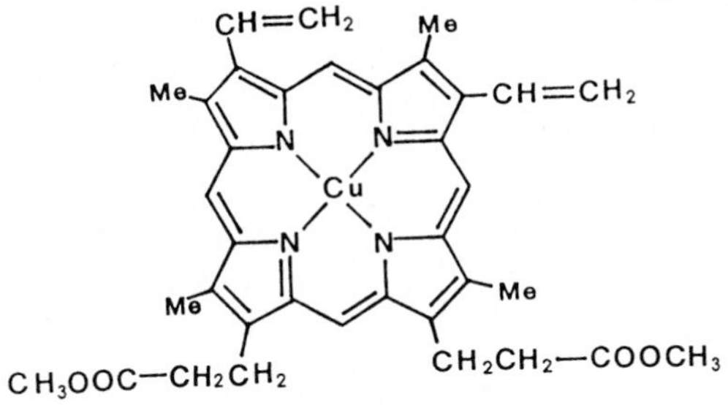

Cu(PPDM)

CHART XLI

As in the case of the preceding porphyrins, free ligands of
the metal complexes shown in Chart XLI are redox active, and the

presence of a central copper(II) ion simply makes easier the sequential removal of two electrons from the ligand, as well as making more difficult the sequential addition of two electrons to the ligand (Table 26).

We do not treat these derivatives any further because no reorganizational effect follows their electron transfers.

TABLE 26

Formal electrode potentials (in Volt) for the redox changes of free porphyrins and their copper(II) complexes schematized in Chart XLI

complex	$E^{o\prime}$ 2+/1+	$E^{o\prime}$ 1+/0	$E^{o\prime}$ 0/1-	$E^{o\prime}$ 1-/2-	solvent	reference
H_2OEP	+1.30	+0.81	–	–	n–BuCN	173
H_2OEP	–	–	-1.46	-1.86	DMSO	173
H_2OEP	+1.40[a]	+0.89	-1.44	-1.90	n–PrCN	174
Cu(OEP)	+1.19	+0.79	–	–	n–BuCN	173
Cu(OEP)	–	–	-1.46	–	DMSO	173
cis-H_2OEC	+1.12[a]	+0.58	-1.43	-1.89	MeCN	175
cis-Cu(OEC)	+1.12[a]	+0.46	-1.51	-2.11[a]	MeCN	175
trans-H_2OEC	+1.18	+0.64	-1.47	-1.95	n–PrCN	174
trans-H_2OEC	+1.10	+0.59	-1.44	-1.89	MeCN	175
trans-Cu(OEC)	+1.09	+0.47	-1.52	-2.15[a]	MeCN	175
H_2OEOx	+1.24	+0.86	-1.23	-1.73	MeCN	176
Cu(OEOx)	+1.27	+0.73	-1.25	-1.85	MeCN	176
H_2Etio I	–	–	-1.34	–	DMF	158
H_2Etio I	–	–	-1.37	-1.80	DMF	162
Cu(Etio I)	–	–	-1.46	-2.05[a]	DMF	158
Cu(Etio IV)	–	–	-1.48	-1.99	DMF	162
H_2PPDM	–	–	-1.24	-1.61	DMF	177
Cu(PPDM)	–	–	-1.37	-1.82	DMF	177

[a]Peak potential value for irreversible process.

4.2.17 <u>Phthalocyanines</u>. Strictly correlated to the porphyrin-copper(II) complexes are the phthalocyanine-copper(II) derivatives shown in Chart XLII.

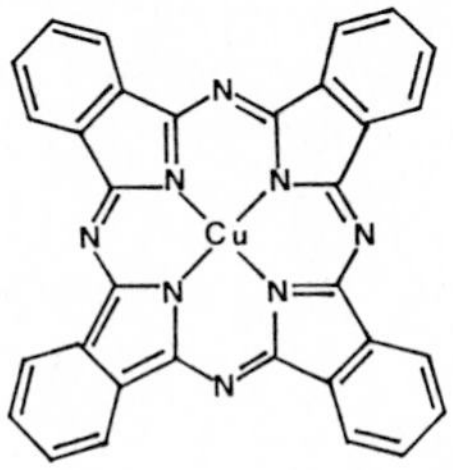

Cu(Pc)

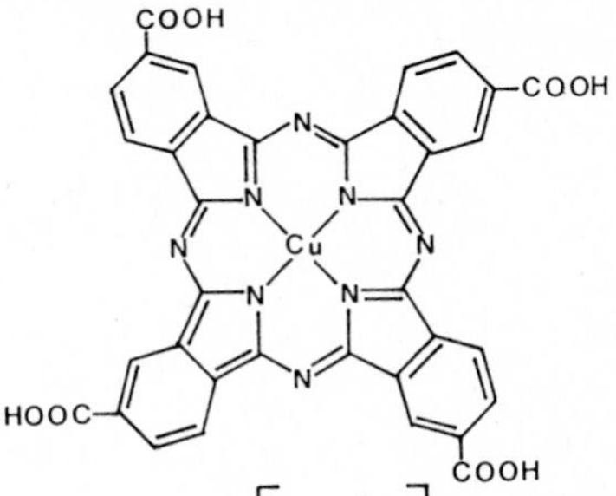

Cu[(COOH)₄Pc]

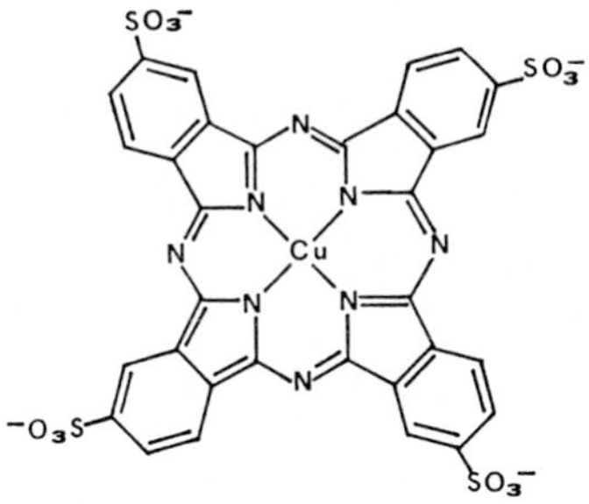

Cu(TSPc)

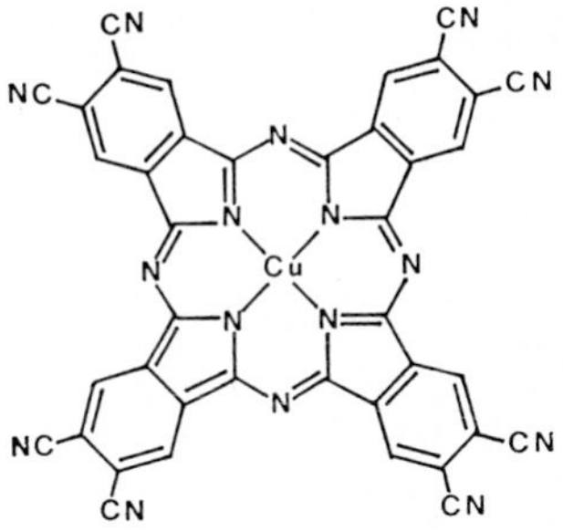

Cu[(CN)₈Pc]

CHART XLII

X-ray investigations have shown that Cu(Pc) is planar with four in-plane Cu-N bonds (1.94 Å) |178|.

As in the case of porphyrins, free base phthalocyanines undergo sequential electron transfers. The presence of the central copper(II) ion produces a negative shift in the electrode potentials of these redox changes (Table 27), and no reduction of copper(II) ion has been observed. Although the importance of these derivatives as synthetic analogues of biological metalloporphyrins is well documented, the lack of significant structural consequences following such redox processes force us to consider them of minor significance for the purposes of the present review.

TABLE 27

Redox potentials (in Volt) for the sequential electron transfers of free phthalocyanines and their copper(II) complexes schematized in Chart XLII

complex	ligand oxidation E°'	ligand reduction				solvent	reference
		E°' 1st	E°' 2nd	E°' 3rd	E°' 4th		
H_2Pc	—	− 0.66	− 1.06	− 1.93	− 2.23	DMF	179
H_2Pc	+ 0.9	—	—	—	—	DMF	180
H_2Pc	+ 1.10	—	—	—	—	CLN	181
H_2Pc	+ 0.73	—	—	—	—	MeCN	182
$Cu(Pc)$	—	− 0.84	− 1.18	− 2.01	− 2.35	DMF	179
$Cu(Pc)$	+ 0.87	—	—	—	—	DMF	180
$Cu(Pc)$	+ 0.98	—	—	—	—	CLN	181
$Cu(Pc)$	+ 0.64	—	—	—	—	MeCN	182
H_2TSPc	+ 0.90	− 0.52	− 0.97	− 1.81	—	DMSO	183
$Cu(TSPc)$	+ 0.87	− 0.73	− 1.11	− 1.89	—	DMSO	183
$Cu(TSPc)$	+ 0.94	—	—	—	—	DMF	184
$Cu|(COOH)_4Pc|$	+ 1.18	—	—	—	—	MeCN	182
$H_2(CN)_8Pc$	—	− 0.10	− 0.45	− 0.88	− 1.50	DMF	185
$Cu|(CN)_8Pc|$	—	− 0.2	− 0.63	− 1.08	− 1.25	DMF	185

4.2.18 <u>Macrocycles</u>. Great attention has been paid to the CuN_4 coordination present in macrocyclic complexes, which in many cases serve as synthetic models of natural products as metalloporphyrins, vitamin B_{12} and chlorophyls.

Let we start with the copper(II) complexes of the saturated tetraaza macrocycles shown in Chart XLIII.

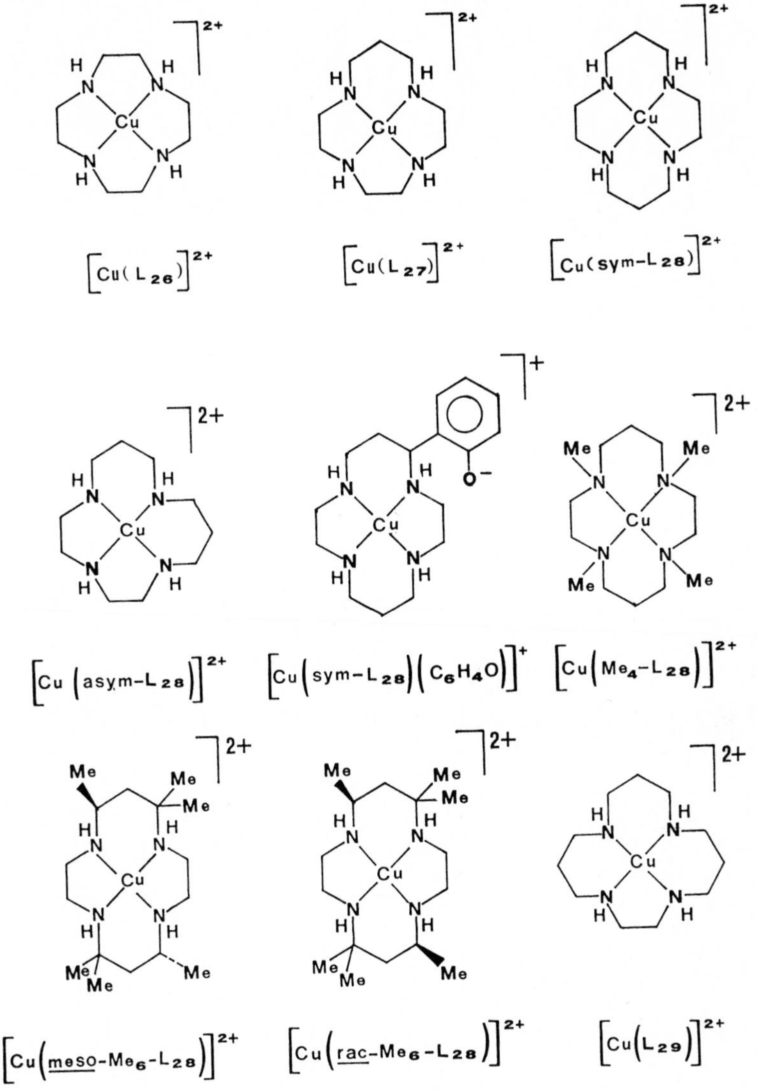

$$\left[Cu(L_{26})\right]^{2+} \qquad \left[Cu(L_{27})\right]^{2+} \qquad \left[Cu(sym-L_{28})\right]^{2+}$$

$$\left[Cu\left(asym-L_{28}\right)\right]^{2+} \qquad \left[Cu\left(sym-L_{28}\right)\left(C_6H_4O\right)\right]^{+} \qquad \left[Cu\left(Me_4-L_{28}\right)\right]^{2+}$$

$$\left[Cu\left(\underline{meso}-Me_6-L_{28}\right)\right]^{2+} \qquad \left[Cu\left(\underline{rac}-Me_6-L_{28}\right)\right]^{2+} \qquad \left[Cu\left(L_{29}\right)\right]^{2+}$$

CHART XLIII

CHART XLIII

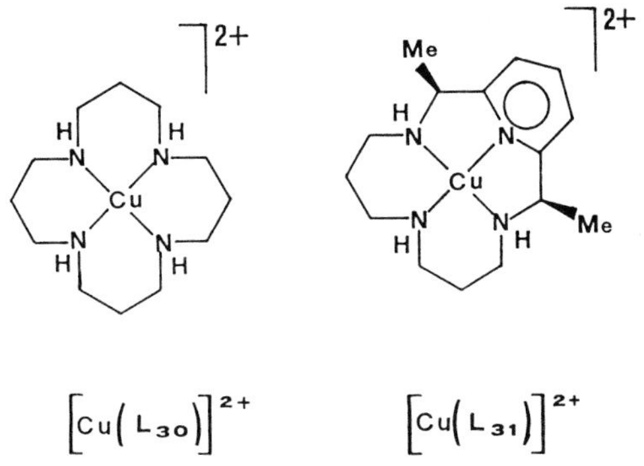

$$\left[Cu(L_{30}) \right]^{2+} \qquad \left[Cu(L_{31}) \right]^{2+}$$

As shown in Figure 49, which refers to $|Cu(sym-L_{28})|^{2+}$ (the well known ligand <u>cyclam</u>), these complexes undergo, at platinum electrodes, two subsequent one-electron reductions and a one-electron oxidation, which are attributed to the metal-centred redox changes $Cu(II)/Cu(I)$, $Cu(I)/Cu(0)$, $Cu(II)/Cu(III)$, respectively.

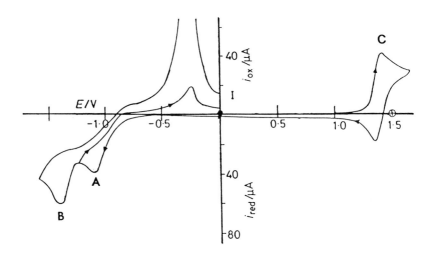

Fig. 49. Cyclic voltammogram recorded at a platinum electrode on a MeCN solution of $|Cu(sym-L_{28})|^{2+}$. $|NEt_4|ClO_4$ (0.1 mol dm^{-3}) supporting electrolyte. Scan rate 0.2 Vs^{-1}.

The copper(II)/copper(I) reduction, in general, is a quasi-reversible process followed by relatively slow decomplexation of the copper(I) ion. The use of high scan rates can prevent the occurrence of such a chemical complication. As a matter of fact, when the copper(II)/copper(I) redox potentials are more negative than the $Cu^{+/0}$ potential, controlled potential electrolysis corresponding to the first step leads to the consumption of $2e^-$/molecule, simply because the newly created copper(I) ion undergoes further reduction to copper metal. Accordingly, the reduction of $|Cu(L_{29})|^{2+}$ and $|Cu(L_{30})|^{2+}$ in acetonitrile, which display redox potentials lower than the reduction potential of free Cu^+ (about -0.6 V in MeCN), which does not even produce the corresponding copper(I) complexes, consumes 1 e^-/molecule.

On the contrary, the second cathodic step results always irreversible, because of the extreme lability of the copper(0) complex. Even at 100 Vs^{-1} no trace of directly associated responses in the reverse scan can be detected. Finally, the copper(II)/copper(III) oxidation proceeds through a reversible or quasireversible process complicated by ligand-centred multielectron removals |189,190,192|. Only in strong acidic solutions ($HClO_4$ 10M) stable copper(III) complexes can be electrogenerated |194|.

In summary the redox behaviour of copper (II) macrocycles can be so schematized:

$$
\begin{array}{cccc}
\text{peak C} & \text{peak A} & \text{peak B} & \\
& +e & +e & +e \\
|Cu^{III}L|^{3+} \underset{-e}{\overset{}{\rightleftarrows}} & |Cu^{II}L|^{2+} \underset{-e}{\overset{}{\rightleftarrows}} & |Cu^{I}L|^{+} \longrightarrow & Cu^0 + L \\
\text{slow} \downarrow & & \downarrow \text{slow} & \\
& & Cu^+ + L \xrightarrow{\;+e\;} & \\
\text{decomposition} & & & \\
\text{products} & & \text{if } E^{\circ\prime}_{Cu^{II}L/Cu^{I}L} < E^{\circ\prime}_{Cu^+/Cu^0} &
\end{array}
$$

At mercury electrodes, the two successive one-electron cathodic steps commonly merge into a single two-electron reduction.

Table 28 summarizes the electrode potentials of such redox changes.

The two redox changes Cu(II)/Cu(I) and Cu(II)/Cu(III) have significant implications concerning the stereochemical reorganizations involved. Even though X-ray data are relatively scarce |188,202,203,204|, there is a body of spectroscopic evidence which allow us to draw some general conclusions.

The most important characteristic in these saturated macrocycles is the size of the central hole of the ligand which must accomodate the copper(II) ion. The best-sized cavity encircles the copper ion in a square planar CuN_4 coordination through strong Cu-N bonds (eventually with two apical solvent molecules or counter anions completing the octahedral geometry). This disfavours the access to the pseudotetrahedral copper(I) geometry, while favouring the access to the planar one of the copper(III) complex. Smaller or larger cavity sizes, weakening the in-plane Cu-N interactions, accomodate the copper(II) ion in a non-planar coordination, so favouring the rearrangement to the copper(I) assembly and disfavouring the rearrangement to the copper(III) assembly |189,201|.

These considerations are congruent with the redox potentials of the unsubstituted macrocycles L_{26}, L_{27}, L_{28}, L_{29}, L_{30}. In fact, sym-L_{28}, which possesses the best-sized cavity to encircle the copper(II) ion |205|, presents the most negative potential for the Cu(II)/Cu(I) step and the less positive one for the Cu(II)/Cu(III) change. L_{26}, L_{27}, which have the smallest cavities, and L_{29}, L_{30}, which have the largest cavities, present the lowest Cu(II)/Cu(I) redox potentials and the highest Cu(II)/Cu(III) ones.

Unfortunately, X-ray structures of $|Cu(sym-L_{28})|^{2+}$ |202,204| versus $|Cu(L_{29})|^{2+}$ |203| do not help in this interpretation, since in both derivatives the trans coordination of two counteranions induced by crystal packing leads to a CuN_4 planarity, Figure 50, not present in solution for $|Cu(L_{29})|^{2+}$.

TABLE 28

Redox potentials (in Volt) for the oxidation and reduction of the copper(II)-macrocyclic complexes schematized in Chart XLIII

complex	$E°'$ Cu^{III}/Cu^{II} [a]	$E°'$ Cu^{II}/Cu^{I} [a]	Ep Cu^{I}/Cu^{0} [a]	E Cu^{II}/Cu^{0} [b,c]	solvent	reference
$|Cu(L_{26})|^{2+}$	-	-0.64	-	-	H_2O (pH 9.6)	186
$|Cu(L_{26})|^{2+}$	-	-	-	-0.33	H_2O (pH 5.0)	187
$|Cu(L_{27})|^{2+}$	+1.41	-0.62	-0.99	-0.66	MeCN	189
$|Cu(L_{27})|^{2+}$	-	-	-	-0.43	H_2O (pH 5.0)	187
$|Cu(sym-L_{28})|^{2+}$	+1.42	-	-	-	MeCN	191
$|Cu(sym-L_{28})|^{2+}$	+1.35	-1.10	-1.34	-0.94	MeCN	192
$|Cu(sym-L_{28})|^{2+}$	-	-0.73	-	-	H_2O (pH 9.6)	186
$|Cu(sym-L_{28})|^{2+}$	+0.76	-	-	-	$HClO_4$ (10M)	194
$|Cu(asym-L_{28})|^{2+}$	+1.40	-0.81	-1.34	-	MeCN	192
$|Cu(\underline{sym}-L_{28})(C_6H_4O)|^{+}$	-	-	-	-0.78	H_2O (pH 10.3)	188
$|Cu(Me_4-L_{28})|^{2+}$	-	-0.32	-0.97	-	MeCN	67
$|Cu(\underline{meso}-Me_6-L_{28})|^{2+}$	-	-	-	-0.62	H_2O (pH 6.0)	195
$|Cu(\underline{meso}-Me_6-L_{28})|^{2+}$	+1.51	-0.74	-1.87	-0.68	MeCN	196
$|Cu(\underline{meso}-Me_6-L_{28})|^{2+}$	+1.60	-0.82	-	-	MeCN	197
$|Cu(\underline{meso}-Me_6-L_{28})|^{2+}$	+0.94	-	-	-	$HClO_4$ (10M)	194
$|Cu(\underline{rac}-Me_6-L_{28})|^{2+}$	-	-0.50	-	-	DMSO	198
$|Cu(\underline{rac}-Me_6-L_{28})|^{2+}$	-	-0.83	-	-	MeCN	199
$|Cu(L_{29})|^{2+}$	-	-	-	-0.51	H_2O (pH 8.7)	200
$|Cu(L_{29})|^{2+}$	+1.56	-0.55	-1.33	-	MeCN	189
$|Cu(L_{30})|^{2+}$	+1.66	-0.31	-1.54	-	MeCN	189
$|Cu(L_{31})|^{2+}$	+1.60	-0.62	-1.50	-	MeCN	192
$|Cu(L_{31})|^{2+}$	-	-	-	-0.58	MeOH	193

[a] Platinum electrode; [b] mercury electrode; [c] peak potential value in cyclic voltammetry, or half wave potential in polarography.

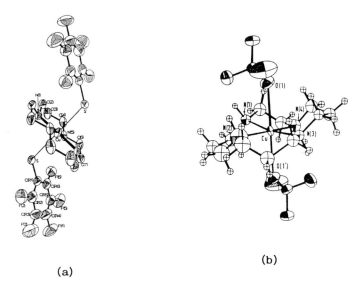

(a) (b)

Fig. 50. Perspective views of: (a) $Cu(sym-L_{28})(SC_6F_5)_2$. $Cu-N_{(mean)}$, 2.01 Å; $Cu-S_{(thiolate)}$, 2.94 Å (from Ref. 204); (b) $Cu(L_{29})(ClO_4)_2$. $Cu-N_{(mean)}$, 2.03 Å; $Cu-O_{(ClO_4)}$, 2.59 Å (from Ref. 203).

The strong in-plane Cu-N coordination in $|Cu(sym-L_{28})|^{2+}$ is further supported by the X-ray structure of $|Cu(sym-L_{28})(C_6H_4O)|^+$, in which a phenolate pendant group, although coordinating in a fifth apical position, does not modify the $Cu-N_4$ planarity, Figure 51.

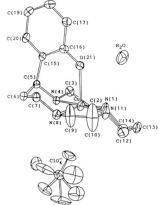

Fig. 51. X-ray structure of $|Cu(sym-L_{28})(C_6H_4O)|(ClO_4) \cdot H_2O$. $Cu-N_{(mean)}$, 2.02 Å; $Cu-O_{(phenolate)}$, 2.14 Å; $Cu \cdots O_{(ClO_4)}$, 3.13 Å (from Ref. 188).

In this connection, indirect evidence that the 12-membered ligand L_{26} cannot planarly encircle the copper(II) ion comes from the X-ray structures of copper complexes with nitrogen substituted analogs |208,209|. In all cases the complexes are five coordinate because of a fifth apical bond with a chloride counteranion; but, as shown in Figure 52, the copper(II) ion is displaced out of the basal plane of the square pyramid (towards the apex) by about 0.5 - 0.6 Å.

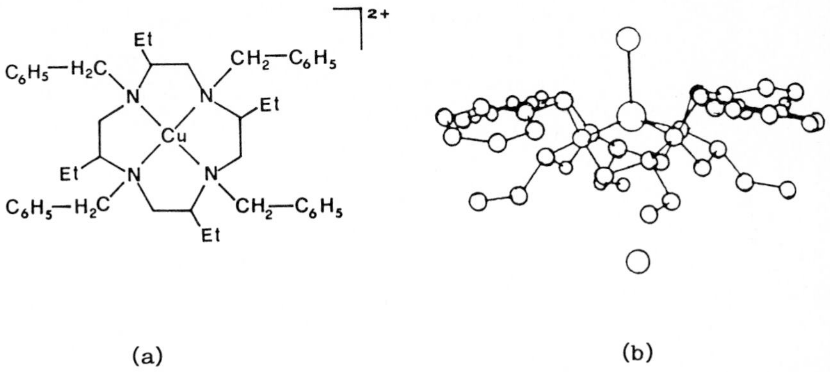

(a)   (b)

Fig. 52. (a) Drawing of $|Cu\{(bz_4et_4)L_{26}\}|^{2+}$. (b) Perspective view of $|Cu\{(bz_4et_4)L_{26}Cl\}|Cl$. $Cu-N_{(mean)}$, 2.06 Å; $Cu-Cl$, 2.43 Å; $Cu\cdots\cdots Cl$, 4.83 Å (from Ref. 209).

The presence of methyl substituents in Me_4-L_{28}, Me_6-L_{28} and L_{31} introduces two additional features - the electron donating effect of the methyl groups disfavour the one-electron addition while favouring the one-electron removal; and, the presence of methyl groups modifies the structure of the copper(II) complex with respect to the unmethylathed one. In this last respect, it must be taken into account that $|Cu(sym-L_{28})|^{2+}$ presents a ring structure of the chair-type $\underline{a}$ in Chart XLIV, whereas $|Cu(Me_4-L_{28})|^{2+}$ is assigned a structure of the boat-type $\underline{b}$ |206|.

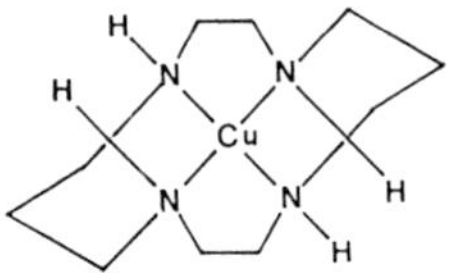

CHART XLIV

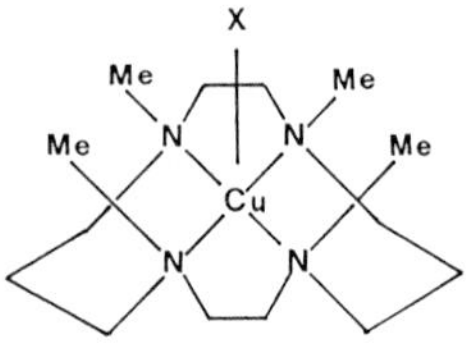

(a)   (b)

Since $|Cu(Me_4-L_{28})|^{2+}$ is notably easier to reduce than $|Cu(sym-L_{28})|^{2+}$, we must conclude that structural aspects overcome the electronic ones. However it seems likely that the determining structural feature is not the chair or boat conformation, but the pentacoordination of the methylated derivative, which inhibiting the square-planar geometry |207| highly favours the reduction step. In a parallel fashion, Me_6-L_{28} complexes are also assigned type _b_ (Chart XLIV) structures |206,207|, since they are more easily reduced than $|Cu(sym-L_{28})|^{2+}$. Within the class of Me_6-L_{28} complexes, electrochemical data are not sufficient to judge if the meso or racemic forms have really different redox abilities.

In order to introduce the wide series of copper(II)- unsaturated macrocycles, we consider first the deprotonated amido complexes schematized in Chart XLV, in which, where possible, it has been maintained a nomenclature similar to that reported in Chart XLIII.

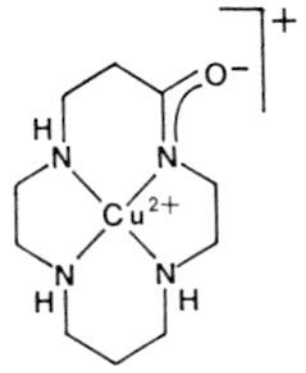

CHART XLV

$[Cu(monox-sym-L_{28})]$ 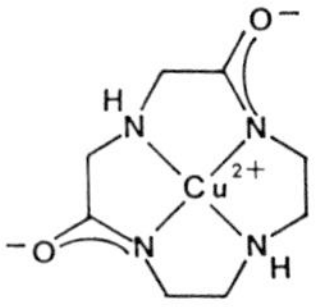 $Cu(diox-L_{26})$

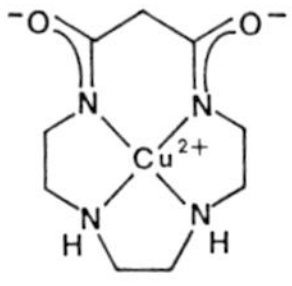

$Cu(diox-L_{27})$ 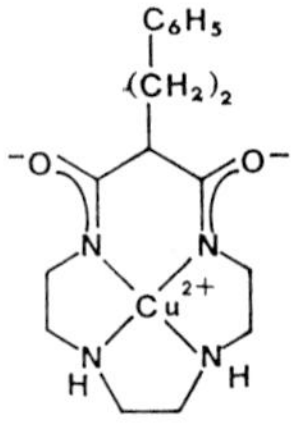 $Cu(L_{32})$ 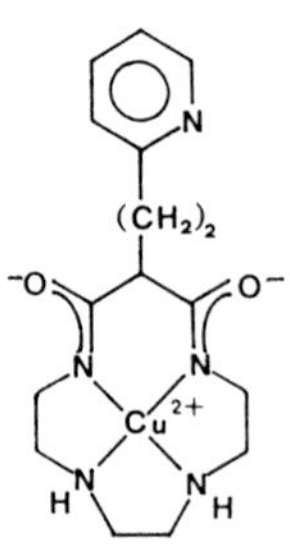 $Cu(L_{33})$

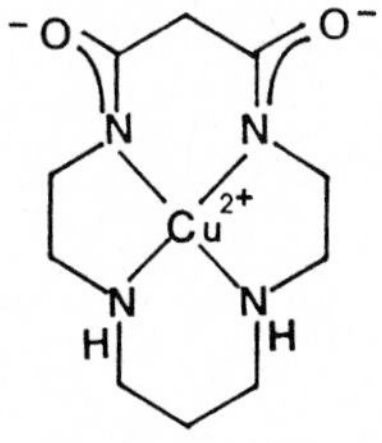

$Cu(diox-sym-L_{28})$ 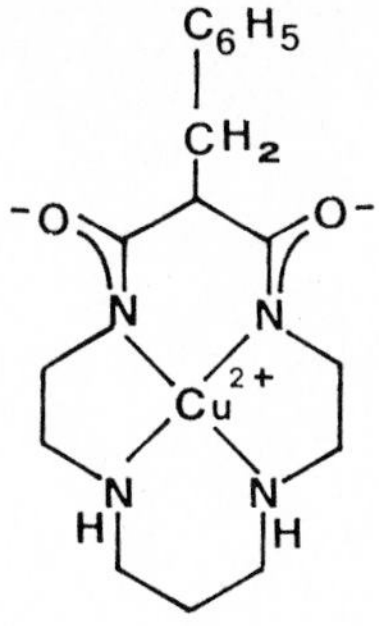 $Cu(L_{34})$ 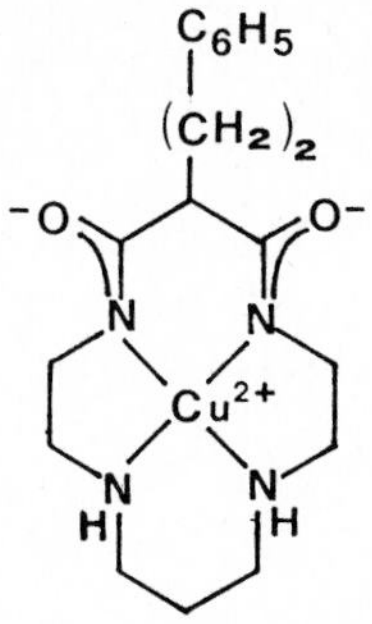 $Cu(L_{35})$ 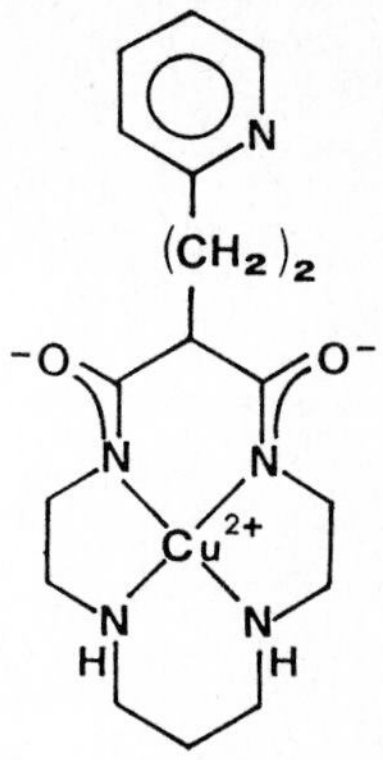 $Cu(L_{36})$

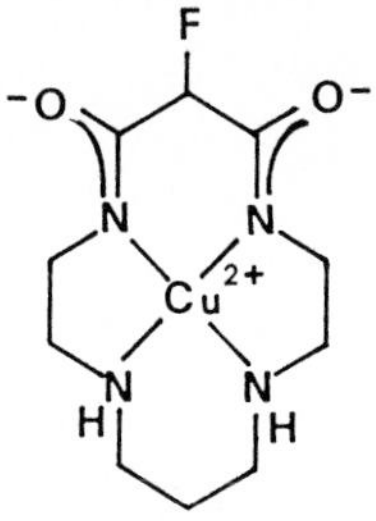

$Cu(L_{37})$ 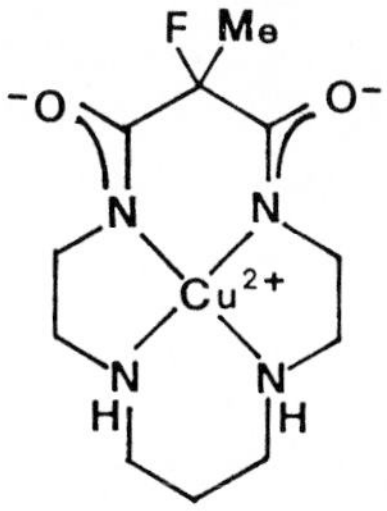 $Cu(L_{38})$ 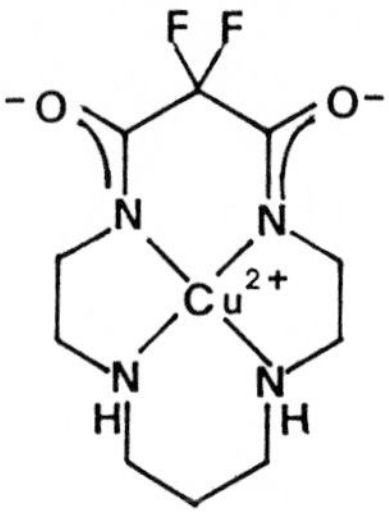 $Cu(L_{39})$

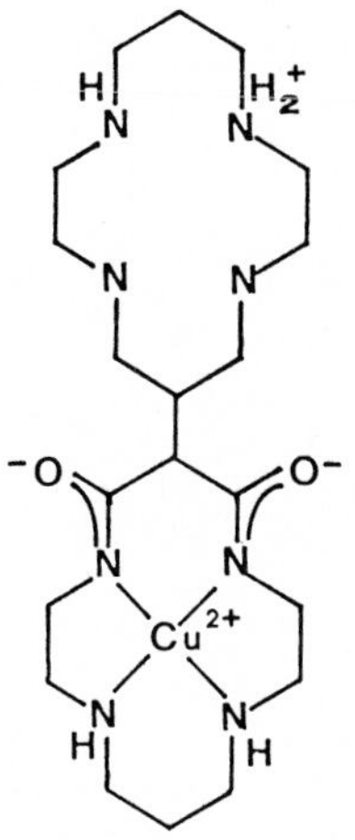

$[Cu(L_{40})]^+$

CHART XLV

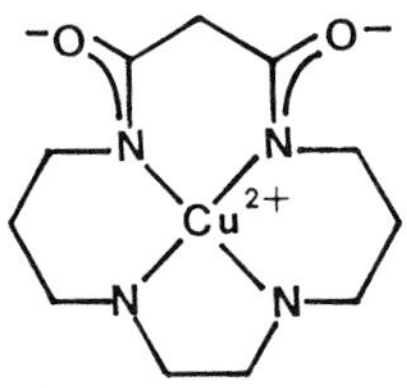

Cu(diox-L₂₉)

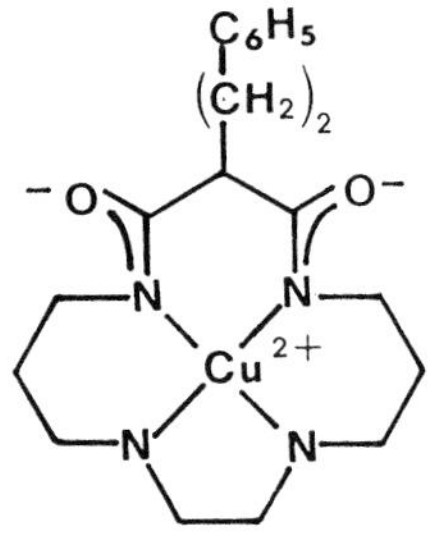

Cu(L₄₁)

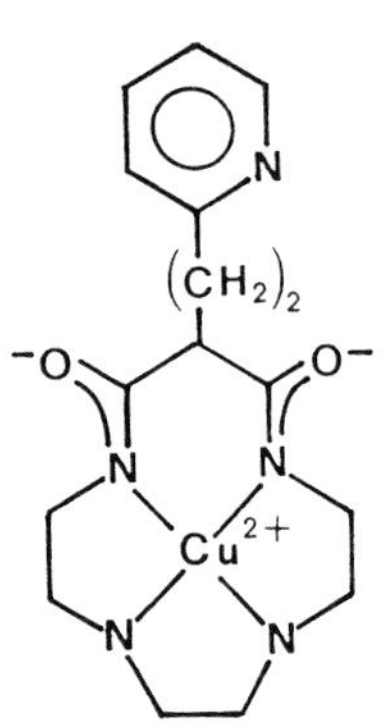

Cu(L₄₂)

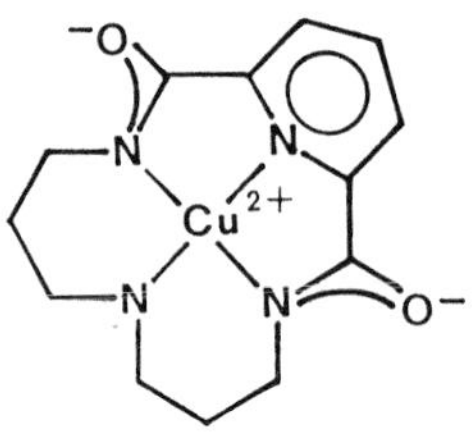

Cu(L₄₃)

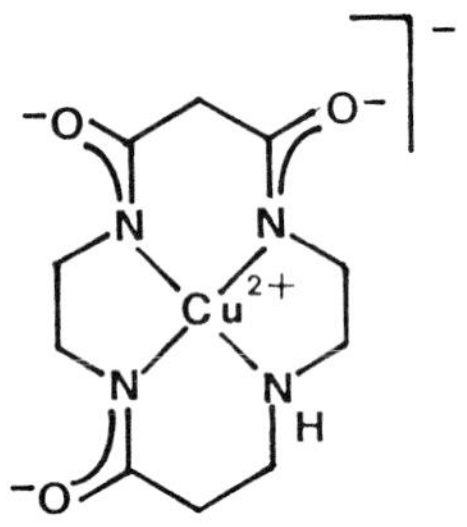

[Cu(triox-sym-L₂₈)]⁻

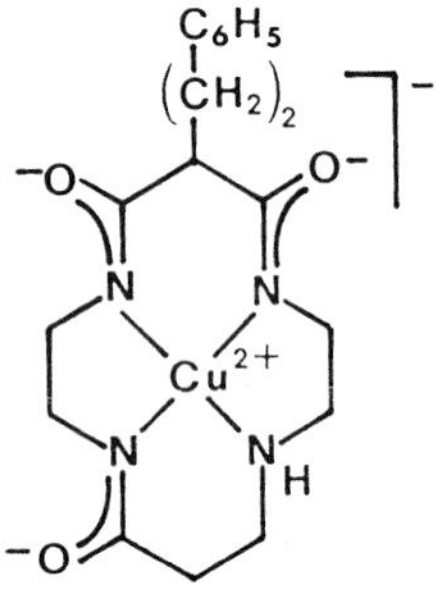

[Cu(L₄₄)]⁻

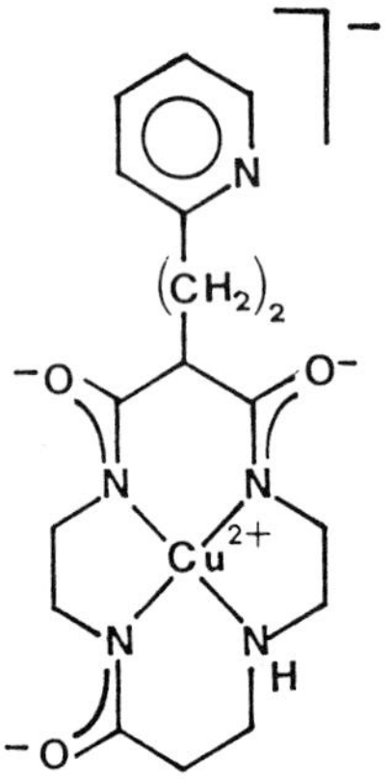

[Cu(L₄₅)]⁻

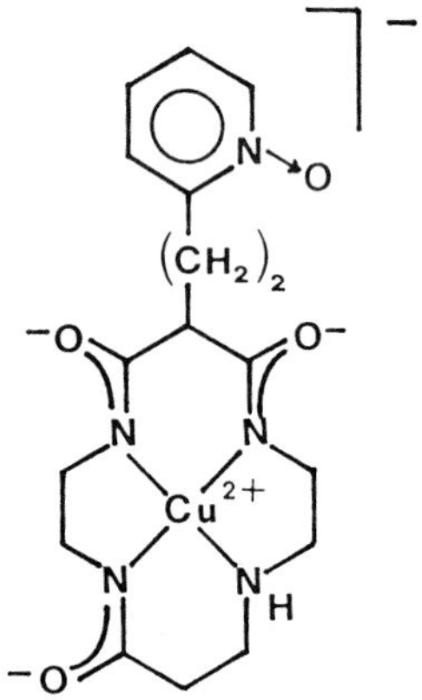

[Cu(L₄₆)]⁻

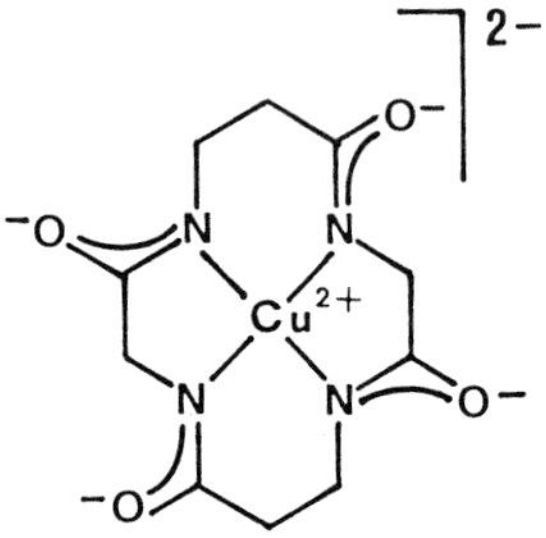

[Cu(tetrox-sym-L₂₈)]²⁻

CHART XLV

As an example, Figure 53 depicts the cyclic voltammogram exhibited by Cu(diox-L$_{26}$) in aqueous solution |97|.

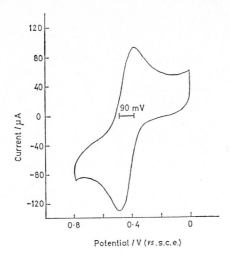

Fig. 53. Cyclic voltammogram recorded at a glassy carbon electrode on an aqueous solution (pH 9.6, Na$_2$SO$_4$ (0.5 mol dm^{-3})) of Cu(diox-L$_{26}$). Scan rate 0.1 Vs^{-1} (from Ref. 87).

As shown, these complexes display a quite easy copper(II)/copper(III) oxidation, commonly quasi-reversible in character. Table 29 summarizes the redox potentials for the complexes schematized in Chart XLV, and compares them with those of the corresponding saturated macrocycles.

As noted also in the case of open-chain polyamino and polypeptide complexes (Sections 4.2.3, 4.2.4, 4.2.15), the presence of deprotonated amido groups exerts a strong beneficial effect on the stabilization of copper(III) species, probably because of the electron donating ability of these negatively charged functions as well as their strong in-plane donor ability which favourably predisposes them to having the planar geometry favoured by copper(III) species.

TABLE 29

Formal electrode potentials (in Volt) for the redox changes displayed by the copper complexes schematized in Chart XLV versus those of the similar complexes reported in Chart XLIII

complex	$E^{\circ\prime}_{Cu^{III}/Cu^{II}}$	$E^{\circ\prime}_{Cu^{II}/Cu^{I}}$	solvent	reference
$\|Cu(monox-sym-L_{28})\|^{+}$	+0.86	–	H_2O (pH 7)	98
$Cu(diox-L_{26})$	+0.42	–	H_2O (pH 9.6)	97
$Cu(diox-L_{26})$	+0.46	–	H_2O (pH 8.5, $NaClO_4$ 3M)	215
$Cu(diox-L_{27})$	+0.56	–	H_2O (pH 9.5)	97,98
$Cu(diox-L_{27})$	–	– 0.74[a]	H_2O (pH 9.6)	186
$Cu(diox-L_{27})$	+0.59	–	H_2O (pH 6.0, $NaClO_4$ 0.1M)	210
$Cu(diox-L_{27})$	+0.51	–	H_2O (pH 6.0, $NaClO_4$ 7M)	210
$\|Cu(L_{27})\|^{2+}$	+1.41	– 0.62	MeCN	189
$Cu(L_{32})$	+0.58	–	H_2O (pH 10)	98
$Cu(L_{33})$	+0.56	–	H_2O (pH 10)	98
$Cu(diox-sym-L_{28})$	+0.64	–	H_2O (pH 9.8)	97
$Cu(diox-sym-L_{28})$	–	– 0.85[a]	H_2O (pH 9.6)	186
$Cu(diox-sym-L_{28})$	+0.64	–	H_2O (pH 7.0)	214
$Cu(diox-sym-L_{28})$	+0.66	–	H_2O (pH 6.0, $NaClO_4$ 0.1M)	210,211
$Cu(diox-sym-L_{28})$	+0.61	–	H_2O (pH 6.0, $NaClO_4$ 7M)	210
$\|Cu(sym-L_{28})\|^{2+}$	–	– 0.73	H_2O (pH 9.6)	186
$\|Cu(sym-L_{28})\|^{2+}$	+1.35	– 1.10	MeCN	192
$Cu(L_{34})$	+0.66	–	H_2O (pH 7.0)	212
$Cu(L_{34})$	+0.53	–	MeCN	212
$Cu(L_{35})$	+0.66	–	H_2O (pH 9.8)	98
$Cu(L_{36})$	+0.66	–	H_2O (pH 9.5)	98
$Cu(L_{37})$	+0.69	–	H_2O (pH 7.0)	214
$Cu(L_{38})$	+0.71	–	H_2O (pH 7.0)	214
$Cu(L_{39})$	+0.83	–	H_2O (pH 7.0)	214

TABLE 29(continued)

complex	$E^{\circ\prime}_{Cu^{III}/Cu^{II}}$	$E^{\circ\prime}_{Cu^{II}/Cu^{I}}$	solvent	reference
$\lvert Cu(L_{40})\rvert^{+}$	+0.60	–	H_2O (pH 7.0)	213
$Cu(diox-L_{29})$	+0.69	–	H_2O (pH 9.0)	97
$\lvert Cu(L_{29})\rvert^{2+}$	+1.56	– 0.55	MeCN	189
$Cu(L_{41})$	+0.72	–	H_2O (pH 9.8)	98
$Cu(L_{42})$	+0.69	–	H_2O (pH 9.8)	98
$Cu(L_{43})$	+0.84	–	H_2O (pH 7.0)	215
$\lvert Cu(triox-sym-L_{28})\rvert^{-}$	+0.43	–	H_2O (pH 12.5)	98
$\lvert Cu(L_{44})\rvert^{-}$	+0.49	–	H_2O (pH 9)	98
$\lvert Cu(L_{45})\rvert^{-}$	+0.49	–	H_2O (pH 9)	98
$\lvert Cu(L_{46})\rvert^{-}$	+0.09	–	H_2O (pH 12.3)	98
$\lvert Cu(tetrox-sym-L_{28})\rvert^{2-}$	+0.23	–	H_2O (pH 12.0)	98,216

[a]Peak potential value for irreversible process.

Also, the higher the number of deprotonated amido groups, the easier the access to the corresponding copper(III) species. This result is well exemplified by the series:

$$Cu(sym-L_{28}), \qquad E^{\circ\prime}_{Cu^{II}/Cu^{III}} = +1.35 \text{ V};$$
$$\lvert Cu(monox-sym-L_{28})\rvert^{+}, \quad E^{\circ\prime}_{Cu^{II}/Cu^{III}} = +0.86 \text{ V};$$
$$Cu(diox-sym-L_{28}), \qquad E^{\circ\prime}_{Cu^{II}/Cu^{III}} = +0.64 \text{ V};$$
$$\lvert Cu(triox-sym-L_{28})\rvert^{-}, \quad E^{\circ\prime}_{Cu^{II}/Cu^{III}} = +0.43 \text{ V};$$
$$\lvert Cu(tetrox-sym-L_{28})\rvert^{-}, \quad E^{\circ\prime}_{Cu^{II}/Cu^{III}} = +0.23 \text{ V}.$$

We note that in the widest series of dioxo complexes, the easiest attainment of the copper(III) complex is reached by the 12-membered $Cu(diox-L_{26})$. This result is in sharp contrast with the situation of the saturated macrocycles, where the 14-membered $\lvert Cu(sym-L_{28})\rvert^{2+}$ was the preferred species forming copper(III) complexes. Attempts to explain this feature on the basis of the fact that one-electron removal causes contraction of the ionic radius of the copper, thus allowing it to accomodate perfectly in the smaller 12-membered cavity have been advanced |186, 201|. However, we do not understand why this should not be equally true for the saturated macrocycles. We think, simply, that the presence of the amido groups modifies the aperture of these macrocycles, allowing the 12-membered ring to reach the best size to encircle, in a perfectly planar fashion, the copper(II) ion. At any rate, differences in redox potentials are rather low ($\sim$0.2 V) passing from the 12-membered Cu(diox-L_{26}) to the 15-membered Cu(diox-L_{29}).

Electronic effects of the substituents present in position 6 of the ring can be noted for the complexes with the ligands L_{34}-L_{40}.

An interesting aspect of the redox changes in dioxo complexes is that the presence of strong supporting electrolyte (NaClO$_4$) concentrations makes the one-electron oxidation easier. This datum has been justified on the basis that the copper(II) complexes are distorted octahedral, with two water molecules apically bound, whereas the copper(III) complexes are square planar. The perchlorate ions, through hydrogen bonding with the ligated water molecules, weakens the axial Cu-O$_{(H_2O)}$ bonds, thereby favouring the rearrangement to the four coordinate copper(III) species |201,210|.

A second series of unsaturated copper(II)-macrocycles is schematized in Chart XLVI.

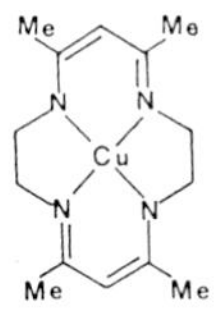

Cu(MeHMe)$_2$(en)$_2$

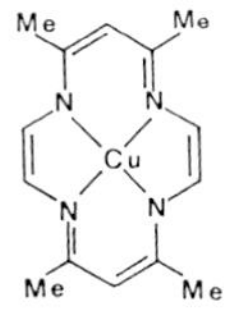

Cu(MeHMe)$_2$(2,9-diene)$_2$

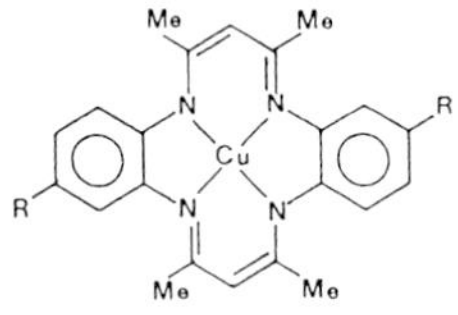

Cu[Me$_4$(R-Bzo)$_2$[14] tetraene N$_4$]

CHART XLVI

We have previously assigned the difficulty of access to copper(I) for Cu(MeHMe)$_2$(en)$_2$ (Section 4.2.13) mainly to the planarity imposed by the macrocycle to the copper(II) species. In principle, the same should have to happen for Cu(MeHMe)$_2$(2,9-diene)$_2$ and Cu|Me$_4$(RBzo)$_2$|14|tetraeneN$_4$|. The only likely difference is that, in the latter derivatives, the presence of two electron reservoirs, such as phenyl groups, may favour the electron addition (see Table 30).

All these derivatives are also characterized by ligand oxidation processes, which, as evident in Figure 54, for Cu(MeHMe)$_2$ (2,9-diene)$_2$, correspond to two reversible one-electron steps.

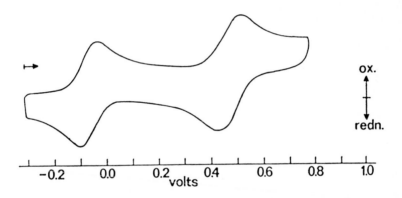

Fig. 54. Cyclic voltammogram exhibited by Cu(MeHMe)$_2$(2,9-diene)$_2$ at a platinum electrode in MeCN solution. Scan rate 0.5 Vs^{-1} (from Ref. 217).

TABLE 30

Redox potentials (in Volt) for the anodic and cathodic processes of the copper(II) complexes schematized in Chart XLVI

complex	ligand oxidation		$E^{o'}$ Cu^{II}/Cu^{I}	solvent	reference
	$E^{o'}$ $2+/1+$	$E^{o'}$ $1+/0$			
Cu(MeHMe)$_2$(en)$_2$	–	+ 0.17	–	DMF	217
Cu(MeHMe)$_2$(en)$_2$	–	–	– 2.26	DMF	120
Cu(MeHMe)$_2$(2,9-diene)$_2$	+ 0.50	– 0.04	–	MeCN	217
Cu|Me$_4$(Bzo)$_2$|14|tetraeneN$_4$|	+ 1.00[a]	+ 0.46[a]	– 1.46	CH$_2$Cl$_2$	218
Cu|Me$_4$(Bzo)$_2$|14|tetraeneN$_4$|	+ 0.87[a]	+ 0.37[a]	– 1.39	MeCN	218
Cu|Me$_4$(CO$_2$CH$_3$Bzo)$_2$|14|tetraeneN$_4$|	+ 1.13[a]	+ 0.61[a]	– 1.22	CH$_2$Cl$_2$	218
Cu|Me$_4$(CO$_2$CH$_3$Bzo)$_2$|14|tetraeneN$_4$|	+ 1.05[a]	+ 0.54[a]	– 1.17	MeCN	218
Cu|Me$_4$(ClBzo)$_2$|14|tetraeneN$_4$|	+ 0.98[a]	+ 0.46[a]	– 1.29	MeCN	219

[a] Peak potential value for irreversible process.

A third series of unsaturated copper(II)-macrocycles is schematized in Chart XLVII.

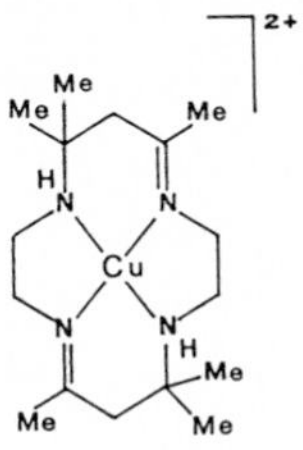

$\left[Cu(\underline{trans}\text{-}[14] \, diene) \right]^{2+}$

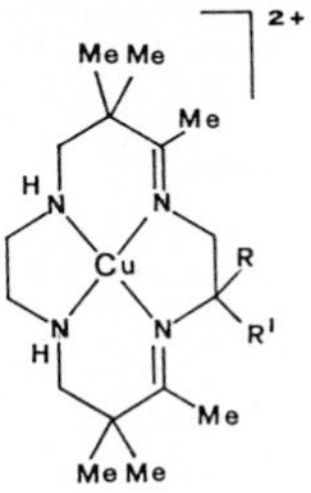

$\left[Cu(R \, R'\text{-} \underline{cis}\text{-}[14] \, diene) \right]^{2+}$

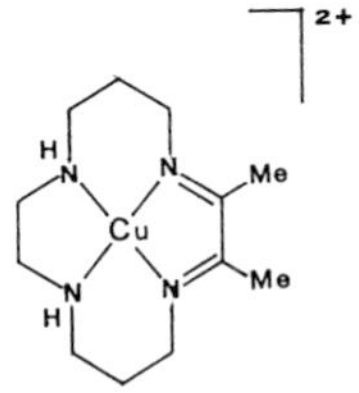

$\left[Cu(L_{47}) \right]^{2+}$

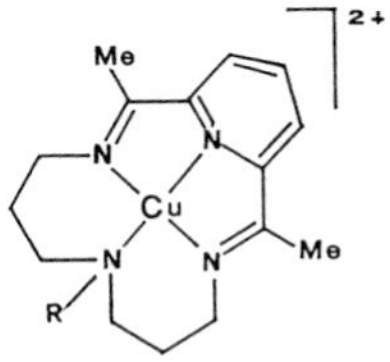

R = H, $\left[Cu(L_{48}) \right]^{2+}$

R = Me, $\left[Cu(L_{49}) \right]^{2+}$

CHART XLVII

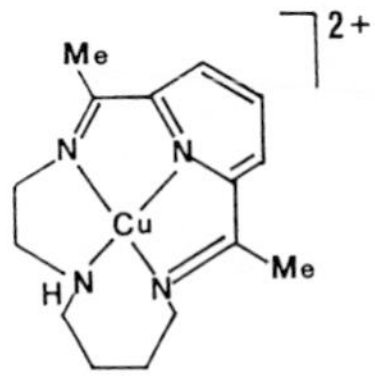

$\left[Cu(L_{50}) \right]^{2+}$

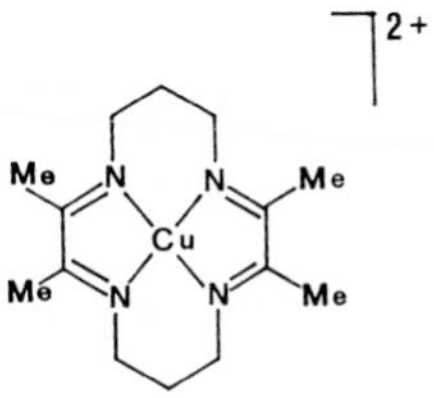

$\left[Cu(TIM) \right]^{2+}$

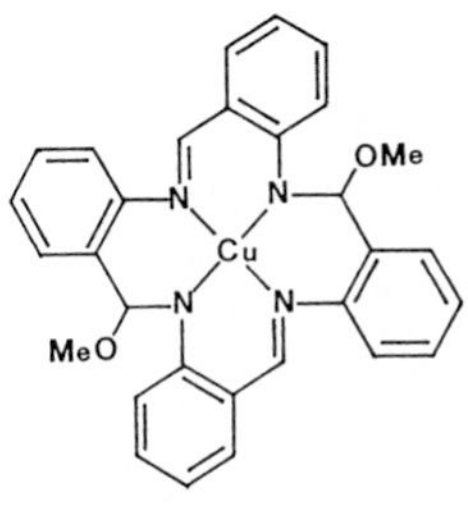

$Cu(TAAB)(OMe)_2$

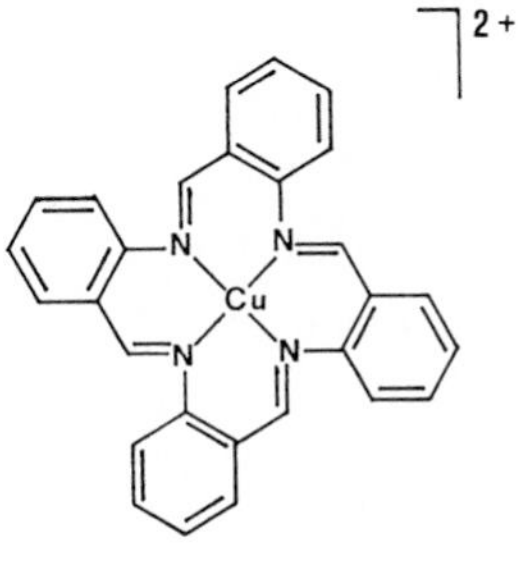

$\left[Cu(TAAB) \right]^{2+}$

284

It is commonly thought that extended, in-plane, unsaturation of the ligand favours reduction of copper(II) complexes through a back-donation mechanism allowing the copper(I) ion to be depleted of charge. A glance at Table 31 soon confirms this hypothesis. All copper(II) complexes schematized in Chart XLVII undergo an easy reduction to copper(I).

$|Cu(trans-|14|diene)|^{2+}$ was the first macrocyclic complex able to be cathodically reduced to the corresponding copper(I) species |195,196,197|. As shown in Figure 55, it exhibits a substantial CuN_4 square-planar coordination, which would disfavour access to copper(I).

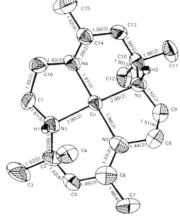

Fig. 55. Crystal structure of $|Cu(dl-trans-|14|diene)|(ClO_4)_2$ with relevant bond distances (from Ref. 229).

It is hence evident that, in this case, easy reduction comes from electronic effects, rather than from favourable stereochemical assembly.

By contrast, spectroscopic evidence assign some degree of tetrahedral distortion to the complexes $|Cu(RR'-\underline{cis}-|14|diene)|^{2+}$. This allows $|Cu(HH-cis-|14|diene)|^{2+}$ to be more easily reducible than $|Cu(trans-|14|diene)|^{2+}$, but methylated derivatives are more difficulty reducible because of their electron donating properties |221|.

Increasing the number of conjugated double bonds up to $|Cu(TIM)|^{2+}$ makes the reduction easier and easier, even if the

TABLE 31

Redox potentials (in Volt) for the electrode activity of the copper(II) complexes schematized in Chart XLVII

complex	$E^{\circ\prime}_{Cu^{III}/Cu^{II}}$	$E^{\circ\prime}_{Cu^{II}/Cu^{I}}$	$E_{p_{Cu^{I}/Cu^{0}}}$	solvent	reference
$\|Cu(\underline{trans}\|14\|diene)\|^{2+}$	–	– 0.49[a]	–	H_2O (pH 6.0)	195
$\|Cu(\underline{trans}\|14\|diene)\|^{2+}$	+ 1.52[a]	– 0.56[a]	– 1.66[a]	MeCN	196
$\|Cu(\underline{trans}\|14\|diene)\|^{2+}$	–	– 0.59[b]	– 1.1[b]	MeCN	196
$\|Cu(\underline{trans}\|14\|diene)\|^{2+}$	–	– 0.58[b]	– 0.99[b]	MeOH	193
$\|Cu(\underline{trans}\|14\|diene)\|^{2+}$	+ 1.56[a]	– 0.63[a,b]	–	MeCN	197
$\|Cu(\underline{trans}\|14\|diene)\|^{2+}$	–	– 0.52[b]	– 1.1[b]	MeCN	221
$\|Cu(\underline{trans}\|14\|diene)\|^{2+}$	+ 0.85[a]	–	–	$HClO_4$ 10M	194
$\|Cu(\underline{trans}\|14\|diene)\|^{2+}$	–	– 0.60[b]	–	DMF	222
$\|Cu(HH-\underline{cis}\|14\|diene)\|^{2+}$	–	– 0.42[b]	– 0.75[b]	MeCN	221
$\|Cu(MeH-\underline{cis}\|14\|diene)\|^{2+}$	–	– 0.61[b]	– 1.0[b]	MeCN	221
$\|Cu(MeMe-\underline{cis}\|14\|diene)\|^{2+}$	–	– 0.81[b]	– 1.1[b]	MeCN	221
$\|Cu(L_{47})\|^{2+}$	–	– 0.62[b]	–	MeOH	220
$\|Cu(L_{48})\|^{2+}$	–	– 0.49[b]	–	MeOH	193, 220
$\|Cu(L_{49})\|^{2+}$	–	– 0.40[b]	–	MeOH	220
$\|Cu(L_{50})\|^{2+}$	–	– 0.40[b]	–	MeOH	220
$\|Cu(TIM)\|^{2+}$	–	– 0.35[b]	–	DMF	222
$\|Cu(TIM)\|^{2+}$	–	– 0.41[b]	–	MeOH	220
$\|Cu(TIM)\|^{2+}$	–	– 0.37[a]	–	MeCN	223
$\|Cu(TIM)\|^{2+}$	–	– 0.36[c]	–	MeCN	226
$\|Cu(TIM)\|^{2+}$	–	– 0.62[c,d]	–	H_2O	226
$\|Cu(TIM)\|^{2+}$	–	– 0.43[c,d]	–	MeOH	226
$\|Cu(TIM)\|^{2+}$	–	– 0.20[c]	–	CH_2Cl_2	226
$Cu(TAAB)(OMe)_2$	–	– 0.21[b]	–	MeOH	227
$\|Cu(TAAB)\|^{2+}$	–	+ 0.06[b]	–	MeOH	227
$\|Cu(TAAB)\|^{2+}$	–	+ 0.11[a]	–	MeOH	228
$\|Cu(TAAB)\|^{2+}$	–	+ 0.05[a,b]	–	DMF	222, 228
$\|Cu(TAAB)\|^{2+}$	–	+ 0.24[a]	–	NB	228
$\|Cu(TAAB)\|^{2+}$	+ 1.59[a]	+ 0.13[a]	– 1.13[a]	MeCN	227
$\|Cu(TAAB)\|^{2+}$	–	+ 0.13[a]	–	MeCN	228
$\|Cu(TAAB)\|^{2+}$	–	+ 0.05[a]	–	H_2O	228
$\|Cu(TAAB)\|^{2+}$	–	+ 0.09[a]	–	EtOH	228
$\|Cu(TAAB)\|^{2+}$	–	– 0.03[a]	–	DMSO	228

[a] Platinum electrode; [b] mercury electrode; [c] glassy carbon electrode; [d] peak potential value for irreversible process.

CuN_4 planarity is maintained $|230|$. Figure 56 shows the cyclic voltammetric response exhibited by $|Cu(TIM)|^{2+}$ in acetonitrile solution.

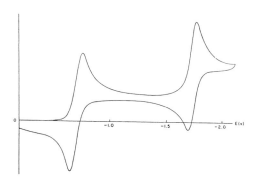

Fig. 56. Cyclic voltammogram recorded at a platinum electrode on a MeCN solution of $|Cu(TIM)|^{2+}$. Potential values vs. Ag/Ag^+ (from Ref. 223).

The first quasireversible reduction step reflects the copper(II)/copper(I) reduction, whereas the second step is assumed to be ligand centred $|223|$.

Please note how sterical effects can add to the electronic ones. In fact, considering the two isomeric complexes $|Cu(L_{48})|^{2+}$ and $|Cu(L_{50})|^{2+}$ having the same ligand atomicity, the presence of a strain inducing tetramethylenic chain in L_{50}, as contrasted to the more symmetrical L_{48}, leads to a significant lowering of the reduction potential of about 0.1 V $|220|$. The non-planarity of the CuN_4 coordination in these complexes is, on the other hand, proved by the X-ray structure of $|Cu(L_{49})|^{2+}$, which, though it crystallizes in a five-coordinate square pyramidal geometry with the fifth apical bond being constituted by one chloride anion, displaces the Cu atom out of the N_4 plane of 0.3 Å $|233|$.

Finally, the porphyrin-like $|Cu(TAAB)|^{2+}$, which has a complete double-bond conjugation, reduces at quite positive potential values, even if both the copper(II) and copper(I) forms are likely to have a substantially square-planar coordination $|231|$. In this connection, it must be recalled that the phenyl ring, methyl substituted |Cu(MeTAAB)|$^{2+}$ affords, during crystallization, the monocation |Cu(MeTAAB)(NO_3)|$^{+}$, in which a nitrate group act as a bidentate ligand. The coordination sphere around the copper centre becomes six-coordinate (N_4O_2), and the overall geometry is bicapped square pyramidal |232|; thus, being barely informative about the real structure of the dication in solution.

Another series of copper(II) macrocycles is shown in Chart XLVIII, in which the macrocycle closure is caused by a BF_2^+ or H^+ fragment.

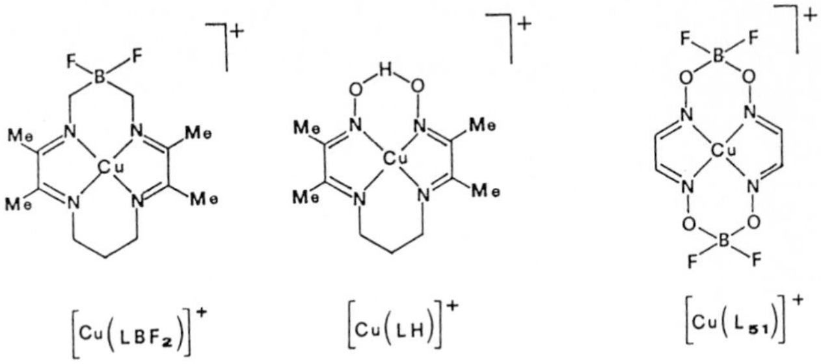

[Cu(LBF2)]+
[Cu(LH)]+
[Cu(L51)]+

CHART XLVIII

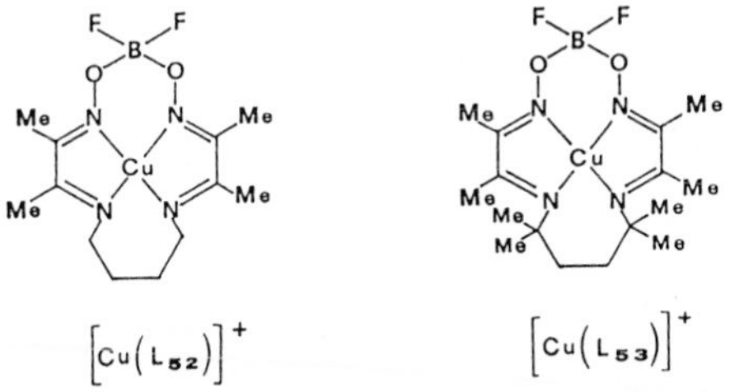

[Cu(L52)]+
[Cu(L53)]+

Figure 57 shows the cyclic voltammetric response exhibited by the copper(II) species |Cu(LBF_2)|$^{+}$ in acetonitrile solution |235|.

ic(μA)
200
100
0
-0.2
-0.4
-0.6
E(v)

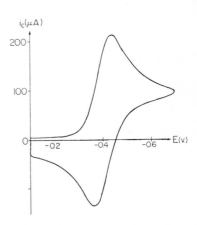

Fig. 57. Cyclic voltammogram recorded at a platinum electrode on a MeCN solution of |Cu(LBF_2)|$^{+}$. Scan rate 0.1 Vs^{-1} (from Ref. 235).

This quasireversible one-electron reduction step $(E^{o'} = -0.40$ V) is assigned to the couple $|Cu(LBF_2)|^+/Cu(LBF_2)$. Both, neither a further one-electron reduction $(E_{1/2} = -1.75$ V) nor a one-electron oxidation $(E_{1/2} = +1.05$ V), have been investigated in detail. The chemical reversibility of the first cathodic step is evident by the fact that constant potential electrolysis affords the crystalline copper(I) derivative $Cu(LBF_2)$ |234|.

Figure 58, which shows the molecular structure of both the copper(II) |237| and the copper(I) |236| complexes of the macro-cycle , in part supports the quasireversibility of the copper(II)/copper(I) reduction connected to the relevant stereochemical change. In fact $|Cu(LBF_2)|^+$, which normally would have to be square-planar $Cu-N_4$ coordinated, because of its crystallization as aquo complex $|Cu(LBF_2)(H_2O)|^+$, displays a square-pyramidal CuN_4O coordination, with the copper atom removed from the N_4 plane by 0.32 $\overset{o}{A}$. The presence of relatively weak strains in the macrocycle seems to be indicated by the fact that, in crystals of this copper(II) complex, both the boat and chair conformations (referred to the relative positions of the boron atom <u>versus</u> the opposite carbon atom C_4) are equally populated |237|. The dihedral angle between the two N_1CuN_2 and N_3CuN_4 planes is less than 4° |237|.

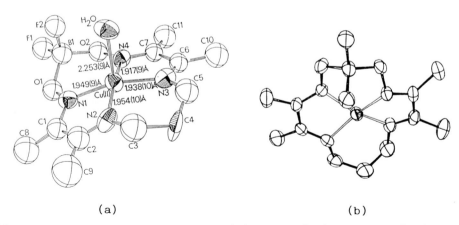

(a)   (b)

Fig. 58. Perspective views of: (a) copper(II) species $|Cu(LBF_2)(H_2O)|^+$. $Cu-N_{(mean)}$,1.94 $\overset{o}{A}$ (from Ref. 237); (b) copper(I) species $Cu(LBF_2)$. $Cu-N$, 1.94 $\overset{o}{A}$ (from Ref. 236).

Reduction to copper(I) is expected to induce some degree of tetrahedral distortion in the CuN_4 coordination. Experimentally, the dihedral angle between the two N_1CuN_2, N_3CuN_4 planes increases to 27°. As a consequence of the additional strain induced by this tetrahedral distortion, the macrocycle assumes the boat conformation, with the boron atom of the BF_2 bridge and the opposite carbon atom bent in the same direction [236]. The constancy of the Cu-N bond length upon one-electron addition suggests that the LUMO of the copper(II) complex possess entirely metal character.

Further support for the planarity of these copper(II) complexes comes from the X-ray structure of $|Cu(LH)|^+$ [237,238]; in it, the Cu atom is displaced out of the N_4 plane only 0.1 Å, probably because of a Cu-O bond with the oxime group of an adjacent molecule (Cu-O, 2.49 Å) [238], Figure 59, or with an apical water molecule (Cu-O, 2.36 Å) [237].

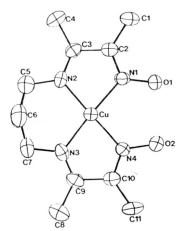

Fig. 59. Perspective view of the copper(II) cation $|Cu(LH)|^+$. Cu-N, 1.95 Å (from Ref. 238).

In comparison with $|Cu(LBF_2)|^+$, $|Cu(LH)|^+$ is more difficult to reduce by about 0.2 V ($E°'_{Cu^{II}/Cu^{I}}= -0.63$ V in MeCN [235]). This datum has been simply attributed to electronic effects, in that the bridging BF_2^+ is more electron withdrawing than H^+ [235].

Table 32 summarizes the redox potentials for the copper(II)/copper(I) couple of the complexes reported in Chart XLVIII.

TABLE 32

Redox potentials (in Volt) for the copper(II)/copper(I) couple of the complexes schematized in Chart XLVIII

complex	$E^{\circ\prime}$ Cu^{II}/Cu^{I}	solvent	reference
$\|Cu(LBF_2)\|^+$	-0.40	MeCN	223, 234, 235
$\|Cu(LBF_2)\|^+$	-0.29	Me_2CO	234
$\|Cu(LBF_2)\|^+$	-0.40	MeOH	235
$\|Cu(LBF_2)\|^+$	-0.51	MeOH	224
$\|Cu(LBF_2)\|^+$	-0.33	$MeNO_2$	235
$\|Cu(LBF_2)\|^+$	-0.40	DMF	222
$\|Cu(LBF_2)\|^+$	-0.16	CH_2Cl_2	224
$\|Cu(LH)\|^+$	-0.63	MeCN	235
$\|Cu(LH)\|^+$	-0.64	MeOH	224
$\|Cu(LH)\|^+$	-0.56	DMF	222
$\|Cu(LH)\|^+$	-0.43	CH_2Cl_2	224
$\|Cu(LH)\|^+$	-0.49	CH_2Cl_2	225
$\|Cu(L_{51})\|^+$	-0.38	DMF	222
$\|Cu(L_{52})\|^+$	-0.21	DMF	222
$\|Cu(L_{53})\|^+$	-0.11	DMF	222

It is clearly evident that the introduction of a planarity-distorting tetramethylenic chain in $\|Cu(L_{52})\|^+$, or tetramethyl substituted tetracarbon chain in $\|Cu(L_{53})\|^+$, makes the reduction more and more facile. Additional evidence for this view is the fact that the copper(II) complexes $\|Cu(TIM)\|^{2+}$ (see Table 31), $\|Cu(LBF_2)\|^{2+}$ and $\|Cu(L_{51})\|^+$ exhibit quite similar redox potentials. It may be thought that both their sterical resemblance and their electronically similar coordination, by four nitrogens, are responsible for this result |222|.

The last member of $Cu-N_4$ macrocycles here considered is the copper(I)-catenate $\|Cu(L_{54})\|^+$ schematized in Chart XLIX.

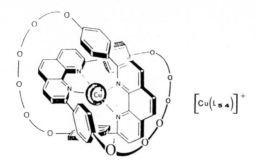

$$\left[Cu(L_{54})\right]^+$$

CHART XLIX

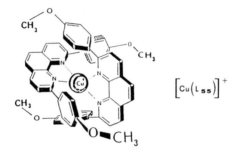

$$\left[Cu(L_{55})\right]^+$$

As shown in Figure 60 (solid line), $|Cu(L_{54})|^+$ undergoes, in DMF solution at a mercury electrode, a chemically reversible one-electron reduction ($E^{o\prime}$= -1.63 V) |239,240|.

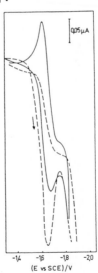

Fig. 60. Cyclic voltammograms recorded at a mercury electrode on DMF solutions of:(——) $|Cu(L_{54})|^+$; (- - -) $|Cu(L_{55})|^+$. Scan rate 0.01 Vs^{-1} (from Ref. 239).

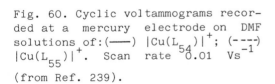

The corresponding, formally copper(0)-catenate (but likely a copper(I) radical anion |240|) resulted stable. By contrast, the cathodic reduction of the open-chain analogue $|Cu(L_{55})|^+$ (dashed line) is complicated by a fast demetallation process which makes irreversible the electrochemical step |239,240|. This example illustrates the stabilizing role of this spectacular catenand arrangement towards electron-rich centers.

As expected for a copper(I) species, the CuN_4 coordination in $|Cu(L_{54})|^+$ is of tetrahedral type (Figure 61); however, the strain produced by the interlocking of the two macrocycles causes a significant distortion, as deducible from a dihedral angle of about 60° between the two phenanthroline planes |241|.

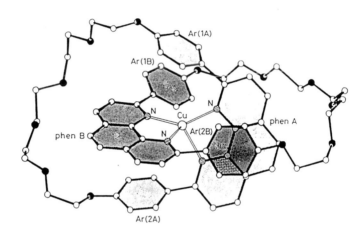

Fig. 61. Perspective view of $|Cu(L_{54})|^+$. Cu-N$_{(averaged)}$, 2.04 Å (from Ref. 241).

We ask ourselves the reasons for which the copper(I)/copper(II) transition has not been observed. We have previously examined the redox behaviour of the somewhat related phenanthroline complexes, together with a precursor of the present macrocycle (Section 4.2.9), both of which display quite high Cu(II)/Cu(I) redox potentials. Since the catenate structure of the complex is suited to force a pseudotetrahedral coordination to the corresponding copper(II) complex, we expect an even more positive redox potential for the $|Cu(L_{54})|^{+/2+}$ couple, likely in the anodic region which could have been obscured by the use of a mercury electrode.

4.3 N_5 donor set

Increasing interest is been devoted to copper complexes displaying a CuN_5 coordination and their redox activity.

4.3.1 Mixed aza functions. The redox behaviour of the copper(I) derivatives schematized in Chart L, in which the N_5 ligand is obtained by condensation of the Schiff-base 2,6-diacetyl-pyridine with histamines and/or 2-(2-aminoethyl)pyridine, has been studied |242-244|.

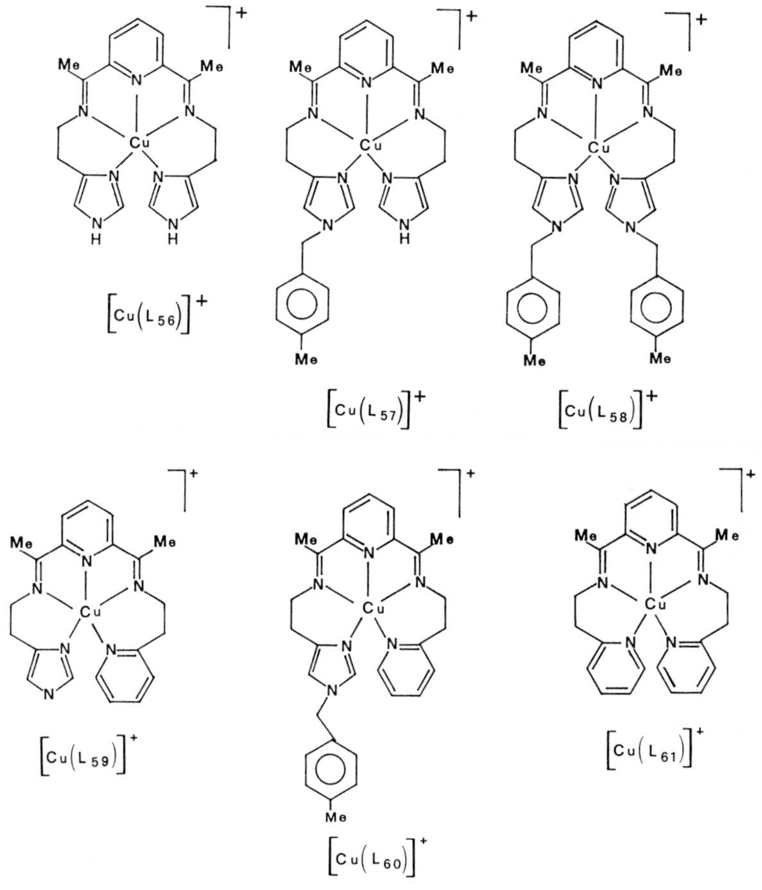

CHART L

These copper(I) complexes undergo an easy, quasi-reversible, one-electron removal to the corresponding copper(II) species.

Figure 62 shows, for instance, the cyclic voltammetric response for the oxidation of $|Cu(L_{61})|^+$ in acetonitrile (lower part), together with that for the reduction of the relevant copper(II) complex $|Cu(L_{61})|^{2+}$ (upper part).

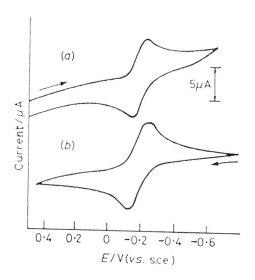

Fig. 62. Cyclic voltammograms recorded at a platinum electrode on a MeCN solution of: (a) $|Cu(L_{61})|^{2+}$; (b) $|Cu(L_{61})|^+$. $|NBu_4|ClO_4$ supporting electrolyte. Scan rate 0.2 Vs^{-1} (from Ref. 242).

Table 33 summarizes the redox potentials of the copper(II)/copper(I) couple for the CuN_5 complexes under discussion. As can be seen, the gradual substitution of imidazoles for pyridines makes the one-electron oxidation relatively more and more difficult. This has to be attributed to the great π-accepting ability of the pyridyl nitrogen, which, depleting the metal centre of electronic density to a higher extent than the imidazolyl nitrogen, thermo-dynamically disfavours the removal of electrons by the electrode.

TABLE 33

Redox potentials (in Volt) for the copper(II)/copper(I) couple of the complexes schematized in Chart L

complex[a]	$E^{\circ\,\prime}_{Cu^{II}/Cu^{I}}$	solvent	reference
$|Cu(L_{56})|^+$	− 0.30	MeCN	242
$|Cu(L_{56})|^+$	− 0.35	MeCN	243
$|Cu(L_{56})|^+$	− 0.27	DMF	242
$|Cu(L_{56})|^+$	− 0.27	DMSO	242
$|Cu(L_{56})|^+$	− 0.20	Py	242
$|Cu(L_{57})|^+$	− 0.33	MeCN	243
$|Cu(L_{58})|^+$	− 0.32	MeCN	243
$|Cu(L_{59})|^+$	− 0.27	MeCN	243
$|Cu(L_{60})|^+$	− 0.25	MeCN	243
$|Cu(L_{61})|^+$	− 0.20	MeCN	243
$|Cu(L_{61})|^+$	− 0.17	MeCN	242
$|Cu(L_{61})|^+$	− 0.13	DMF	242
$|Cu(L_{61})|^+$	− 0.15	DMSO	242

[a] some minor differences in redox potentials have been observed depending upon the counterion (ClO_4^- or BF_4^-) of the starting complexes |242, 243|.

It is now useful to offer a brief discussion of the structural aspects of the copper(I)/copper(II) redox change. The only X-ray structural informations available concern the copper(II) complex $|Cu(L_{56})|^{2+}$, which is isormophous with the zinc(II) species $|Zn(L_{56})|^{2+}$ shown in Figure 63 |245|.

296

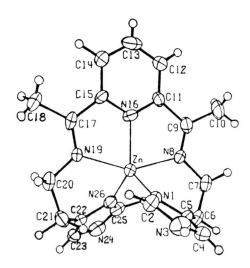

Fig. 63. Stereoscopic view of $|Zn(L_{56})|^{2+}$ (from Ref. 245).

The metal(II) coordination (zinc or copper) is described as intermediate between a trigonal bipyramid and a square pyramid $(Cu-N_{(mean)}, 2.02 \text{ Å}) |245|$. The near reversibility of the copper(I)/copper(II) electrochemical step allows us to foresee that a substantially similar geometry holds as far as the corresponding copper(I) complex is concerned. As a matter of fact, five-coordinate copper(I) species, even if rare, proved to exist, and the X-ray structures of copper(I) complexes such as $Cu(LBF_2)(CO) |234|$, $|Cu(dien)(C_7H_{10})|^+$ (dien = diethylenetriamine)$|246|$, as well as a somewhat related quinquedentate N_3S_2 copper(I) species $|247|$ have been reported. Indeed it is worth recalling here, that, as we shall elaborate in Section 7.5, in this latter complex the copper(II)/copper(I) reduction is accompanied by a square-pyramidal/trigonal-bipyramidal reorganization.

Another series of CuN_5 derivatives comes from complexes able to model the copper(II) complex of the antitumor antibiotic bleomycin (BLM) $|248,249|$. Chart LI just schematizes bleomycin and some related ligands.

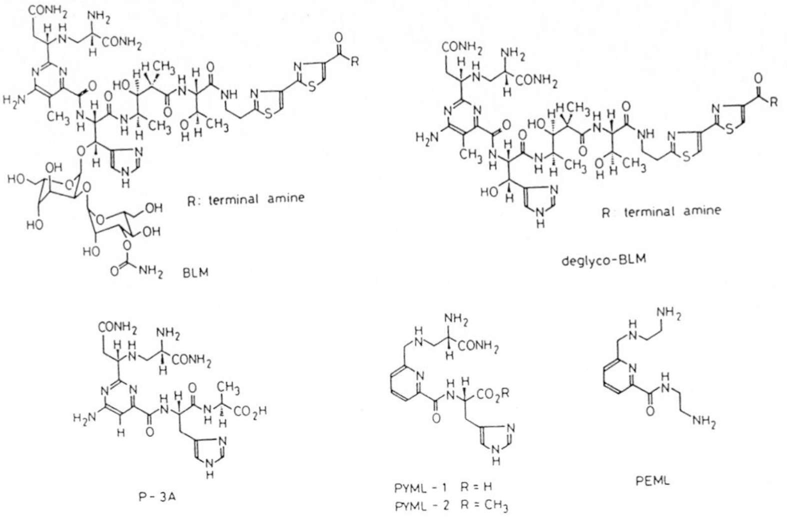

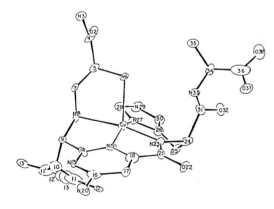

CHART LI

On the basis of the X-ray structure of the copper(II) complex
of P-3A, shown in Figure 64, a distorted square pyramidal CuN_5
coordination with the copper atom displaced out of the basal plane,
towards the apical nitrogen, is assigned to all the copper(II)
complexes of the ligands in Chart LI. This geometry is supported
also by the crystal structure of a somewhat similar synthetic
analogue of copper(II) bleomycin |249|.

Fig. 64. X-ray structure of Cu-P-3A. Cu-N$_{(averaged)}$ in the basal
plane, 2.04 Å; Cu-N$_{(apical)}$, 2.31 Å. The Cu atom is displaced out
of the basal plane of 0.20 Å (from Ref. 250).

All these complexes undergo a quasireversible copper(II)/ copper(I) reduction at moderately negative potential values (Table 34). On this basis, it seems not ventured to think a primary square-pyramidal/trigonal-bypiramidal conversion may accompany the one-electron addition, even if deeper reorganizations have been proposed on the basis of spectroscopic data (namely it seems that the deprotonated amide nitrogen of the ß-hydroxyhistidine residue is no longer metal bonding) |251|.

TABLE 34

Redox potentials (in Volt) for the copper(II)/copper(I) reduction of the copper(II) complexex schematized in Chart LI

complex	$E^{\circ\prime}_{Cu^{II}/Cu^{I}}$	solvent	reference
BLM - Cu(II)	- 0.57	H_2O (pH 7.2)	248
BLM - Cu(II)	- 0.58	H_2O (pH 8.0)	252
deglyco-BLM - Cu(II)	- 0.42	H_2O (pH 7.2)	248
P-3A - Cu(II)	- 0.38	H_2O (pH 7.2)	248
PYML-1 - Cu(II)	- 0.56	H_2O (pH 7.2)	248
PEML - Cu(II)	- 0.30	H_2O (pH 7.2)	248

4.3.2 Tripodal ligands. As discussed in Section 4.2.14 one-electron oxidation of the copper(I) complex with a tripodal ligand, $|Cu(L_{17})|^+$, favours the coordination of a fifth ligand. As examples, the two five-coordinate copper(II) complexes schematized in Chart LII, displaying a CuN_5 coordination, have been synthesized by introducing a methylimidazole molecule in the coordination of copper(II) with a tripodal N_4 ligand |253|.

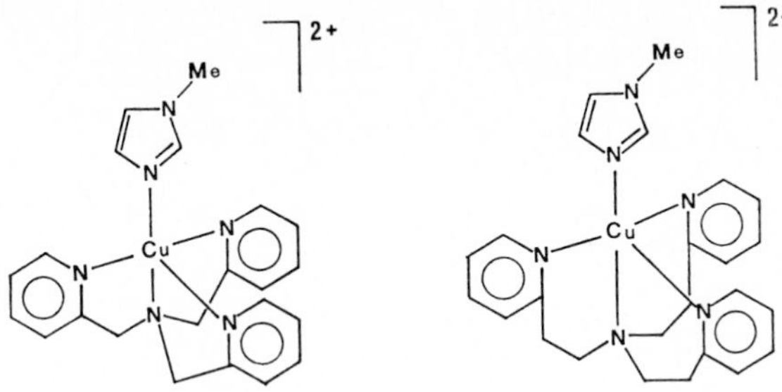

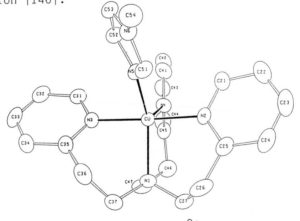

CHART LII

$$\left[Cu(L_{62})\right]^{2+} \qquad \left[Cu(L_{63})\right]^{2+}$$

It has been briefly reported that both the complexes undergo, in dimethylformamide solution, the copper(II)/copper(I) reduction at -0.46 V for $|Cu(L_{62})|^{2+}$, and at +0.02 V for $|Cu(L_{63})|^{2+}$ |146, 253|.

Figure 65 shows the square pyramidal geometry of the copper centre in $|Cu(L_{63})|^{2+}$ with the imidazolyl nitrogen occupying a basal position |146|.

Fig. 65. X-ray structure of $|Cu(L_{63})|^{2+}$. Cu-N$_{(mean)}$, 2.07 Å (from Ref. 146).

Even if electrochemical information is scanty, the quasi-reversibility of the reduction step allows us to hypothesize that the corresponding five-coordinate copper(I) complex may be stable, at least in the time scale of cyclic voltammetry. We presume, in fact, that the instantaneous release of the methylimidazole molecule upon one-electron addition, to give the previously discussed, well characterized, $|Cu(L_{17})|^+$, should have imparted instrumental responses characteristic of charge transfers complicated by subsequent reactions.

The strong impact on the redox potentials caused by the presence of lateral methylenic chains in $|Cu(L_{62})|^{2+}$ vs. lateral ethylenic chains in $|Cu(L_{63})|^{2+}$ must be emphasized. The origin of the difference of about 0.5 V may likely be found in the geometrical differences between the two copper(II) compounds; in fact, five-coordinate complexes of the tripodal N_4 ligand present in $|Cu(L_{62})|^{2+}$ are commonly trigonal bypyramidal, whereas those of the N_4 ligand present in $|Cu(L_{63})|^{2+}$ are, as illustrated, square-based pyramidal |254|.

4.3.3 Macrocycles. Chart LIII shows a series of five-coordinate CuN_5 copper(II)- macrocycles of known redox behaviour.

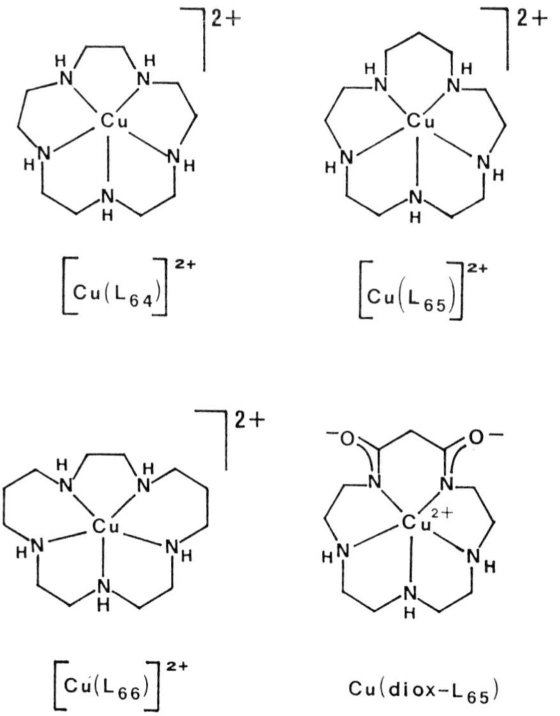

$$\left[Cu(L_{64})\right]^{2+} \qquad \left[Cu(L_{65})\right]^{2+}$$

$$\left[Cu(L_{66})\right]^{2+} \qquad Cu(diox\text{-}L_{65})$$

CHART LIII

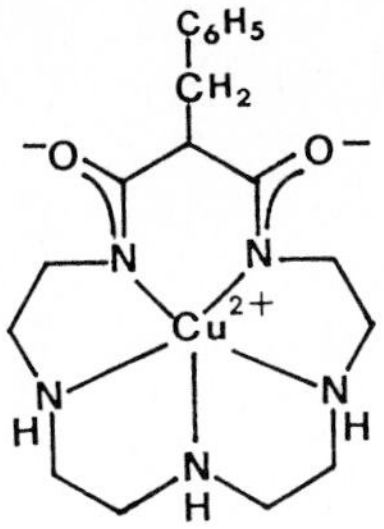

CHART LIII

Cu(L$_{67}$)

As in the case of CuN$_4$ macrocycles, these complexes undergo both a Cu(II)/Cu(I) reduction and a Cu(II)/Cu(III) oxidation. Table 35 reports the relevant redox potentials. Also in this case, the presence of two deprotonated amido groups significantly favours access to copper(III) with respect to saturated macrocycles. Provided that a comparison with the somewhat similar CuN$_4$ complexes can be made (see Tables 28 and 29), one can see that the likely square-pyramidal structure of the present five-coordinate copper(II) derivatives |256| does not appreciably improve the attainment of either copper(I) or copper(III) species.

4.4. N$_6$ donor set

The CuN$_6$ cromophore has been the object of intensive spectroscopic and structural investigations, but decidedly lacks redox information.

Copper(II) complexes having a CuN$_6$ coordination have essentially an octahedral geometry, either elongated or compressed as a consequence of dynamic/static Jahn-Teller distortions |see for instance Refs. 257-260|.

Electrochemical behaviour of the copper(II) complexes schematized in Chart LIV has been reported.

TABLE 35

Redox potentials (in Volt) for the redox changes of the copper(II) complexes schematized in Chart LIII

complex	$E°'_{Cu^{III}/Cu^{II}}$	$E°'_{Cu^{II}/Cu^{I}}$	$E_{p\,Cu^{I}/Cu^{0}}$	$E^{a,b}_{Cu^{II}/Cu}$	solvent	reference
$|Cu(L_{64})|^{2+}$	–	–	–	– 0.68	H_2O (pH 9.5)	255
$|Cu(L_{64})|^{2+}$	–	– 0.57[b]	–	–	H_2O (pH 9.6)	186
$|Cu(L_{64})|^{2+}$	+ 1.50[c]	– 0.73[c]	– 1.35[c]	–	MeCN	189
$|Cu(L_{65})|^{2+}$	–	–	–	– 0.65	H_2O (pH 9.5)	255
$|Cu(L_{65})|^{2+}$	–	– 0.56[b]	–	–	H_2O (pH 9.5)	186
$|Cu(L_{65})|^{2+}$	+ 1.56[c]	– 0.78[c]	– 1.45[c]	– 0.82	MeCN	189
$|Cu(L_{66})|^{2+}$	–	–	–	– 0.57	H_2O (pH 9.6)	255
$|Cu(L_{66})|^{2+}$	+ 1.62[c]	– 1.01[c]	– 1.31[c]	– 0.80	MeCN	189
$Cu(diox-L_{65})$	+ 0.68[c]	–	–	–	H_2O (pH 9.0)	256
$|Cu(L_{67})|^{2+}$	+ 0.74[c]	–	–	–	H_2O (pH 10.5)	256

[a]Peak potential value in cyclic voltammetry of half-wave potential value in polarography for irreversible processes; [b]mercury electrode; [c]platinum electrode.

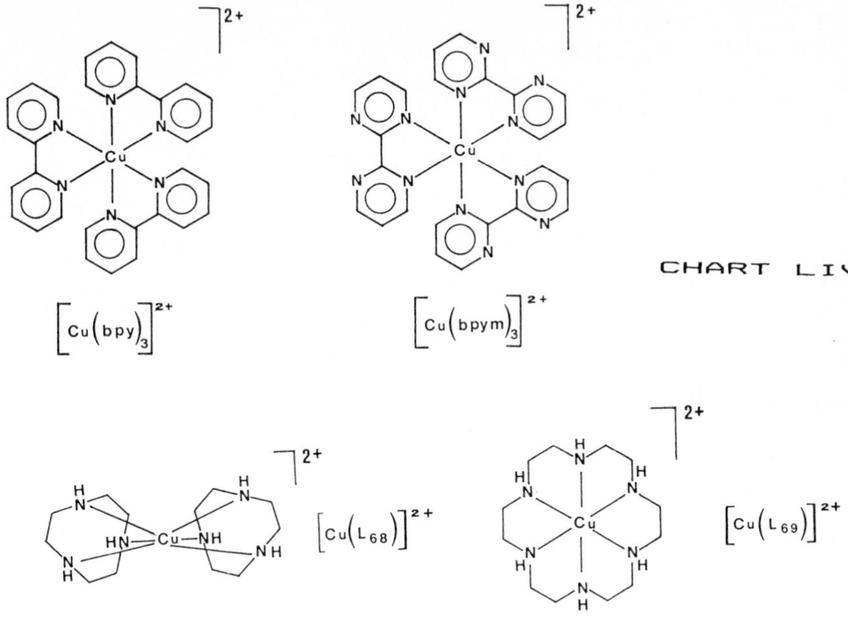

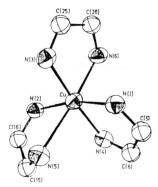

$$\left[Cu(bpy)_3 \right]^{2+} \qquad \left[Cu(bpym)_3 \right]^{2+}$$

CHART LIV

$$\left[Cu(L_{68}) \right]^{2+} \qquad \left[Cu(L_{69}) \right]^{2+}$$

Figure 66 shows the CuN_6 coordination polyhedron in the tris-bipyridil complex $|Cu(bpy)_3|^{2+}$.

Fig. 66. Perspective view of the core of $|Cu(bpy)_3|^{2+}$. Peripheral carbon atoms have been omitted (from Ref. 261).

The copper(II) site can be regarded as having an elongated octahedral geometry with four equivalent equatorial Cu–N bonds (Cu–N1, Cu–N2, Cu–N3, Cu–N4) of mean length 2.03 Å, and two longer but unequal apical Cu–N bonds (Cu–N5, 2.45 Å; Cu–N6, 2.23 Å) |261|.
$|Cu(bpy)_3|^{2+}$ is redox unstable. In aqueous solution the

copper(II)/copper(I) reduction is followed by fast ligand loss |262,263| according to:

$$|Cu(bpy)_3|^{2+} \xrightarrow{+e} |Cu(bpy)_2|^+ + bpy$$

The half-wave potential of this irreversible step ranges from -0.2 V at pH 7 to - 0.17 V at pH 5. In acetonitrile solution the decomposition of the tris-chelate complex is perhaps less quick, and an E°' value of -0.03 V is assigned to the couple $|Cu(bpy)_3|^{2+/+}$ |264|.

The very similar tris-bipyrimidine complex $|Cu(bpym)_3|^{2+}$ behaves essentially in the same manner, and in acetonitrile solution an E°' value of +0.24 V is assigned to the couple $|Cu(bpym)_3|^{2+/+}$. The more favourable reduction with respect to the dipyridine complex is attributed to an increased electron delocalization over the organic moiety |264|.

In this class of compounds the best redox characterization belongs to the bis-sandwich complex $|Cu(L_{68})|^{2+}$, formed by copper(II) with the tridentate macrocycle 1,4,7-triazacyclononane.

Two X-ray structures are available for this cation, with $|Cu(CN)_3|^{2-}$ |260| and $|ClO_4|^-$ |86| counteranions, respectively. Figure 67 just shows the structure of $|Cu(L_{68})|^{2+}$.

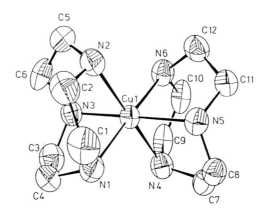

Fig. 67. Perspective view of the cation $|Cu(L_{68})|^{2+}$ at 293°K. Counteranion $|Cu(CN)_3|^{2-}$ (from Ref. 260).

It consists roughly of an elongated distorted octahedron, with the four equatorial Cu-N bonds having an averaged length of 2.11 Å |260| or 2.13 Å |86|, and the two apical Cu-N bonds of mean length 2.23 Å |86,260|.

$|Cu(L_{68})|^{2+}$, in acetonitrile, undergoes irreversibly the copper(II)/copper(I) reduction ($E°' = -1.03$ V). However, quite interestingly, it also undergoes quasi-reversibly the copper(II)/copper(III) oxidation at $E°' = +1.28$ V. This anodic step is complicated by successive decomposition of the copper(III) species ($t_{1/2} = 0.8$ sec at 19°C). The kinetic complication associated with the unstability of the copper(III) complex can be prevented either by the use of high sweep rates, or reducing the temperature. In this connection Figure 68 shows the anodic behaviour of $|Cu(L_{68})|^{2+}$ at -19°C |86|.

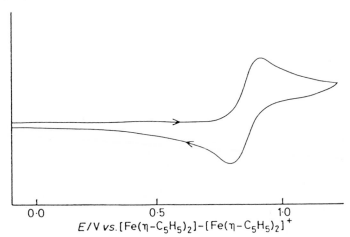

E/V vs. $[Fe(\eta-C_5H_5)_2]-[Fe(\eta-C_5H_5)_2]^+$

Fig. 68. Cyclic voltammogram of $|Cu(L_{68})|^{2+}$ in acetonitrile solution at 254°K. Platinum working electrode. Scan rate 0.26 Vs^{-1} (from Ref. 86).

The full chemical reversibility of this anodic process ($i_{pc}/i_{pa} = 1$) throws light upon the fact that octahedral geometries are not at all unaccessible to copper(III) species.

Finally, as far as the hexaaza-18-crown-6 complex $|Cu(L_{69})|^{2+}$ is concerned, it has been briefly reported that, at variance with the well defined redox changes occurring for the corresponding hexathia complex (see Section 3.4), in aqueous solution it undergoes irreversibly the copper(II)/copper(I) reduction |80|. Concerning this CuN_6 coordination, it is useful to note that the copper(II) complexes of the hexaazamacrocycle shown in Chart LV are thought to be coordinated only to the four secondary amine nitrogens, so that they behave electrochemically just like the tetraaza 14-membered $|Cu(sym-L_{28})|^{2+}$ |265|.

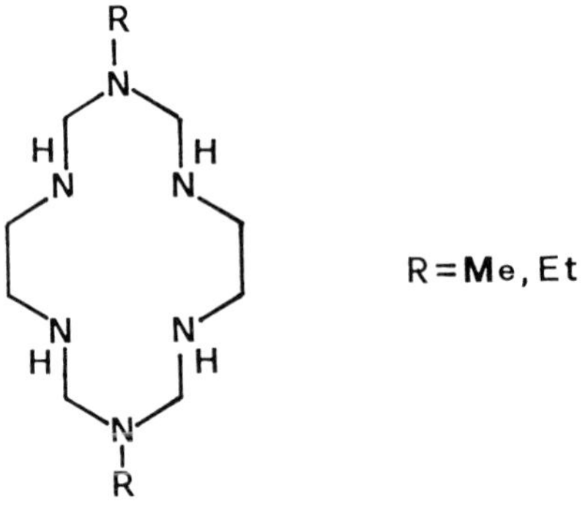

R = Me, Et

CHART LV

5. MIXED OXYGEN-SULFUR LIGANDS

5.1 O_2S_2 donor set

A series of copper(II)-thiohydroxamates, schematized in Chart LVI, have been investigated from the redox viewpoint.

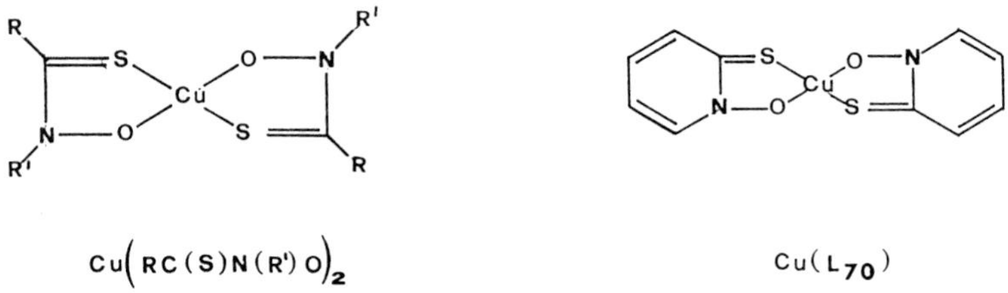

$Cu\left(RC(S)N(R')O\right)_2$

$Cu(L_{70})$

CHART LVI

As shown in Figure 69 for the complex with R = H, R'= Me, the structure of these complexes is substantially square planar, with the copper atom slightly displaced from the O_2S_2 plane |266|.

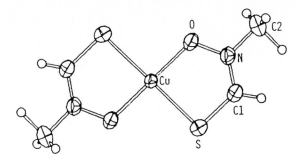

Fig. 69. X-ray structure of $Cu(HC(S)N(Me)O)_2$. Cu–S, 2.27 Å; Cu–O, 1.91 Å. Copper ion displaced out of the S_2O_2 plane of 0.13 Å (from Ref. 266).

These complexes undergo, in dimethylformamide solution, a reversible copper(II)/copper(I) reduction, and an irreversible copper(II)/copper(III) oxidation |266|. The relevant redox potentials are reported in Table 36.

TABLE 36

Redox potentials (in Volt) for the one-electron reduction and one-electron oxidation of the thiohydroxamate - copper(II) complexes schematized in Chart LVI in dimethylformamide solution |266|

complex	$E_{P_{Cu^{III}/Cu^{II}}}$	$E^{\circ'}_{Cu^{II}/Cu^{I}}$
$Cu(RC(S)N(R')O)_2$		

R	R'		
H	Me	+ 1.01	− 0.63
Ph	H	+ 0.97	− 0.56[a]
Ph	Me	+ 0.93	− 0.72
Ph	Ph	+ 1.04	− 0.60
$Cu(L_{70})$		+ 0.89	− 0.52

[a] Peak potential value for irreversible processes.

Given these results, some discrepancies exist between the electrochemical data and the structural features of these complexes. In fact, taken for granted the planarity of such species, the full electrochemical reversibility of the copper(II)/copper(I) reduction is puzzling in view of the planar ⇌ tetrahedral conversion, which should induce some degree of departure from reversibility to the heterogeneous electron transfer.

In addition the copper(II)/copper(III) oxidation should be favoured in view of the likely maintainement of the planar stereochemistry.

It must, however, be recognized that the reduced number of linkages within the ligand probably make these molecules reasonably flexible, so that tetrahedral distortions require minimal reorganization energies. This could favour reduction to copper(I), as well as disfavour stabilization of copper(III).

6.   MIXED OXYGEN-NITROGEN LIGANDS

6.1 ON$_2$ donor set

A limited number of copper(II) complexes, exhibiting a CuN$_2$O coordination, have been studied from the electrochemical viewpoint. Chart LVII schematizes such derivatives [267].

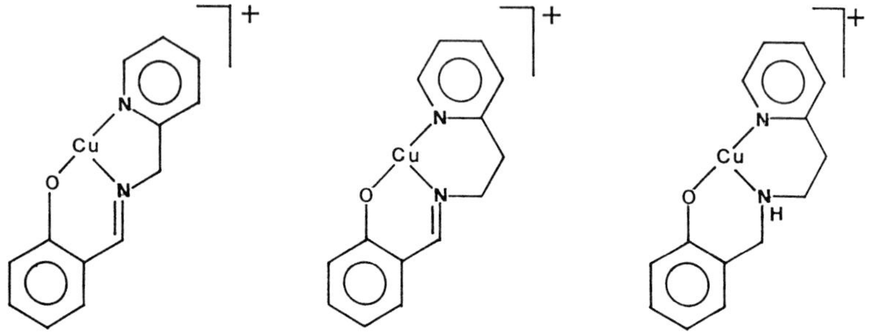

$$[Cu(SALIMP)]^+ \qquad [Cu(SALIEP)]^+ \qquad [Cu(SALAEP)]^+$$

CHART LVII

In dimethylformamide solution, these species undergo a quasireversible copper(II)/copper(I) reduction, complicated by following chemical reactions, indicating that the corresponding copper(I) complexes are unstable |267|. This redox change occurs at moderately negative potential values (|Cu(SALIMP)|$^+$, E_p = -0.56 V; |Cu(SALIEP)|$^+$, E_p = -0.36 V; |Cu(SALAEP)|$^+$, E_p = -0.38 V).

Figure 70 shows the X-ray structure of Cu(SALIMP)(NO$_3$).

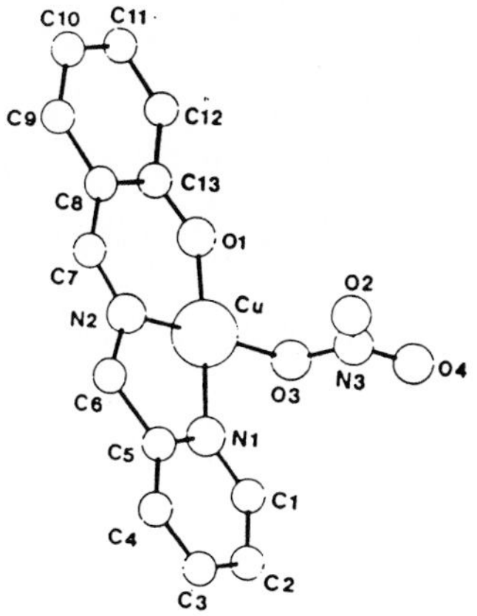

Fig. 70. Crystal structure of Cu(SALIMP)(NO$_3$). Cu-N(mean), 1.96 Å; Cu-O1, 1.87 Å; Cu-O3, 2.01 Å (from Ref.267).

The coordination to a NO$_3^-$ counteranion leads to an essentially planar CuN$_2$O$_2$ assembly. In solution however the NO$_3^-$ ion is not coordinated; instead it is thought that solvent molecules may complete the coordination sphere |267|. This copper(II) geometry experiences some difficulty in rearranging to the copper(I) tetrahedral one (note however that trigonal planar copper(I) complexes are stable |268|). Increasing the N   N carbon chain length to |Cu(SALIEP)|$^+$ and/or |Cu(SALAEP)|$^+$ likely causes tetrahedral distortion in the copper(II) coordination, thereby favouring thermodynamically access to the still kinetically labile copper(I) complexes.

6.2 O$_2$N$_2$ donor set

CuN$_2$O$_2$ coordination is very common in copper chemistry, and the redox behaviour of a wide series of mononuclear copper(II)-Schiff base complexes has been pioneeringly reviewed some years ago in connection with their structural changes |120|.

Let us start with the bis-chelate copper(II) complexes of salicylaldimines and ß-ketoamines schematized in Chart LVIII.

Cu(R-sal)$_2$ Cu(R-PhHMe)$_2$

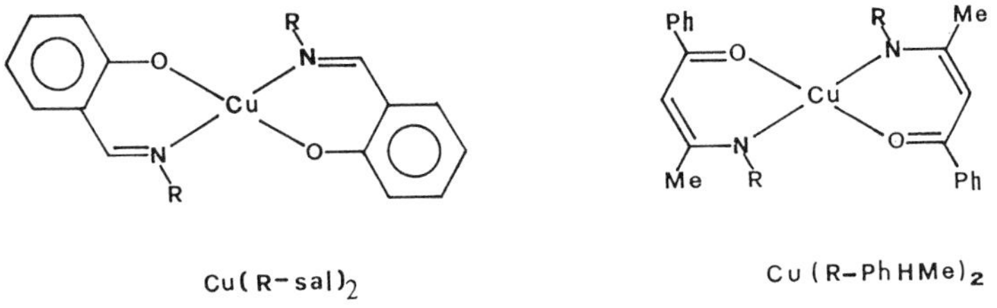

CHART LVIII

These complexes undergo quasi-reversibly the copper(II)/copper(I) redox change; in some cases, the relevant copper(I) species are unstable and subsequently decompose to copper(0) metal [120]. Table 37 summarizes the redox potentials for this one-electron reduction.

TABLE 37

Redox potentials (in Volt) for the copper(II)/Cu(I) reduction of the copper(II) complexes schematized in Chart LVIII in dimethyl-formamide solution [120]

complex	$E^{\circ\,\prime}_{Cu^{II}/Cu^{I}}$
Cu(R-sal)$_2$	
R	
Me	− 0.90
Et	− 0.86
i-Pr	− 0.74
t-Bu	− 0.66
Cu(R-PhHMe)$_2$	
R	
Me	− 0.96
i-Pr	− 0.78

As it can be seen, notwithstanding the increase of electron donating ability of N-substituents, in the order Me<Et<i-Pr<t-Bu, the redox process becomes easier and easier. It has been recognized that this is the result of favourable stereochemical assemblies. In fact, when the alkyl substituent is of minor steric bulk (Me, Ph, n-Bu) the CuN_2O_2 coordination is substantially square planar. Increasing the bulkiness of the substituents (i-Pr, t-Bu) is accompanied by significant tetrahedral distortions. For instance, in the $Cu(R-sal)_2$ complexes, the dihedral angle between the two CuNO planes is: 0° for R = Me; 35.6° for R = Et (monoclinic form); 60° for R = i-Pr, and 62° for R = t-Bu. The same tetrahedral distortion holds as far as the $Cu((i-Pr)-PhHMe)_2$ versus the planar $Cu(Me-PhHMe)_2$ (and the X-ray determined $Cu(H-PhHMe)_2$ |277|) are concerned |120|.

The polarographic behaviour of many other complexes of these types has been examined in pyridine-H_2O solution |13|, but the results are decidedly less straightforward.

Consider now the tetradentate copper(II) complexes of salicylaldimines and related ligands schematized in Chart LIX.

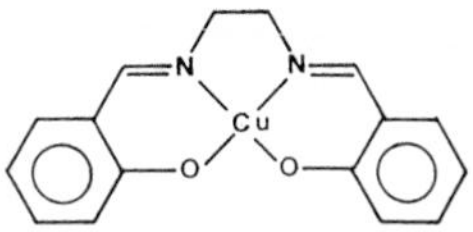

Cu(sal)₂ en

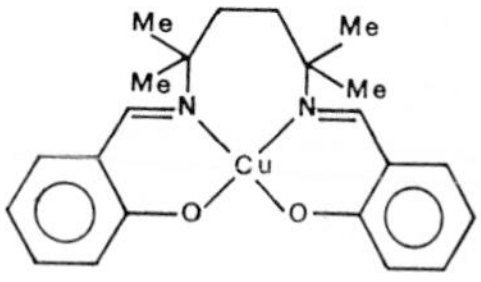

Cu(sal)₂ tmput

CHART LIX

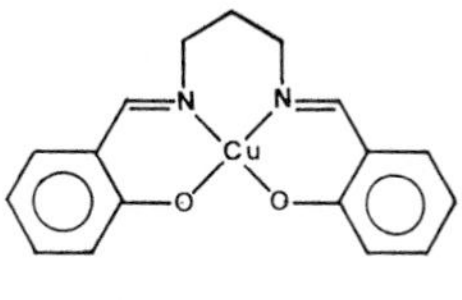

Cu(sal)₂-1,3-pn

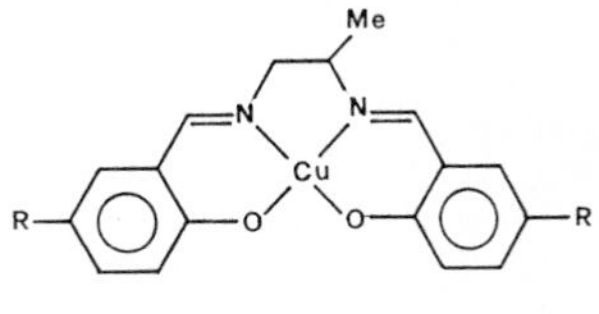

Cu(R-sal)₂-1,2 pn

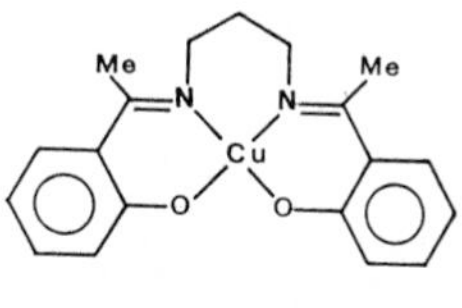

Cu(apim)₂-1,3-pn

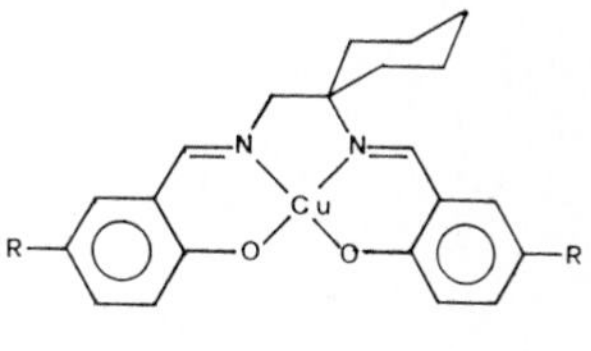

Cu(R-sacda)

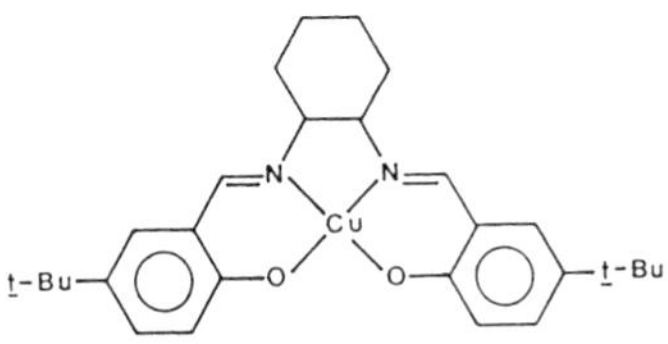

312

Cu(bstc)

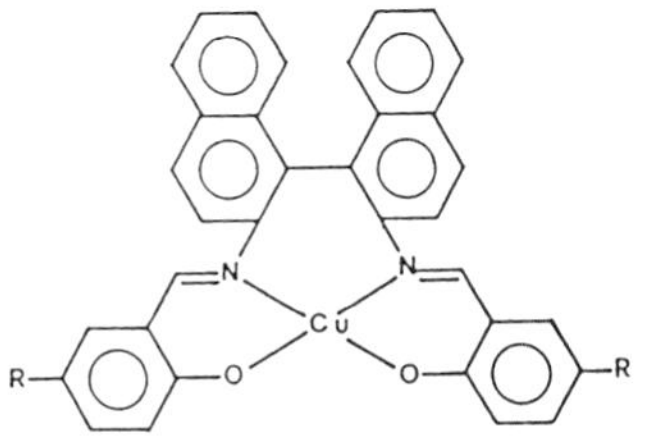

Cu(R-sal)₂(dinaph)

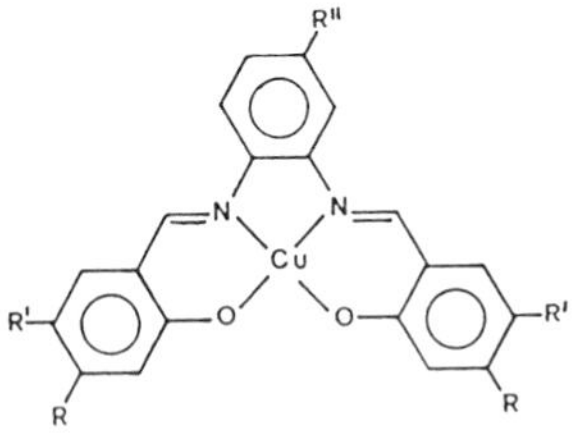

Cu(R-sal)$_2$(dinaph)

Cu(R R'-sal)$_2$(o-phen-R'')

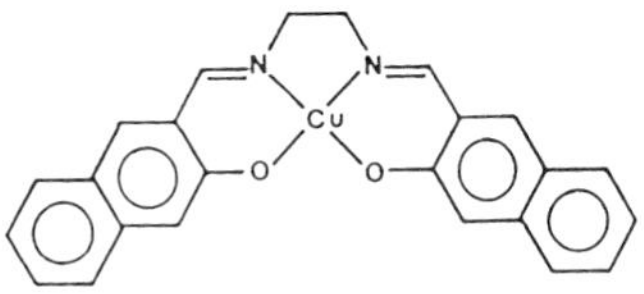

Cu(naph)$_2$ en

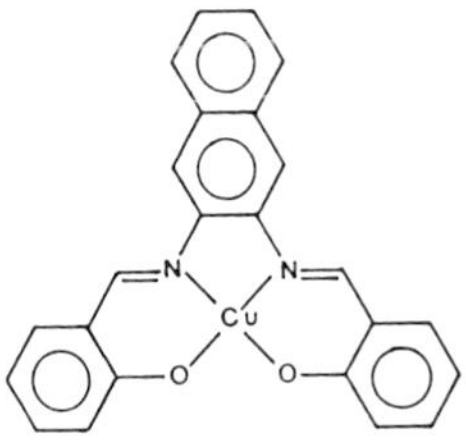

Cu(sal)$_2$- 2,3-naphten

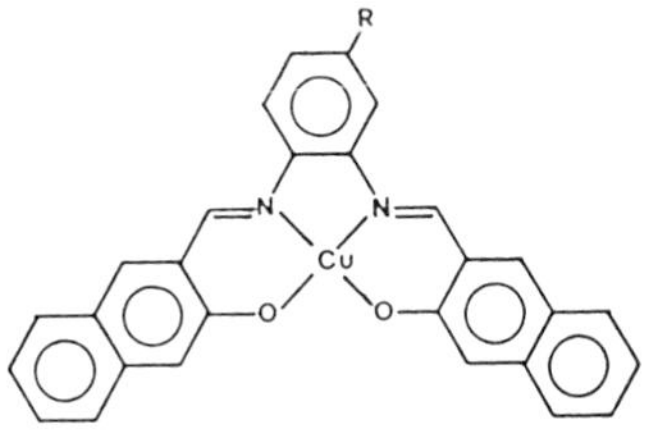

Cu(naph)$_2$(o-phen-R)

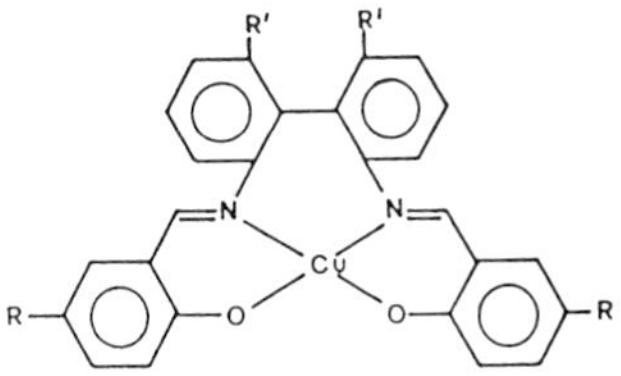

Cu(R-sal)$_2$(diphen- R')

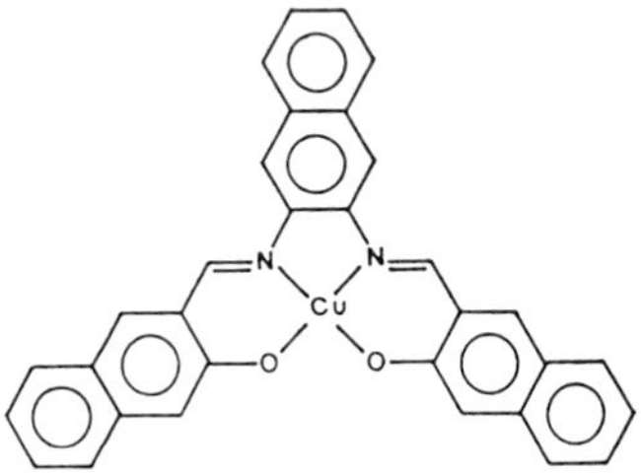

Cu(naph)$_2$(2,3-naphten)

CHART LIX

As usual, the most significant redox change is the copper(II)/ copper(I) reduction, which proceeds quasireversibly. As a typical example, Figure 71 shows the cyclic voltammogram exhibited by Cu(H-sal)$_2$(diphen-H).

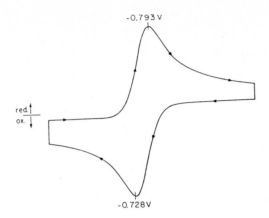

Fig. 71. Cyclic voltammetric response recorded on a DMF solution of Cu(H-sal)$_2$(diphen-H). Scan rate 0.1 Vs^{-1} (from Ref. 120).

In some cases, displayed in Table 38, the corresponding copper(I) complexes are unstable for longer times of electrolysis.

As shown in Table 38, the copper(II)/copper(I) redox change is generally located at rather negative potential values (more negative than -1 V). This datum, together with the quasireversibility of the electron transfer, agrees with the planar-to-tetrahedral reorganization following the reduction of such species. Cu(sal)$_2$en can be assumed as an example of these planar complexes |278|. Two exceptions are quite evident in Table 38; namely, Cu(sal)$_2$tmput and Cu(R-sal)$_2$(diphen-R') |and its analogue (R-sal)$_2$(dinaph)|; they are more easily reduced by at least 0.3 V. Also, this result nicely agrees with the geometrical features of these derivatives. In fact, X-ray structures of both Cu(sal)$_2$tmput |279| and Cu(H-H-sal)$_2$(diphen-H) |280| prove they have a tetrahedrally distorted CuN$_2$O$_2$ coordination.

314

TABLE 38

Redox potentials (in Volt) for the copper(II)/copper(I) reduction of the copper(II) complexes schematized in Chart LIX

complex	$E^{\circ\prime}$ Cu^{II}/Cu^{I}	solvent	reference
Cu(sal)$_2$en	-1.21^a	DMF	120
Cu(sal)$_2$en	-1.15^a	DMF	269
Cu(sal)$_2$en	-1.18^b	DMSO	270
Cu(sal)$_2$-1,3-pn	-1.04	DMF	271
Cu(sal)$_2$-1,3-pn	-0.99	DMF	130
Cu(sal)$_2$-1,3-pn	-1.06	DMF	273
Cu(sal)$_2$-1,3-pn	-1.3^c	DMSO	272
Cu(apim)-1,3-pn	-1.09	MeCN	274
Cu(sal)$_2$tmput	-0.67	DMF	130
Cu(H-sal)$_2$-1,2-pn	-1.22	CH$_2$Cl$_2$	275
Cu(Me-sal)$_2$-1,2-pn	-1.28	CH$_2$Cl$_2$	275
Cu($\underline{t}$-Bu-sal)$_2$-1,2-pn	-1.32	CH$_2$Cl$_2$	275
Cu(H-sacda)	-1.32	CH$_2$Cl$_2$	275
Cu(Me-sacda)	-1.34	CH$_2$Cl$_2$	275
Cu($\underline{t}$-Bu-sacda)	-1.35	CH$_2$Cl$_2$	275
Cu(bstc)	-1.37	CH$_2$Cl$_2$	276
Cu(H-H-sal)$_2$($\underline{o}$-phen-H)	-1.11^a	DMF	120
Cu(H-H-sal)$_2$($\underline{o}$-phen-H)	-1.08	DMF	269
Cu(H-H-sal)$_2$($\underline{o}$-phen-H)	-0.96	DMF	273
Cu(H-H-sal)$_2$($\underline{o}$-phen-H)	-1.12^b	DMSO	270
Cu(MeO-H-sal)$_2$($\underline{o}$-phen-H)	-1.19^b	DMSO	270
Cu(H-MeO-sal)$_2$($\underline{o}$-phen-H)	-1.14^b	DMSO	270
Cu(H-Cl-sal)$_2$($\underline{o}$-phen-H)	-1.03^a	DMSO	270
Cu(H-H-sal)$_2$($\underline{o}$-phen-Cl)	-1.08^a	DMSO	270
Cu(H-H-sal)$_2$($\underline{o}$-phen-OMe)	-1.15^b	DMSO	270
Cu(H-H-sal)$_2$($\underline{o}$-phen-Et)	-1.08^a	DMF	269
Cu(sal)$_2$-2,3-naphten	-1.06^a	DMF	269
Cu(H-sal)$_2$(diphen-H)	-0.75^a	DMF	120

TABLE 38 (continued)

complex	$E^{o'}_{Cu^{II}/Cu^{I}}$	solvent	reference
Cu(H-sal)$_2$(diphen-Me)	- 0.78^a	DMF	120
Cu($\underline{t}$-Bu-sal)$_2$(diphen-Me)	- 0.83	DMF	120
Cu($\underline{t}$-Bu-sal)$_2$(dinaph)	- 0.78	DMF	120
Cu(naph)$_2$en	- 1.16	DMF	269
Cu(naph)$_2$($\underline{o}$-phen-H)	- 1.05	DMF	269
Cu(naph)$_2$($\underline{o}$-phen-Et)	- 1.05	DMF	269
Cu(naph)$_2$-2,3-naphten	- 1.04	DMF	269

[a]Stable or [b]unstable in macroelectrolysis time scale; [c]peak potential value for irreversible process.

Figure 72 shows the molecular structure of Cu(sal)$_2$tmput.

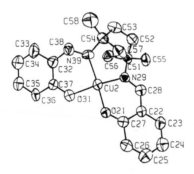

Fig. 72. Perspective view of Cu(sal)$_2$tmput. Cu-O, 1.90 Å; Cu-N, 1.98 Å (from Ref. 279).

The dihedral angle between the two phenyl wings is about 49°. In Cu(H-H-sal)$_2$(diphen-H) the angle is 37° (and, speculatively, this complex is slightly less easily reduced than Cu(sal)$_2$tmput).

Also, for a number of derivatives schematized in Chart LIX, less significant polarographic data in pyridine-H$_2$O solution |13, 281| have been reported.

We now continue our examination with the tetradentate

copper(II) complexes of ß-ketoiminates (and related ligands) schematized in Chart LX.

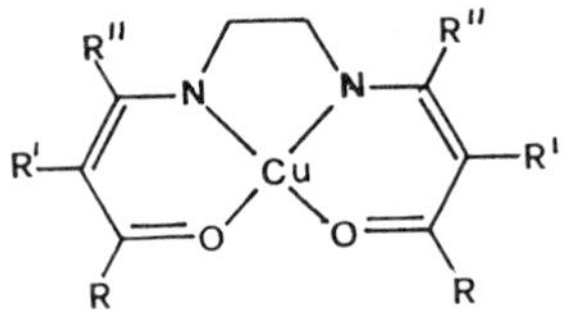

$Cu(R-R'-R'')_2 en$

CHART LX

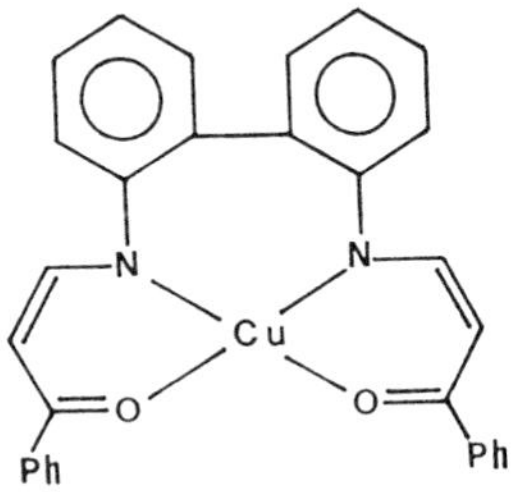

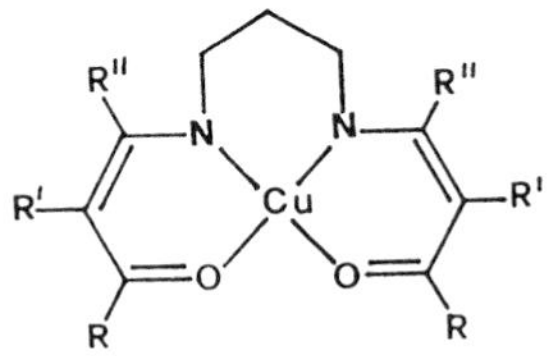

$Cu(R-R'-R'')-1,3-pn$

$Cu(Ph-H-H)_2(diphen)$

The redox behaviour of these complexes qualitatively parallels that of related ones in Chart LIX. Their planarity, typified by $Cu(Me-H-Me)_2en$ |285|, makes difficult access to copper(I), Table 39. In addition, the presence of peripheral electron donating groups, instead of the electron withdrawing aryl groups, present in the complexes shown in Chart LIX, makes even more difficult such reduction step. Further, these electronic effects are evident in the CF_3-substituted derivatives, in which the presence of the electron withdrawing trifluoromethyl groups dramatically lowers the reduction potential.

Passing from the ethylene bridged to the corresponding propylene bridged species, a significant lowering of the reduction potential occurs. It is likely that the increasing of the bridging carbon chain induces some distortion in the CuN_2O_2 coordination.

As expected, the non-planar complex $Cu(Ph-H-H)_2(diphen)$ exhibits the lowest reduction potential.

A commonly neglected point, which indeed deserves some attention, is the fact that the planarity of these complexes should favour access to copper(III) species.

TABLE 39

Redox potentials (in Volt) for the redox changes exhibited by the copper(II) complexes schematized in Chart LX

complex	$E^{\circ\prime}_{Cu^{III}/Cu^{II}}$	$E^{\circ\prime}_{Cu^{II}/Cu^{I}}$	solvent	reference
$Cu(Me-H-Me)_2en$	–	– 1.54[a]	DMF	120
$Cu(Me-H-Me)_2en$	–	– 1.58[b]	DMF	269
$Cu(Me-H-Me)_2en$	+ 0.80	– 1.60[b]	MeCN	282
$Cu(Ph-H-Me)_2en$	–	– 1.41	DMF	120
$Cu(Ph-H-Me)_2en$	–	– 1.40[b]	DMF	269
$Cu(CF_3-H-Me)_2en$	–	– 1.08	DMF	120
$Cu(Ph-H-H)_2en$	–	– 1.26	DMF	120
$Cu(Me-COCH_3-H)_2en$	–	– 1.09	DMF	283
$Cu(Me-H-Me)_2-1,3-pn$	–	– 1.22	DMF	120
$Cu(Ph-H-Me)_2-1,3-pn$	–	– 1.14	DMF	120
$Cu(CF_3-H-Me)_2-1,3-pn$	–	– 0.74	DMF	120
$Cu(Ph-H-H)_2-1,3-pn$	–	– 1.08	DMF	120
$Cu(Me-COOH_3-H)_2-1,3-pn$	–	– 0.98	DMF	284
$Cu(Ph-H-H)_2diphen$	–	– 0.82	DMF	120

[a]Stable electrogenerated copper(I) complex; [b]peak potential value for irreversible process.

Figure 73 shows the redox changes exhibited by $Cu(Me-H-Me)_2en$ in acetonitrile solution.

While the copper(II)/copper(I) reduction never shows an associated reoxidation response (even at 100 Vs^{-1}), indicating that the copper(I) complex is extremely labile, the copper(II)/copper(III) oxidation displays an associated rereduction response, which suggests that the copper(III) complex, even if labile, is able to exist ($t_{1/2} \cong 0.4s$) |282|.

318

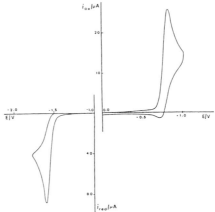

Fig. 73. Cyclic voltammograms recorded on a MeCN solution of Cu(Me-H-Me)$_2$en. Cathodic scan: mercury working electrode. Anodic scan: platinum working electrode. Scan rate: 0.2 Vs^{-1} (from Ref. 282).

Another series of CuN$_2$O$_2$ complexes is constituted by Schiff base ligands able to give potentially binucleating compartmental ligands |1|, partially discussed in Section 2.1. These copper(II) complexes are schematized in Chart LXI.

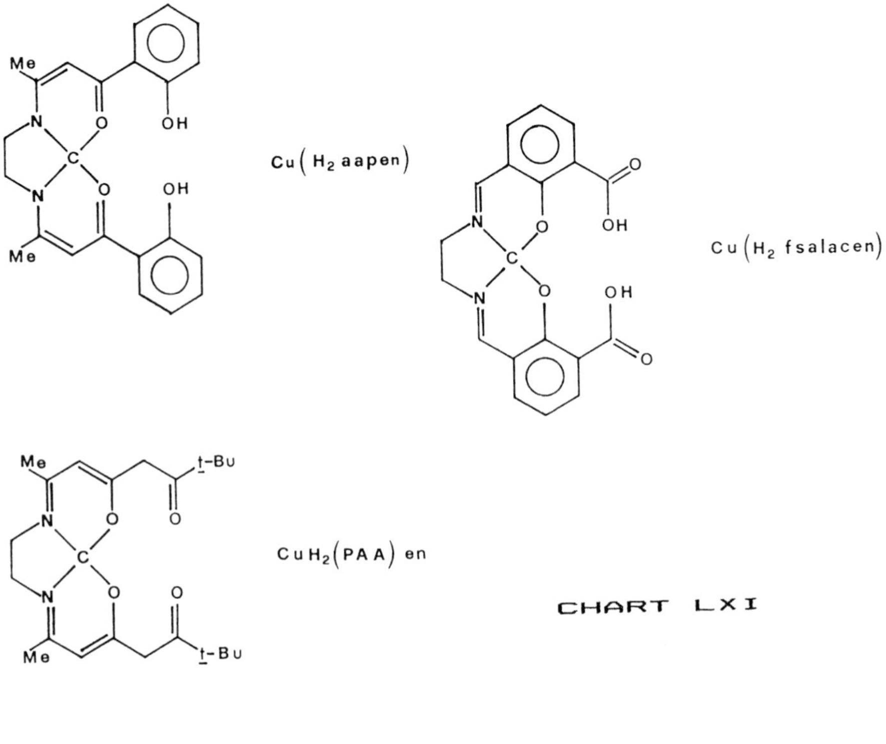

Cu(H$_2$ aapen)

Cu(H$_2$ fsalacen)

CuH$_2$(PAA) en

CHART LXI

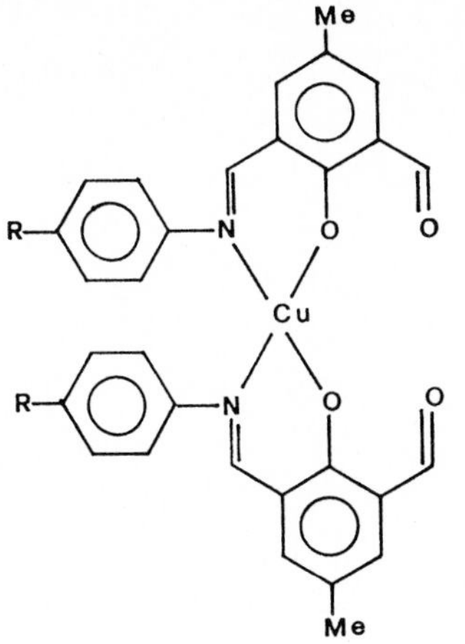

$Cu\left(R-L_{71}\right)$

CHART LXI

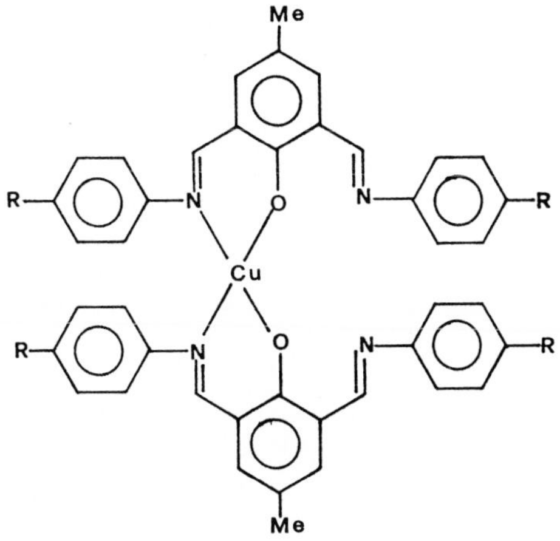

$Cu\left(R-L_{72}\right)$

The first three specimens of this series display a planar CuN_2O_2 coordination. In partial support of this hypothesis, the X-ray structure of binuclear Cu_2(aapen) shows a square planar CuN_2O_2 geometry (whereas the CuO_2O_2 chamber is square pyramidal because of a fifth coordination to a water molecule) |288|. This planar configuration is expected to disfavour the copper(II)/copper(I) reduction. Accordingly, Figure 74 shows the cathodic process of $Cu(H_2fsalacen)$.

320

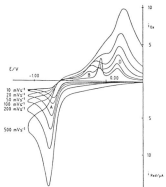

Fig. 74. Cyclic voltammograms at different scan rates recorded at a platinum electrode on a DMSO solution of Cu(H$_2$fsalacen). |NEt$_4$|ClO$_4$ (0.1 mol.dm^{-3}) supporting electrolyte (from Ref.286).

These redox profiles are quite common to copper complexes generating short-lived copper(I) species, and are indicative of an e.c.e. mechanism of the type discussed also in Section 4.2.18:

$$
\begin{array}{ccccc}
& \text{peak A} & & & Cu^{2+} \\
& & & & {-e \uparrow \text{peak D}} \\
Cu^{II}(H_2\text{fsalacen}) & \underset{-e}{\overset{+e}{\rightleftharpoons}} & |Cu^{I}(H_2\text{fsalacen})|^{-} & \longrightarrow & Cu^{+} + |H_2\text{fsalacen}|^{2-} \\
& & & & {+e \updownarrow -e \quad \text{peak C}} \\
& \text{peak B} & & & Cu^{0}
\end{array}
$$

At slow scan rates, the reduction of released copper(I) ions occurs before their reoxidation; at high scan rates the reoxidation of the released copper(I) ions is kinetically predominant over their deposition on the electrode surface.

The large departure from electrochemical reversibility, and hence the high molecular reorganization involved in the one-electron reduction, becomes evident by the separation (about 0.55 V) between peaks A and B. A half-life of about 4s |286| has been computed for |Cu^I(H$_2$fsalacen)|$^-$. In the case of Cu(H$_2$aapen) and CuH$_2$(PAA)$_2$en, their copper(I) complexes are much more unstable, and because of the presence of electron donating alkyl substituents they reduce with even more difficulty |18,28|.

Table 40 summarizes the relevant redox potentials.

TABLE 40

Redox potentials (in Volt) for the redox changes exhibited by the copper(II) complexes schematized in Chart LXI

complex	$E^{\circ\prime}_{Cu^{III}/Cu^{II}}$	$E^{\circ\prime}_{Cu^{II}/Cu^{I}}$	solvent	reference
$Cu(H_2aapen)$	+ 0.88[b]	- 1.19[a]	DMSO	28
$CuH_2(PAA)_2en$	-	- 1.2[c]	DMF	19
$Cu(H_2fsalacen)$	-	- 0.47	DMSO	286
$Cu(H-L_{71})$	-	- 0.52	DMF	287
$Cu(Me-L_{71})$	-	- 0.54	DMF	287
$Cu(MeO-L_{71})$	-	- 0.54	DMF	287
$Cu(Cl-L_{71})$	-	- 0.50	DMF	287
$Cu(COCH_3-L_{71})$	-	- 0.49	DMF	287
$Cu(NO_2-L_{71})$	-	- 0.46	DMF	287
$Cu(H-L_{72})$	-	- 0.56	DMF	287
$Cu(Me-L_{72})$	-	- 0.58	DMF	287
$Cu(MeO-L_{72})$	-	- 0.58	DMF	287
$Cu(Cl-L_{72})$	-	- 0.54	DMF	287
$Cu(COCH_3-L_{72})$	-	- 0.50	DMF	287

[a] Measured at a mercury electrode at 100 Vs^{-1}; [b] measured at a platinum electrode at 100 Vs^{-1}; [c] peak potential value for irreversible process.

In this case also, the planarity of the CuN_2O_2 moiety allows access to copper(III) species, even if it is transient.

Finally, as far as the complexes $Cu(R-L_{71})$ and $Cu(R-L_{72})$ are concerned, even though X-ray structures are not available, it is amply evident that steric repulsions between pendant aryl groups induce appreciable tetrahedral distortions in the CuN_2O_2 geometry, thus significantly stabilizing the copper(I) species. Indeed, the copper(II)/copper(I) reduction is described as a simple quasireversible step, and it occurs at moderately negative potential values |287|. Electronic effects from R-substituents linearly correlate with redox potentials, indicating full conjugation of the aryl groups with the metal centre.

A final CuN_2O_2 complex showing an unusual redox behaviour is the o-quinone monooxime-copper(II) species, one limiting form of which is schematized in Chart LXII.

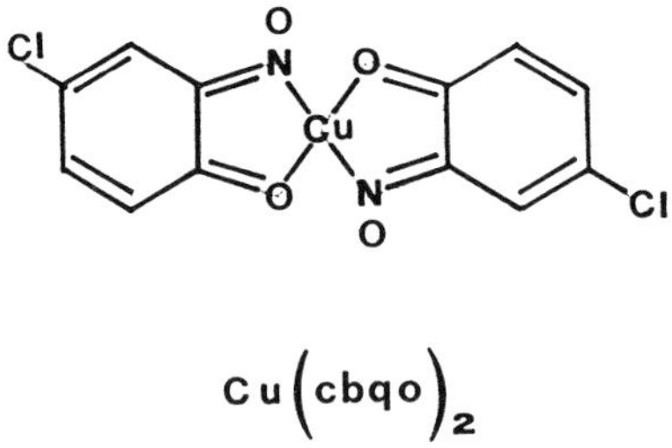

$$Cu\left(cbqo\right)_2$$

CHART LXII

In solution this complex likely displays a planar CuN_2O_2 assembly. In the solid state it easily forms adducts with Lewis bases. Figure 75 shows the X-ray crystal structure of $Cu(cbqo)_2 \cdot$ MeOH |289|.

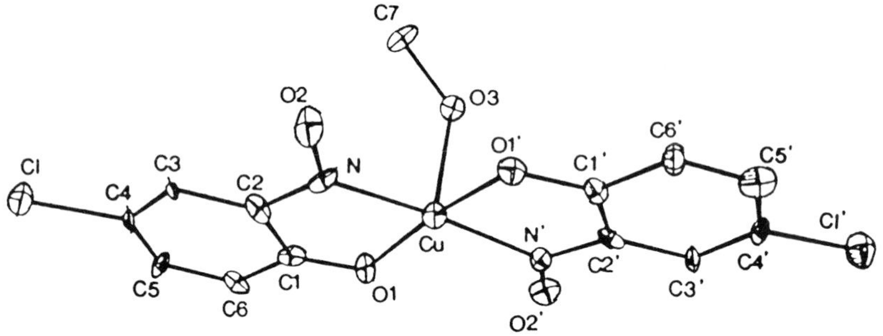

Fig. 75. Perspective view of $Cu(cbqo)_2 \cdot$ MeOH. Cu-N, 2.00 Å; Cu-O, 1.96 Å; Cu-O$_{(MeOH)}$, 2.27 Å (from Ref. 289).

In acetonitrile solution, $Cu(cbqo)_2$ undergoes in the cyclic voltammetric time scale two quasireversible, one-electron reduction processes at + 0.01 V and - 0.74 V, respectively |290|. While in the longer time of electrolysis, the most cathodic step revealed to be complicated by following chemical reactions (leading to unidentified products), the first one-electron reduction proved to be ligand centred. Spectroscopic data strongly support the formation of a quinone-semiquinone complex of the type schematized (one limiting form) in Chart LXIII |290|.

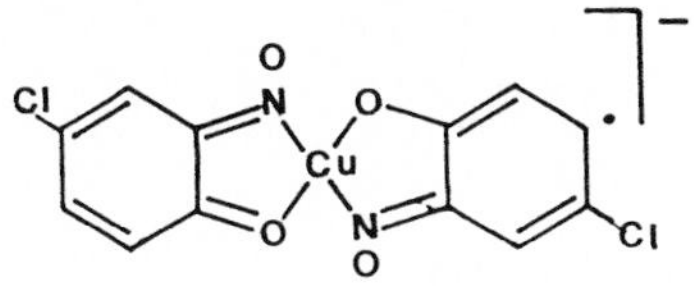

$$\left[Cu(cbqo)(cbsqo)\right]^-$$

CHART LXIII

The likely planar $(Cu^{II}(cbqo)_2)$ to planar $(|Cu^{II}(cbqo)$ $(cbsqo)|^-)$ redox change agrees well with the relatively high redox potential (the planar to tetrahedral copper(II)/copper(I) conversion should have been disfavoured), even if the quasireversibility of the electrochemical step could be indicative of some significant structural variation.

6.3 ON_3 donor set

The only report dealing with copper(II) complexes displaying a CuN_3O coordination concerns the derivatives schematized in Chart LXIV |130|.

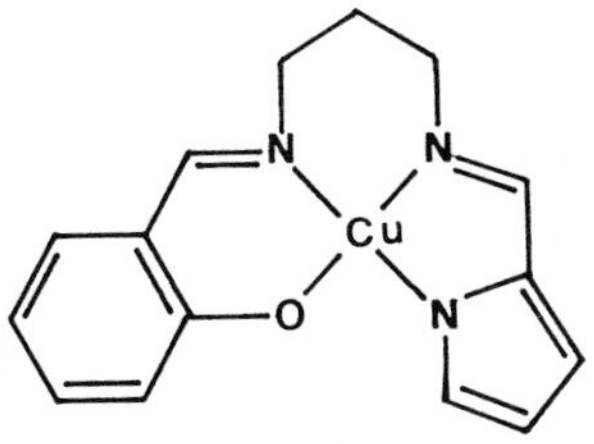

$Cu(slppl)$

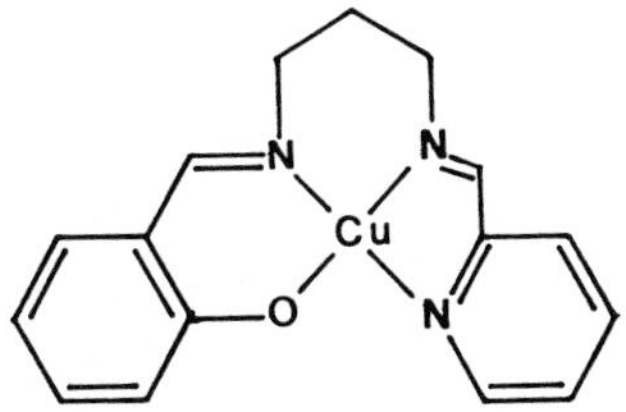

CHART LXIV

$$\left[Cu(slppy)\right]^+$$

324

As an example, Figure 76 shows the cyclic voltammetric response of the copper(II)/copper(I) reduction of |Cu(slppy)|$^+$.

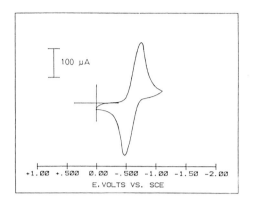

Fig. 76. Thin-layer cyclic voltammogram of |Cu(slppy)|$^+$ in dimethylformamide solution (from Ref. 130).

Even if the large peak-to-peak separation is in good part attributable to uncompensated solution resistance in the optically transparent thin-layer cell, the process is likely quasireversible, and the corresponding copper(I) species results quite stable |130|.

The following formal electrode potentials for the copper(II)/copper(I) couple have been computed:

$$E^{\circ\prime}{}_{Cu(slppl)/|Cu(slppl)|^-} = -0.98 \text{ V}$$
$$E^{\circ\prime}{}_{|Cu(slppy)|^+/Cu(slppy)} = -0.62 \text{ V}$$

On the basis of spectroscopic data, these copper(II) complexes are thought to be substantially square planar |291|. It has hence been remarked that, while the presence of a deprotonated pyrrolic nitrogen in Cu(slppl) or a deprotonated phenolic oxygen in Cu(sal)$_2$-1,3-pn (see Table 38) involve roughly the same energetic reorganizational barrier to the copper(II)/copper(I) reduction, the presence of a pyridine nitrogen in |Cu(slppy)|$^+$ notably lowers such a barrier. The electronic factor favouring the reduction step has been attributed to the ability of the pyridine nitrogen to stabilize copper(I) through back-bonding |130|.

6.4 O_2N_3 donor set

A large part of the copper complexes with potentially penta-dentate N_3O_2 ligands have been not characterized from the redox viewpoint |292|. Those studied electrochemically are schematized in Chart LXV.

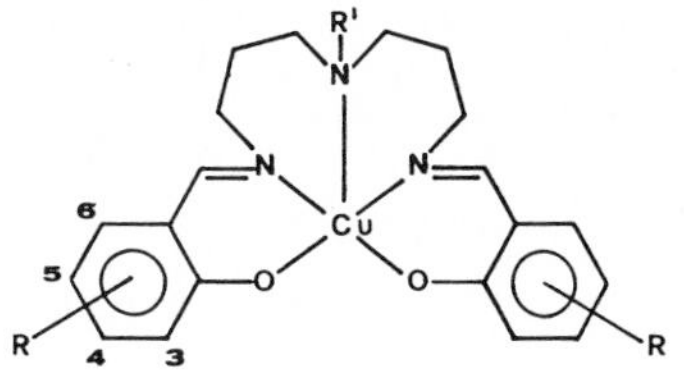

Cu(R-SAL-R¹-DPT)

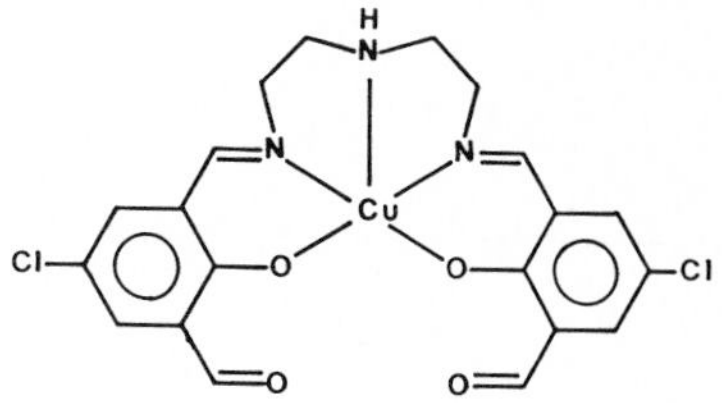

Cu(L₇₄)

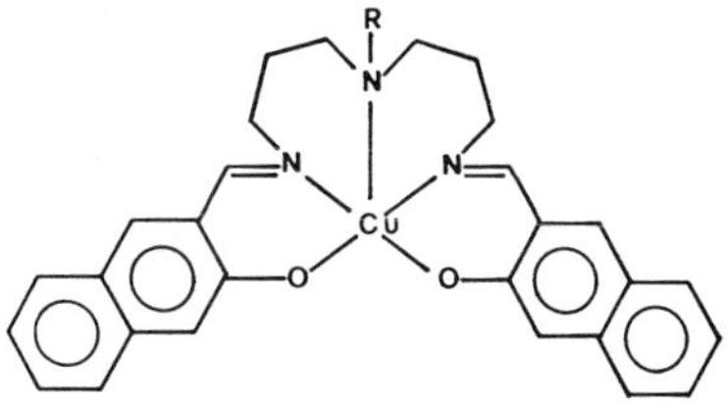

Cu(NAPH-R-DPT)

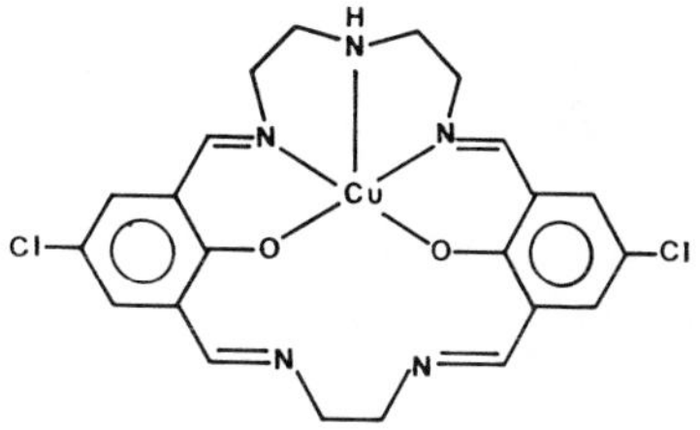

Cu(L₇₅)

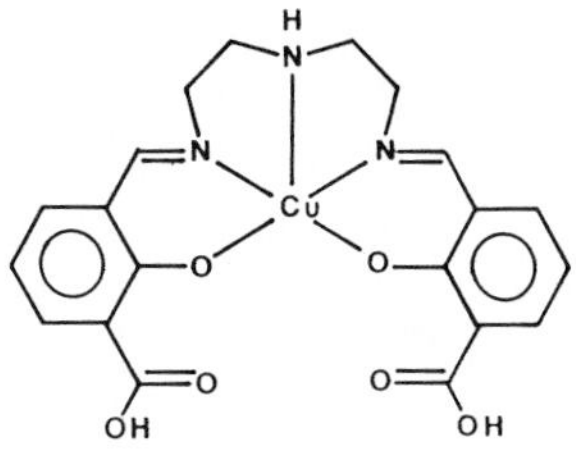

Cu(L₇₃)

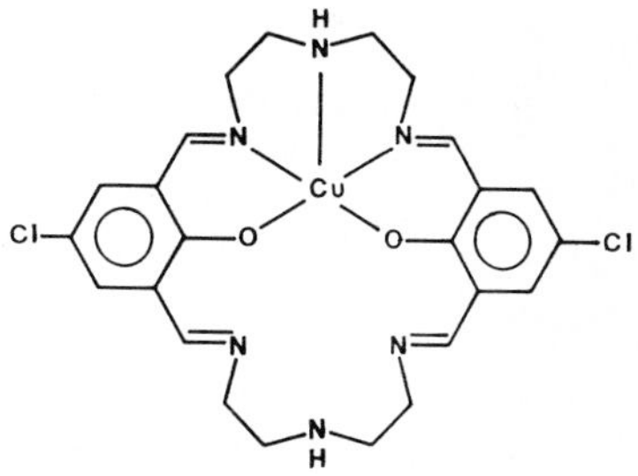

Cu(L₇₆)

CHART LXV

As shown in Figure 77, which refers to the cyclic voltammetric response of Cu(3MeO-SAL-H-DPT), these complexes are rather easily oxidizable, as well as exhibiting difficult reducibility. Both the copper(II)/copper(III) oxidation, and the copper(II)/copper(I) reduction are quasireversible processes complicated by subsequent chemical reactions, _i.e._ neither the corresponding copper(III) complexes, nor the copper(I) ones are fully stable |293-296|.

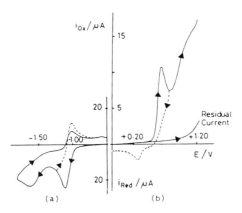

Fig. 77. Cyclic voltammetric responses exhibited by Cu(3MeO-SAL-H-DPT) in DMSO solution. (a) Mercury electrode; (b) platinum electrode. Scan rate 0.2 Vs^{-1} (from Ref. 293).

Table 41 summarizes the redox potentials for the complexes schematized in Chart LXV. Electronic effects from peripheral substituents account for the differences in electrode potentials (it has been shown that in complexes Cu(R-SAL-R'-DPT), depletion or enchancement of the charge on the metal by aryl substituents follows the way Ph-O-Cu-O-Ph |293|).

Significant exceptions are illustrated by $Cu(L_{73})$, $Cu(L_{74})$ and $Cu(L_{76})$ which reduce at potentials notably less negative than all other species. We have not sufficient stereochemical information to explain this datum. Copper(II) complexes of Cu(R-SAL-R'-DPT) type (and likely their naphthylideneimine analogues) have a five coordinate geometry intermediate between trigonal bipyramidal and square pyramidal |297,298|. The coordination of the copper(II) ion in the complexes $Cu(L_{73})$-$Cu(L_{76})$ should be the same. However, it

TABLE 41

Redox potentials (in Volt) for the electrode processes exhibited by the copper(II) complexes schematized in Chart LXV

complex	$E°'_{Cu^{III}/Cu^{II}}$	$E°'_{Cu^{II}/Cu^{I}}$	$E_{p\,Cu^{I}/Cu^{0}}$	solvent	reference
Cu(H-SAL-H-DPT)	+ 0.84 [a,b]	− 1.06 [c]	− 1.50 [c]	DMSO	293
Cu(H-SAL-H-DPT)	−	− 1.08 [d]	−	DMF	283
Cu(3MeO-SAL-H-DPT)	+ 0.43 [a]	− 1.05 [c]	− 1.62 [c]	DMSO	293
Cu(5MeO-SAL-H-DPT)	+ 0.47 [a]	− 1.07 [c]	− 1.55 [c]	DMSO	293
Cu(5Cl-SAL-H-DPT)	+ 0.69 [a]	− 0.96 [b,c]	− 1.35 [c]	DMSO	293
Cu(5NO$_2$-SAL-H-DPT)	−	− 0.80 [b,c]	− 1.05 [c]	DMSO	293
Cu(H-SAL-Me-DPT)	+ 0.71 [a]	− 1.02 [c]	− 1.46 [c]	DMSO	293
Cu(3MeO-SAL-Me-DPT)	+ 0.55 [a]	− 1.00 [c]	− 1.65 [c]	DMSO	293
Cu(5MeO-SAL-Me-DPT)	+ 0.54 [a]	− 1.02 [c]	− 1.50 [c]	DMSO	293
Cu(5Cl-SAL-Me-DPT)	+ 0.80 [a]	− 0.94 [c]	− 1.55 [c]	DMSO	293
Cu(5NO$_2$-SAL-Me-DPT)	−	− 0.73 [b,c]	−	DMSO	293
Cu(NAPH-H-DPT)	+ 0.75 [a,b]	− 1.11 [c]	− 1.30 [c]	DMSO	294
Cu(NAPH-Me-DPT)	+ 0.86 [a,b]	− 1.16 [c]	− 1.24 [c]	DMSO	294
Cu(L$_{73}$)	−	− 0.42 [c]	−	DMSO	295
Cu(L$_{74}$)	−	− 0.76 [c]	−	DMSO	296
Cu(L$_{75}$)	−	− 1.08 [c]	−	DMSO	296
Cu(L$_{76}$)	−	− 0.77 [c]	−	DMSO	296

[a] Platinum electrode; [b] peak potential value for irreversible process; [c] mercury electrode; [d] glassy carbon electrode.

cannot be excluded that in these complexes the ligand can act as tetradentate with a N_2O_2 coordination. This being true, $Cu(L_{73})$, $Cu(L_{74})$, and $Cu(L_{76})$ could be tetrahedrally distorted, whereas the less flexible nature of the macrocycle $Cu(L_{75})$ should impart the metal a planar coordination. The relative reorganizations to tetrahedral copper(I) species would hence be obvious. In this connection, it should be taken into account that the geometry of the copper(II) complex with the ligand L_{77} (Chart LXVI), which can be roughly considered a precursor of these complexes, is just N_2O_2 four coordinate and essentially planar |299|.

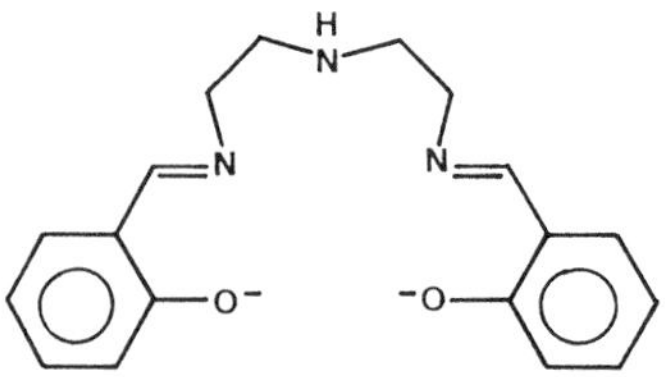

CHART LXVI

L_{77}

This speculative argumentation may account for the cited differences in redox potentials.

6.5 $\underline{O_4N_2}$ donor set

The only electrochemically studied derivative having a formal CuN_2O_4 coordination is the copper(II) complex of the macrocycle schematized in Chart LXVII.

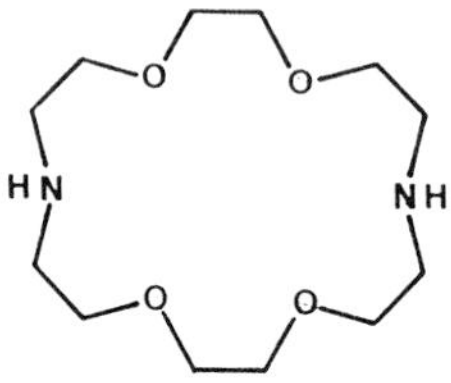

CHART LXVII

L_{78}

Like the corresponding hexaazamacrocyclic complex $|Cu(L_{69})|^{2+}$ (see Section 4.4), $|Cu(L_{78})|^{2+}$ undergoes, in propylene carbonate, a single copper(II)/copper(0) reduction ($E_p = -0.95$ V), coupled to a fast demetallation process, which makes irreversible the two-electron addition |300|.

This quite disfavoured reduction process may likely be accounted for considering the geometry of the complex $Cu(L_{78})Cl_2$, shown in Figure 78 |301|.

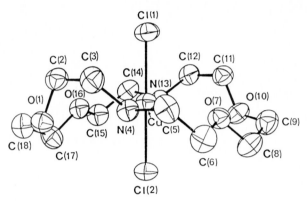

Fig. 78. Perspective view of $Cu(L_{78})Cl_2$. Cu-N, 2.03 Å; Cu-O, 2.73 Å; Cu-Cl, 2.31 Å; Cu⋯⋯O (O1, O10), 3.65 Å (from Ref. 301).

Even if crystal packing makes the chloride anions bonding, it seems plausible that in solution $|Cu(L_{78})|^{2+}$ may be planarly CuN_2O_2 coordinated, the two oxygens O1 and O10 being non-bonding because of the strain imposed by the macrocycle. As oft times cited, the square planar copper(II) geometry is not apt to reorganize to the tetrahedral copper(I) geometry.

7. MIXED SULFUR-NITROGEN LIGANDS

The fact that copper complexes coordinated to both nitrogen and sulfur atoms are the active sites of a number of important biological functions |2-4| makes this type of complexes of major interest. Unfortunately, although the relevant biomolecules are largely associated with electron-transfer processes, most of the synthetic analogs lack redox characterization.

7.1 $\underline{SN_2 \text{ donor set}}$

The redox behaviour of the formally tridentate complexes schematized in Chart LXVIII has been studied.

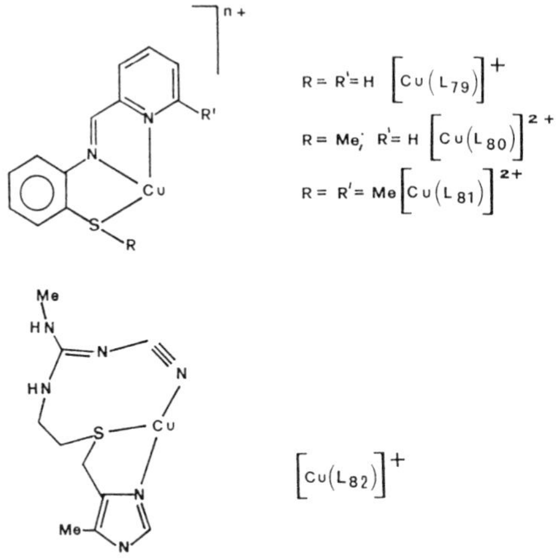

$$R = R' = H \quad \left[Cu\left(L_{79}\right)\right]^{+}$$

$$R = Me, \ R' = H \quad \left[Cu\left(L_{80}\right)\right]^{2+}$$

$$R = R' = Me \left[Cu\left(L_{81}\right)\right]^{2+}$$

CHART LXVIII

$$\left[Cu\left(L_{82}\right)\right]^{+}$$

The solid state structure of the copper(II) complexes $|Cu(L_{79})|^{+}$ and $|Cu(L_{81})|^{2+}$ can be deduced from that of $Cu(L_{80})|^{2+}$ |141|.

As shown in Figure 79, the coordination geometry around the copper(II) ion can be summarily described as elongated octahedral, both perchlorate ions (in axial positions) and one water molecule (in an equatorial position) being coordinated in addition to the pyridyl and imine nitrogens and to the thioether sulfur of the ligand. This indicates that the real core of the complex is CuN_2SO. Spectroscopic data seem to indicate that such a coordination is also maintained in solution.

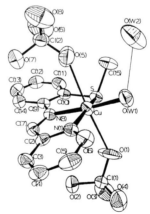

Fig. 79. X-ray structure of $Cu(L_{80})(ClO_4)_2 \cdot 2H_2O$. Cu-N$_{(mean)}$, 1.97 Å; Cu-S, 2.32 Å; Cu-O$_{(ClO_4)}$, 2.53 Å; Cu-O$_{(H_2O)}$, 1.93 Å (from Ref.141).

As shown in Table 42, these copper(II) complexes display an easy copper(II)/copper(I) reduction. The quasireversibility of the cathodic step suggests the occurrence of some significant reorganization in the structure of the relevant copper(I) complexes; however, information is lacking not only on the structure but also on the stability of these electrogenerated copper(I) species at times longer than those typical of cyclic voltammetry.

TABLE 42

Formal electrode potentials (in Volt) for the copper(II)/copper(I) couple of the copper complexes schematized in Chart LXVIII

complex	$E°'_{Cu^{II}/Cu^{I}}$	solvent	reference
$|Cu(L_{79})|^+$	-0.25	DMF	141
$|Cu(L_{80})|^{2+}$	$+0.44$	$MeNO_2$	141
$|Cu(L_{81})|^{2+}$	$+0.41$	MeCN	141
$|Cu(L_{81})|^{2+}$	$+0.53^a$	$MeNO_2$	141
$|Cu(L_{82})|^+$	$+0.16$	H_2O (pH 7.0)	302

[a] Peak potential value for irreversible process.

The rather easy access to copper(I) species, starting from tetragonal copper(II) complexes, seems apparently unusual. However, it must be taken into account that the presence of soft sulfur atoms in the donor atom set always stabilizes the copper(I) state because of their high π-acceptor ability.

In particular, comparison of the reduction potential of $|Cu(L_{79})|^+$ with those of $|Cu(L_{80})|^{2+}$ and $|Cu(L_{81})|^{2+}$ supports the concept that thioether sulfur atoms favour the copper(II)/copper(I) reduction to an extent greater than thiolate sulfur atoms do |141|.

As far as the copper(I) complex $|Cu(L_{82})|^+$ (the copper(I) complex of the pharmacologically important ligand cimetidine) is concerned, even if its low solubility did not allow extensive characterization, it is supposed to be four-coordinate with a fourth bond to the oxygen of a water molecule, giving rise to a square planar geometry |302|. Assuming this description is correct, its rather positive potential value seems unexpected considering the presence of one thioether sulfur atom.

7.2 SN_3 donor set

The two complexes schematized in Chart LXIX, displaying a CuN_3S coordination, have been investigated by electrochemical methods.

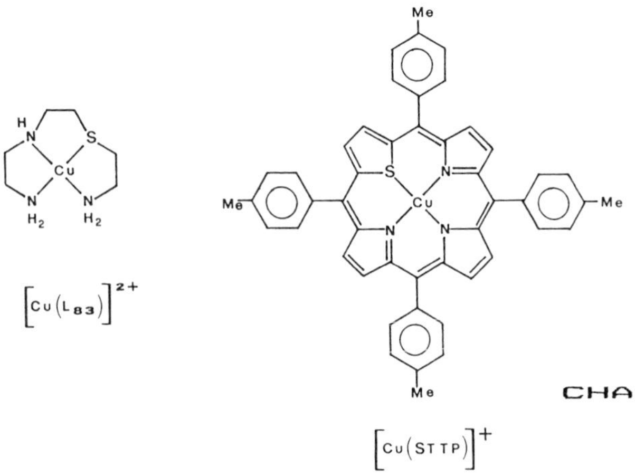

$$\left[Cu(L_{83})\right]^{2+}$$

$$\left[Cu(STTP)\right]^{+}$$

CHART LXIX

The complex $|Cu(L_{83})|^{2+}$ shows a copper(II)/copper(I) reduction at $E°' = +0.12$ V in aqueous-80% methanol |29|. This rather positive potential value, once again, indicates that the presence of a sulfur atom in the ligand significantly stabilizes the copper(I) state; in fact, the corresponding tetraaza complex $|Cu(L_3)|^{2+}$ reduces at potentials 0.6 V more negative (see Table 17).

As far as the copper(II)-thiatetra(p-tolyl)porphyrin complex Cu(STTP)Cl is concerned, it has been reported that it undergoes a copper(II)/copper(I) reduction at $E°' = -0.1$ V (in CH_2Cl_2 solution ?)|303|. Recalling that, usually, copper(II)-porphyrin complexes do not exhibit a copper(II)/copper(I) reduction, except for N-substituted porphyrins (see Section 4.2.16), easy access to copper(I) must arise from a favourable geometrical assembly. As a matter of fact, the related complex Cu(STTP)(CO_3H) exhibits a non-planar CuN_4 coordination, Figure 80, even if crystal packing affords a fifth Cu-O bond with the HCO_3^- anion |304|.

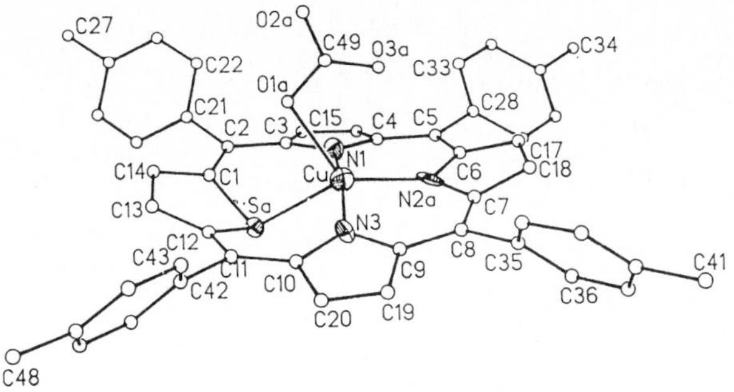

Fig. 80. X-ray structure of Cu(STTP)(CO₃H). $Cu-N_{(mean)}$, 2.05 Å; Cu-S, 2.28 Å; $Cu-O_{(CO_3H)}$, 2.25 Å (from Ref. 304).

The non-planarity of porphyrin ring is clearly evident. The thiophene plane gives rise to a dihedral angle of 27.5° with the CuN_3 plane, and the $S-Cu-N_2$ angle is 148° (in comparison with 180° for a planar assembly).

7.3 S_3N donor set

One potential CuS_3N coordination may be present in the copper(I) complex of the tripodal ligand schematized in Chart LXX.

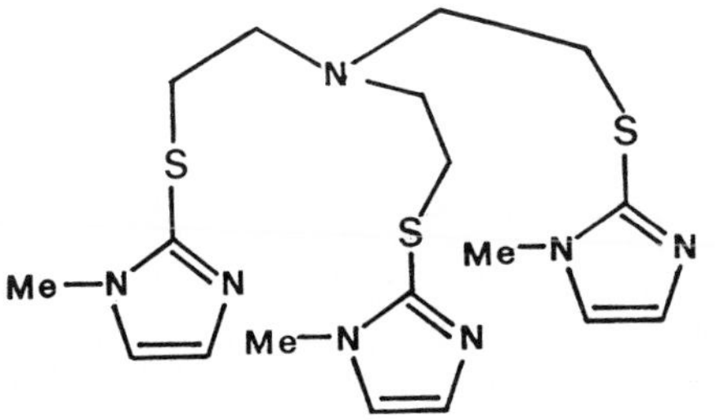

CHART LXX

teSI

$|Cu(teSI)|^+$ undergoes, in dimethylformamide solution, the copper(I)/copper(II) oxidation at $E°'= +0.07$ V. This relatively low potential value for a NS_3 donor set makes uncertain the real coordination in this complex |305|.

7.4 S_2N_2 donor set

The CuS_2N_2 coordination is by far the most important in bioinorganic copper chemistry, since the active site of the best structurally characterized "blue" protein poplar plastocyanin possesses such a coordination both in the oxidized (Cu(II)) and reduced (Cu(I)) forms. As shown in Figure 81, the metal core of the copper(II) protein undergoes a relatively minor geometrical reorganization upon one-electron reduction; in fact, the geometry remains pseudotetrahedral even if the two Cu-N bonds lengthen of about 0.1 Å. Only at low pH values (3.8) a deep reorganization occurs accompanied by redox-inactivity. In fact, one Cu-N bond is broken, and the copper(I) becomes trigonally coordinated |306|.

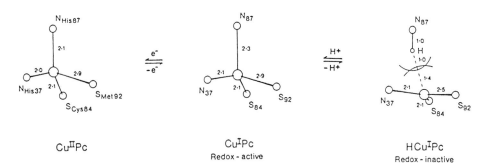

Fig. 81. Geometrical reorganizations of the active site accompanying the one electron reduction of copper(II) poplar plastocyanin (from Ref. 306).

Both the minor stereochemical reorganization upon one-electron reduction, and the presence of soft sulfur atoms, cooperate in making the redox potential of the copper(II)/copper(I) couple of plastocyanin rather high (E°'= +0.13V, in neutral aqueous solution).

Since we have just recently reviewed the electrochemistry of mononuclear CuN_2S_2 complexes and the relevant stereochemical implications |307|, we limit here our discussion to a brief updating.

In the preceding review article we have subdivided the copper complexes according to the following ligand designs (L = N or S):

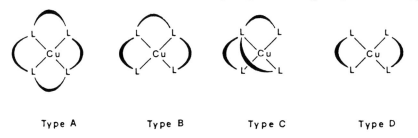

Type A   Type B   Type C   Type D

Two series of compounds, Chart LXXI, belonging to the "Type B" complexes have recently been investigated.

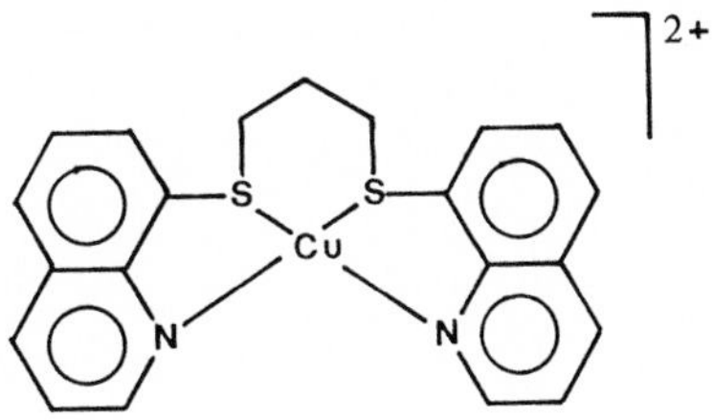

$$\left[Cu\left(L_{84}\right)\right]^{2+}$$

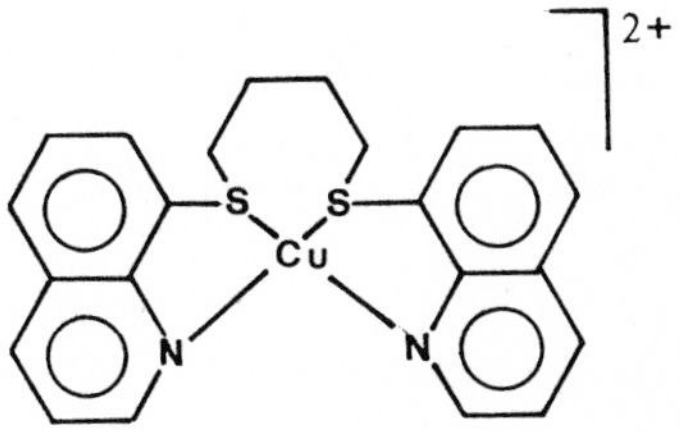

$$\left[Cu\left(L_{85}\right)\right]^{2+}$$

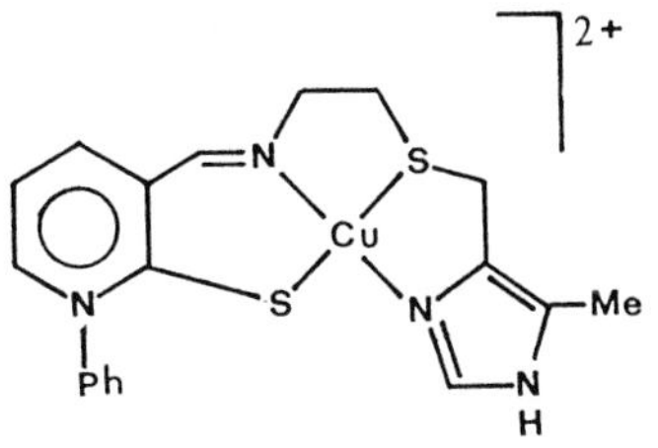

$$\left[Cu\left(L_{86}\right)\right]^{2+}$$

CHART LXXI

The mercaptoquinoline derivatives $|Cu(L_{84})|^{2+}$ and $|Cu(L_{85})|^{2+}$ undergo, in acetonitrile solution, a copper(II)/copper(I) reduction electrochemically quasireversible but chemically reversible |308|.

As shown in Table 43 the redox potential of this reduction step is notably high in both cases. The most positive potential value for $|Cu(L_{85})|^{2+}$ comes from the more marked tetrahedral distortion operated by the bridging tetracarbon chain, with respect to the tricarbon chain of $|Cu(L_{84})|^{2+}$, which obviously facilitates the reorganization to the tetrahedral geometry of the corresponding copper(I) species.

336

TABLE 43

Formal electrode potentials (in Volt) for the copper(II)/copper(I) reduction of the complexes schematized in Chart LXXI in acetonitrile solution.

complex	$E^{\circ\prime}$ Cu^{II}/Cu^{I}	reference
$|Cu(L_{84})|^{2+}$	+ 0.40	308
$|Cu(L_{85})|^{2+}$	+ 0.66	308
$|Cu(L_{86})|^{2+}$	+ 0.26	309

The methylimidazole-pyridinethione derivative $|Cu(L_{86})|^{2+}$ also displays a quasireversible copper(II)/copper(I) reduction generating the corresponding, stable, copper(I) complex |309|. Since the CuN_2S_2 coordination is substantially planar (in a tetragonal environment completed by solvent molecules |310|), we once again point out that the relatively high redox potential (Table 43) derives from the presence of sulfur donor atoms rather than from an appropriate stereochemical assembly.

Stereochemical effects on redox potentials in these types of complexes are evident in the homologues depicted in Chart LXXII, having variable length of the bridging carbon chains |307, 311|.

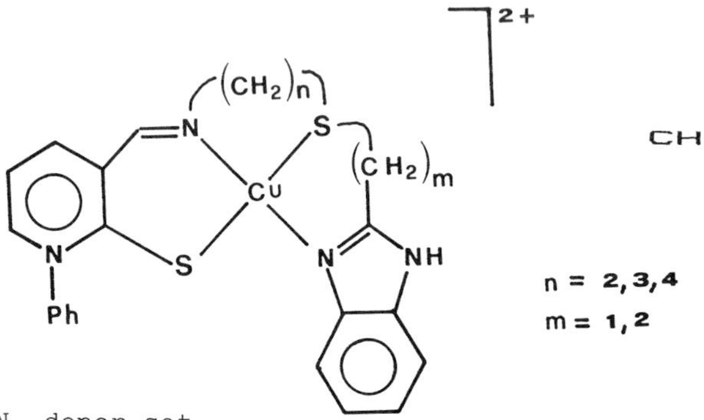

CHART LXXII

n = 2,3,4
m = 1,2

7.5 $\underline{S_2N_3}$ donor set

The copper(II)/copper(I) redox couple of the ligand shown in Chart LXXIII has been structurally characterized |247, 312|.

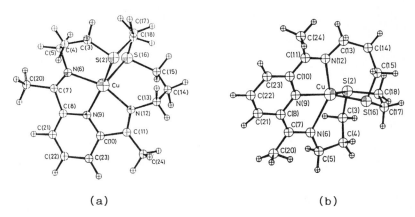

CHART LXXIII

L_{87}

Figure 82 shows the geometrical arrangement of the CuN_3S_2 coordination in both the oxidized, a, and reduced, b, forms.

(a)   (b)

Fig. 82. X-ray structures of: (a) $|Cu(L_{87})|(ClO_4)_2$; (b) $|Cu(L_{87})|(ClO_4)$ (from Refs. 312, 247).

The copper(II) complex can be described as square-pyramidal with a sulfur atom in axial position. The averaged distances are: Cu-N, 2.01 Å; Cu-S, 2.42 Å |312|.

The copper(I) complex can be thought as trigonal bipyramidal, with the two sulfur atoms in equatorial positions. The averaged distances are: Cu-N, 2.30 Å; Cu-S, 2.33 Å |247|.

As it can be seen, upon one-electron reduction the geometrical reorganization is accompanied by an enhancement of the Cu-S bonds, as well as by an weakening of the Cu-N bonds.

Unfortunately, we do not know the electrochemical charac-teristics of this structurally appealing redox couple, even if the lack of reactivity of the copper(I) complex towards dioxygen roughly indicates a relatively high redox potential.

The CuN_3S_2 complexes in Chart LXXIV have been electrochemically studied.

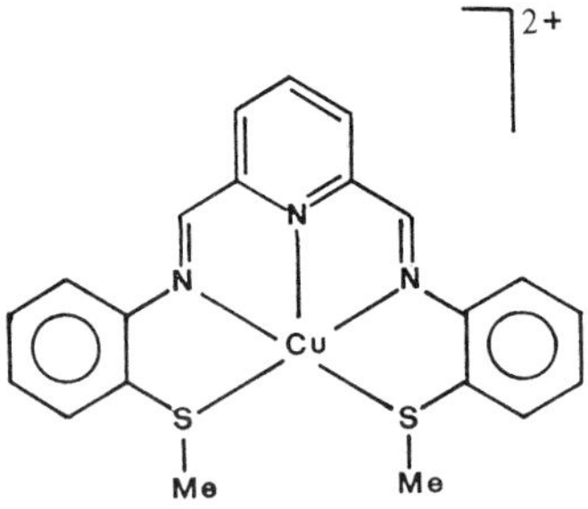

$$\left[Cu\left(L_{88}\right)\right]^{2+}$$

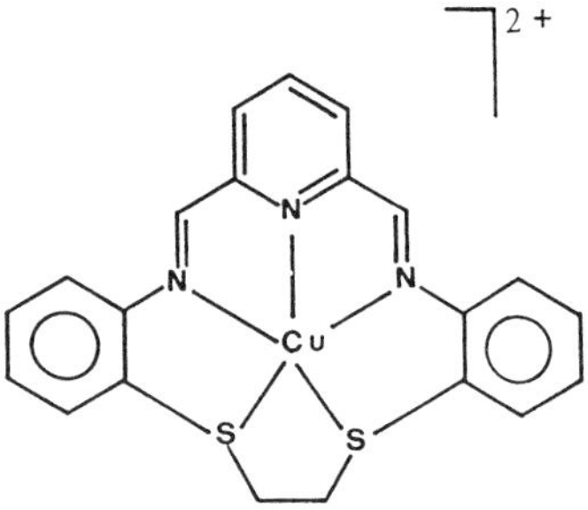

$$\left[Cu\left(L_{89}\right)\right]^{2+}$$

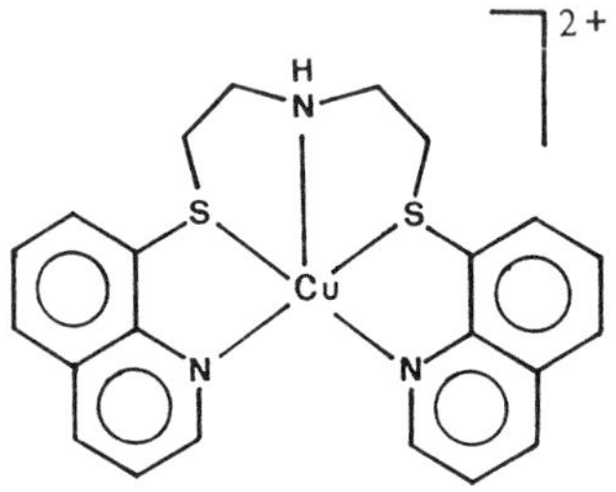

$$\left[Cu\left(L_{90}\right)\right]^{2+}$$

CHART LXXIV

These complexes undergo, quasireversibly, the copper(II)/copper(I) reduction at quite positive potential values [141, 308] (see Table 44). X-ray structural data are not available; however, both the spectral evidence [141] and the chemical reversibility of the quasireversible cathodic reduction [308] strongly support the idea that the copper(II)/copper(I) redox couple of the ligands L_{88}-L_{90} may have a geometrical coordination essentially similar to that of L_{87}, even if the presence of side-arm aryl rings makes these assemblies less flexible.

TABLE 44

Standard electrode potentials (in Volt) for the copper(II)/
copper(I) reduction in acetonitrile solution of the complexes
schematized in Chart LXXIV

complex	$E^{\circ\prime}$ Cu^{II}/Cu^{I}	reference
$|Cu(L_{88})|^{2+}$	+ 0.47	141
$|Cu(L_{89})|^{2+}$	+ 0.48	141
$|Cu(L_{90})|^{2+}$	+ 0.48	308

7.6 S_3N_2 donor set

The propensity of copper(I) to assume a pentadentate coordination is evidenced by the redox behaviour of the macrocyclic
copper(II) complexes schematized in Chart LXXV, which display a
CuN_2S_3 coordination.

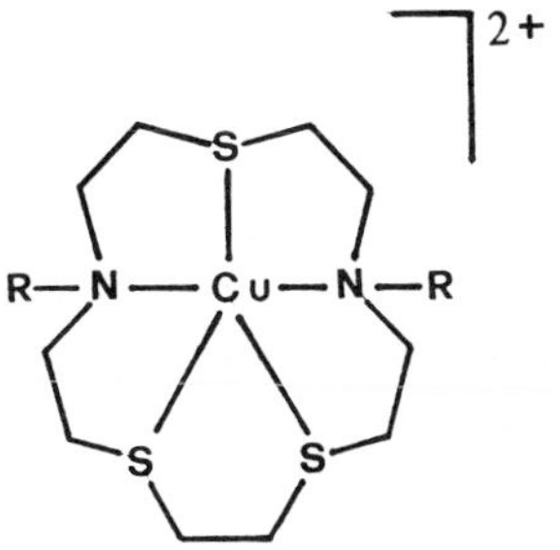

$$R = H, \left[Cu(L_{91})\right]^{2+}$$
$$R = Me, \left[Cu(L_{92})\right]^{2+}$$

CHART LXXV

As an example, Figure 83 shows the cyclic voltammetric
response exhibited by $|Cu(L_{91})|^{2+}$ in the copper(II)/copper(I)
reduction |300|.

An almost reversible electron transfer occurs at positive
potential values, which testifies to a minor geometrical
reorganization accompaning the copper(II)/copper(I) redox changes.

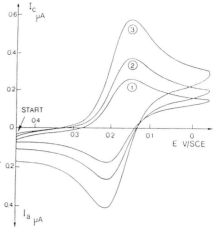

Fig. 83. Cyclic voltammograms recorded at a platinum electrode on a PC solution of $|Cu(L_{91})|^{2+}$. Scan rates: 1, 0.01 Vs^{-1}; 2, 0.02 Vs^{-1}; 3, 0.05 Vs^{-1} (from Ref.300).

It is not unlikely that the relatively unflexible geometry of the macrocycle forces both the copper(II) and copper(I) to assume an essentially similar geometry, intermediate between the square pyramidal and the trigonal bipyramidal illustrated previously for the copper(II)/copper(I) couple of the CuN_3S_2 assembly.

This consideration may be speculatively enforced by the fact that in $|Cu(L_{92})|^{2+}$, in which the presence of methyl groups disfavour electronically the reduction step, is significantly easier to reduce (see Table 45), and this reduction step is fully reversible, which means that bulky methyl groups even more markedly force the redox couple to assume the same geometry.

TABLE 45

Redox potentials (in Volt) for the reduction pathway of the copper(II) complexes schematized in Chart LXXV |300|

complex	$E^{\circ\prime}_{Cu^{II}/Cu^{I}}$	$E_{p_{Cu^{I}/Cu^{0}}}$	solvent
$|Cu(L_{91})|^{2+}$	+ 0.18	− 1.46	PC
$|Cu(L_{92})|^{2+}$	+ 0.32	−	PC
$|Cu(L_{92})|^{2+}$	+ 0.14	−	H_2O

7.7 $\underline{SN}_4$ donor set

Copper complexes displaying CuN_4S coordination have been prepared through coordination of different thiolates to copper(II)-tetraazamacrocycles. A series of stable copper(II)-mercaptide complexes make use of the CuN_4 moiety of $Cu(\underline{rac}-Me_6-L_{28})$ discussed in Section 4.2.18. They commonly have a trigonal-bipyramidal CuN_4S assembly, with the thiolate sulfur atom occuping an equatorial position |313|. Unfortunately, their redox behaviour is still unknown. On the other hand, the electrochemistry of the copper(II) complexes schematized in Chart LXXVI, based on some tetraaza macrocycles reported in Chart XLVIII, has been investigated.

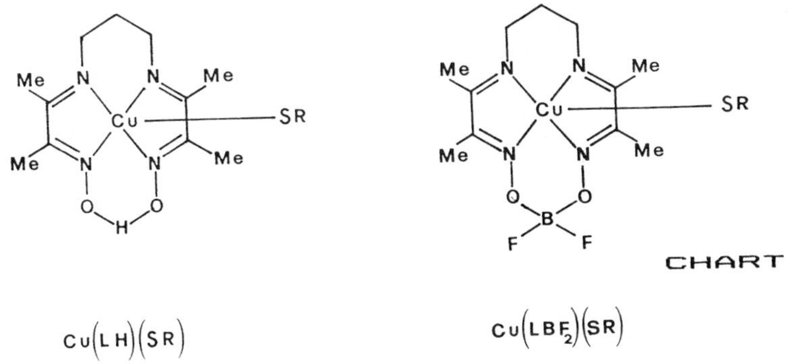

CHART LXXVI

$$Cu(LH)(SR) \qquad Cu(LBF_2)(SR)$$

Figure 84 shows the typical crystal structure of Cu(LH)(SR) derivatives. It refers to $Cu(LH)(S-C_6H_4-\underline{p}-Cl)$.

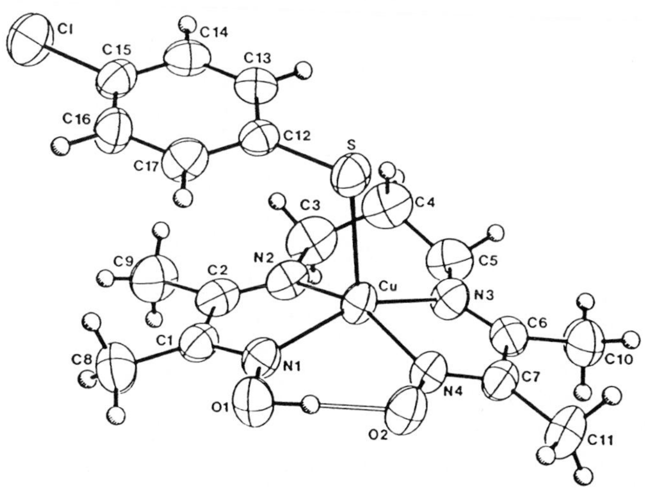

Fig. 84. X-ray structure of $Cu(LH)(S-C_6H_4-\underline{p}-Cl)$. Cu-N$_{(averaged)}$, 1.98 Å; Cu-S, 2.42 Å (from Ref. 225).

The N_4S donor set assumes a square-pyramidal geometry, with the sulfur atom occupying the apical position. The copper atom is displaced out of the N_4 plane (towards the sulfur atom) by 0.43 Å |225|.

This same geometry is thought to hold for the $Cu(LBF_2)(SR)$ complexes |224|.

As illustrated in Figure 85 for $Cu(LH)(S-C_6H_4-\underline{p}-Cl)$, these CuN_4S derivatives undergo a copper(II)/copper(I) reduction at potentials significantly more negative than their CuN_4 precursors (compare Table 46 with Table 32).

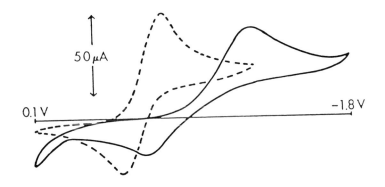

Fig. 85. Cyclic voltammograms recorded at 0°C, at a glassy-carbon electrode in CH_2Cl_2 solutions of: (-------), $|Cu(LH)|^+$; (———) $Cu(LH)(S-C_6H_4-\underline{p}-Cl)$. Scan rate 0.1 Vs^{-1} (from Ref. 225).

Also, the extent of electrochemical reversibility of the redox change (as measured by the peak-to-peak separation) is notably lower. This suggests that deeper stereochemical reorganizations occur upon one-electron reduction of the square pyramidal CuN_4S complexes when compared to the corresponding square-planar CuN_4 complexes. No data (spectroscopic or crystallographic) are available concerning these copper(I)-N_4S complexes.

TABLE 46

Redox potentials (in Volt) for the copper(II)/copper(I) reduction of the complexes schematized in Chart LXXVI in dichloromethane solution

complex	$E^{\circ\prime}_{Cu^{II}/Cu^{I}}$	reference
$Cu(LH)(S-C(C_6H_5)_3)$	-0.54	224
$Cu(LH)(S-C_6H_4-\underline{p}-Cl)$	-0.86	225
$Cu(LH)(S-C_6H_5)$	-0.85	225
$Cu(LBF_2)(S-C(C_6H_5)_3)$	-0.46	224
$Cu(LBF_2)(S-C_6H_4-\underline{p}-Cl)$	-0.38	224
$Cu(LBF_2)(S-C_{10}H_7)$	-0.39	224

Stabilization of the copper(II) oxidation state in the CuN_4S assembly, over the parent CuN_4 assembly, is mostly attributed to electronic effects in that the electron donating ability of the apical thiolate ligand should make more difficult the one-electron addition [224,225].

7.8 $\underline{S}_2\underline{N}_4$ donor set

The 1:2 copper(II) complex of the cimetidine ligand, previously cited as forming a 1:1 complex (see Section 7.1), crystallizes in the polymeric form shown in Figure 86.

As can be seen, an octahedral CuN_4S_2 geometry is achieved through coordination of two imidazoles and two thioethers in the equatorial plane (Cu-N, 1.97 Å; Cu-S, 2.70 Å), and two cyanoguanidines (from adjacent molecules) in axial positions (Cu-N, 2.04 Å) [314].

344

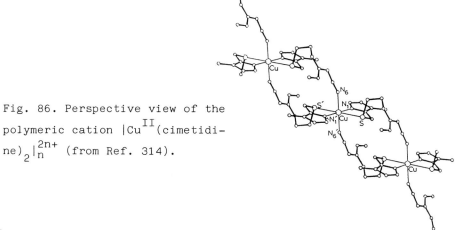

Fig. 86. Perspective view of the polymeric cation $|Cu^{II}(cimetidine)_2|_n^{2n+}$ (from Ref. 314).

In aqueous solution, at pH 7.8, this bis-ligand copper(II) complex undergoes a substantially reversible one-electron reduction, Figure 87, $(E^{\circ\prime}{}_{Cu}{}^{II}{}_{/Cu}{}^{I} = +0.11\ V)$; this datum suggests that the copper(I) complex may assume either an octahedral CuN_4S_2 coordination |302|, as a tetrahedral CuN_2S_2 geometry (see the subsequent paragraph). Unfortunately, precipitation of the insoluble 1:1 copper(I) complex, previously discussed, during the chemical or electrochemical reduction |302| prevents the obtainment of such 1:2 copper(I) species.

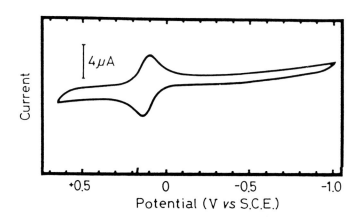

Fig. 87. Cyclic voltammetric response of the copper(II)/copper(I) reduction of $|Cu(cimetidine)_2|^{2+}$ in aqueous solution (pH 7.8). Glassy-carbon working electrode (from Ref. 302).

7.9 $\underline{S_4N_2}$ donor set

The copper(II) complexes of the N_2S_4 macrocycle shown in Chart LXXVII display chemically reversible reduction to the corresponding copper(I) complexes |300, 315|. Again, in confirmation of the stability of these copper(I) species, $|Cu(L_{94})|^{2+}$ undergoes self-reduction to $|Cu(L_{94})|^+$ |300|.

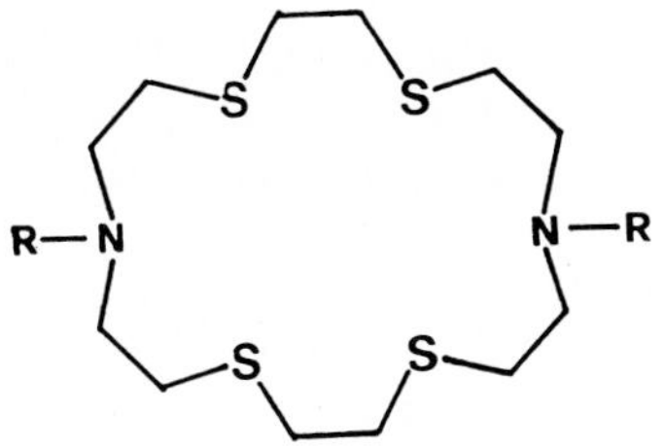

R = H , L93

R = Me , L94

CHART LXXVII

The electrochemical step is substantially reversible, and the relevant redox potentials are reported in Table 47.

TABLE 47

Redox potentials (in Volt) for the redox changes exhibited by the copper(II) complexes of the ligands schematized in Chart LXXVII |300|

complex	$E^{\circ\prime}_{Cu^{II}/Cu^{I}}$	$E_{p Cu^{I}/Cu^{0}}$	solvent
$|Cu(L_{93})|^{2+}$	− 0.10	−	H_2O
$|Cu(L_{93})|^{2+}$	+ 0.04	−	PC
$|Cu(L_{94})|^{2+}$	+ 0.31	−	H_2O
$|Cu(L_{94})|^{2+}$	+ 0.44	− 1.25	PC

This redox behaviour, together with the donor atoms composition, are quite reminiscent of the redox-to-structural changes observed in the similar macrocyclic complex $|Cu(|18|-ane-S_6|^{2+}$ (see Section 3.4). As a consequence, it does not seem ventured to think that, upon reduction the octahedral $Cu^{II}S_4N_2$ complex may assume a tetrahedral geometry leaving the two nitrogen atoms unbonding.

As deducible from Table 48, in the 18-membered macrocycles having six donor atoms, increasing the number of sulfur atoms leads to increasing stabilization of copper(I) complexes.

TABLE 48

Electrochemical characteristics of the copper(II)/copper(I) transition in different $|18|$-membered copper(II) macrocyclic complexes

complex	$E^{\circ\prime}$ Cu^{II}/Cu^I (volt)	extent of reversibility	solvent	reference
$|Cu(|18|-ane-N_6)|^{2+}$	—	irreversible	H_2O	80
$|Cu(|18|-ane-O_4N_2)|^{2+}$	− 0.95	irreversible	PC	300
$|Cu(|18|-ane-S_4N_2)|^{2+}$	+ 0.04	quasireversible	PC	300
$|Cu(|18|-ane-S_6)|^{2+}$	+ 0.72	quasireversible	$MeNO_2$	80

8. APPENDIX

Some pertinent papers have appeared just after the completion of this review article.

1. Addendum to Section 4.2.7

The planarity of the bis(2,2'-biimidazole)copper(II) complex $|Cu(H_2bim)_2|^{2+}$, has now been demonstrated by an X-ray analysis $|316|$. Figure 88 just shows the molecular structure of $|Cu(H_2bim)_2|^{2+}$.

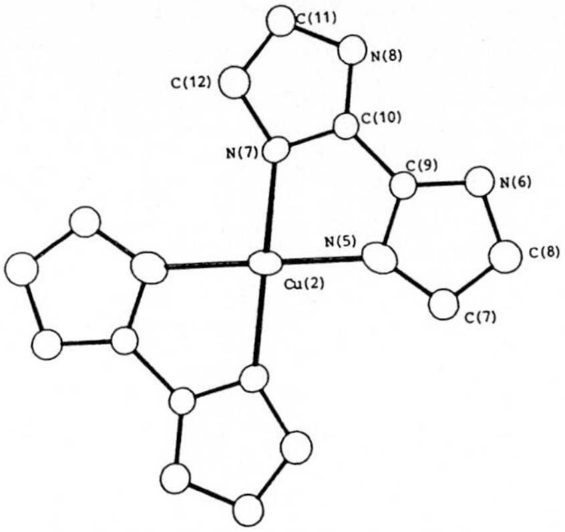

Fig. 88. Perspective view of $|Cu(H_2bim)_2|^{2+}$. Cu-N mean distance, 2,01 Å (from Ref. 316).

2. Addendum to Section 4.2.18

A series of saturated tetraaza macrocycles. Schematized in Chart LXXVIII, has been characterized |317|.

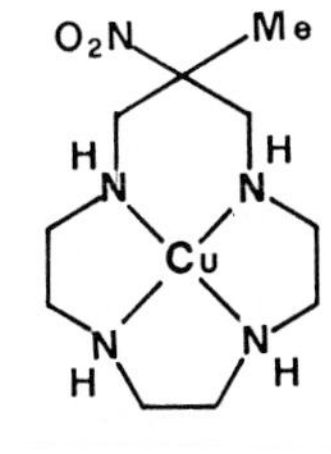

$$\left[Cu(L_{95})\right]^{2+}$$

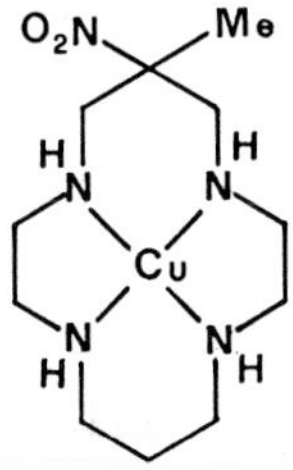

$$\left[Cu(L_{96})\right]^{2+}$$

CHART LXXVIII

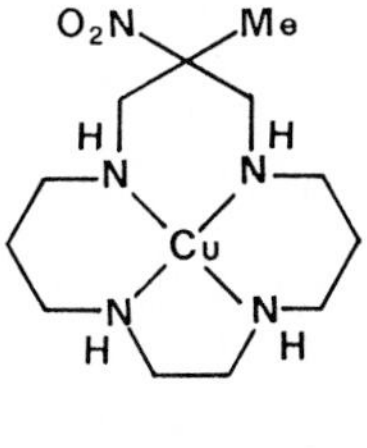

$$\left[Cu(L_{97})\right]^{2+}$$

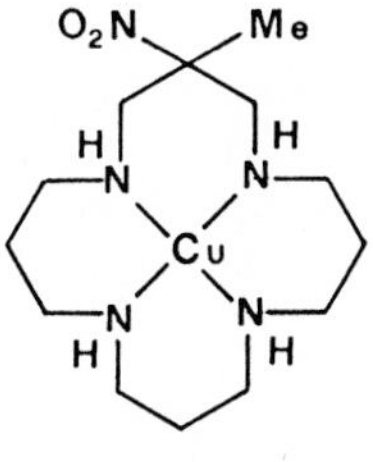

$$\left[Cu(L_{98})\right]^{2+}$$

They are formally related to the species $|Cu(L_{27})|^{2+}$, $|Cu(sym-L_{28})|^{2+}$, $|Cu(L_{29})|^{2+}$ and $|Cu(L_{30})|^{2+}$, discussed in Section 4.2.18.

Even if the electrochemistry of these complexes has been only briefly reported, the copper(II)/copper(I) redox change, the potentials of which are reported in Table 49, conforms to what was said in previous discussions. In fact, the 14-membered macrocycle L_{96}, which is best suited to encircle a planar copper(II) ion, undergoes irreversible reduction to copper(I). By contrast, the 13-membered L_{95}, as well as the 15- and 16-membered L_{97}, L_{98} cycles, disfavour the planar CuN_4 coordination because their smallest and largest cavities, respectively, favour access to copper(I).

TABLE 49

Redox potentials (in Volt) for the copper(II)/copper(I) reduction of the copper(II) complexes schematized in Chart LXXVIII in aqueous solution ($NaClO_4$ 0.1 M)[317]

complex x	$E^{o'}_{Cu^{II}/Cu^{I}}$
$|Cu(L_{95})|^{2+}$	- 0.61
$|Cu(L_{96})|^{2+}$	- 0.81[a]
$|Cu(L_{97})|^{2+}$	- 0.69
$|Cu(L_{98})|^{2+}$	- 0.45

[a] Peak potential value for irreversible process.

As represented in Figure 89, X-ray investigations show that $|Cu(L_{95})|^{2+}$ and $|Cu(L_{96})|^{2+}$ have the usual trans-axial coordination by oxygen atoms of perchlorate anions (Figure 89a), while $|Cu(L_{97})|^{2+}$ and $|Cu(L_{98})|^{2+}$ display axial Cu-O bonds constituted from one side by perchlorate oxygen and from the other side by one oxygen of the pendant nitro group (Figure 89b).

The chemical reversibility of the copper(II)/copper(I) redox change suggests that, as is frequently observed, six-coordination is induced by crystal packing, but it does not persist in solution.

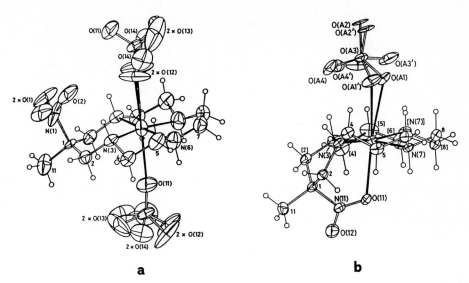

a

b

Fig. 89. Perspective view of the copper(II) environment in : (a) $|Cu(L_{95})|^{2+}$ (from Ref. 318); (b) $|Cu(L_{97})|^{2+}$ (from Ref. 317).

Bond lengths and angles in these complexes are reported in Figure 90.

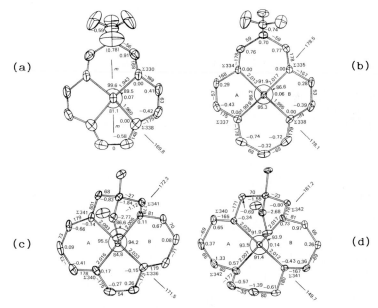

Fig. 90. Bond lengths and angles in: (a) $|Cu(L_{95})|^{2+}$; (b) $|Cu(L_{96})|^{2+}$; (c) $|Cu(L_{97})|^{2+}$; (d) $|Cu(L_{98})|^{2+}$ (from Ref. 317).

Finally, with reference to the unsaturated copper(II)-macrocycles schematized in Chart XLVI, the new member of tetraaza|14| annulene-copper(II) complexes schematized in Chart LXXIX has been recently prepared |319|.

CHART LXXIX

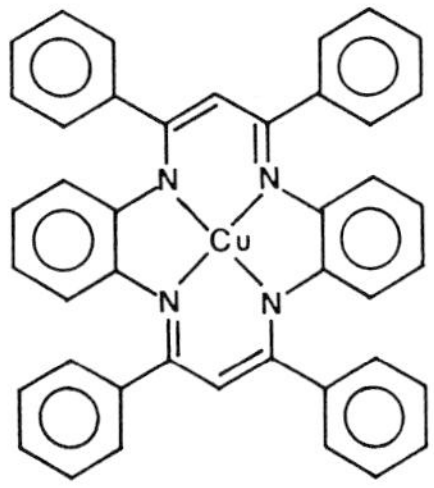

$$Cu\left[Ph_4\left(Bzo\right)_2[14]\,tetraene\,N_4\right]$$

We previously underlined that reduction to copper(I) was favoured by the presence of phenyl groups in $Cu|Me_4(R-Bzo)_2|14|$ tetraeneN$_4$| with respect to $Cu(MeHMe)_2(en)_2$ and $Cu(MeHMe)_2(2,9-diene)_2$. It is hence evident that the further presence of four electron withdrawing phenyl rings in $Cu|Ph_4(Bzo)_2|14|tetraeneN_4|$ must make more and more facile the electron addition. This is really the case, and the redox potential for the copper(II)/copper(I) couple significantly shifts towards less negative potential values (in dichloromethane solution, $E°'_{Cu^{II}/Cu^{I}}(Cu|Me_4(Bzo)_2|14|tetraeneN_4|) = -1.46$ V; $E°'_{Cu^{II}/Cu^{I}}(Cu|Ph_4(Bzo)_2|14|tetraeneN_4|) = -1.00$ V) |319|.

3. Addendum to Section 4.3.1

The X-ray structure of $|Cu(L_{56})|^{2+}$, Figure 91, is now available thus allowing us a brief discussion of the stereochemical rearrangement following the copper(II)/copper(I) redox change |320|.

As shown, in nice agreement with electrochemical inferences, the copper(I) complex maintains a five-coordination, roughly similar to that of the corresponding copper(II) complex illustrated in Figure 63. In both cases, pentacoordination does not induce a precise geometry (as previously mentioned, the coordination is intermediate between square-pyramidal and trigonal-bipyramidal); however, as limiting forms, the copper(I) complex is closer to a trigonal-bipyramidal description (the two imino nitrogens constituting the two axial vertices), whereas the copper(II) complex is closer to a square-pyramidal description (one imidazole nitrogen constituting the orbital vertex).

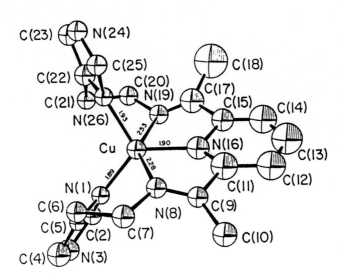

Fig. 91. ORTEP drawing of $|Cu(L_{56})|^+$, with relevant bond distances (in Å) (from Ref. 320).

It has been pointed out that the copper(II)/copper(I) reduction involves an averaged elongation of the Cu-N bonds from 2.02 Å to 2.11 Å; in particular however, upon reduction, there is a significant weakening of the copper-imino nitrogens bonds, which change from a mean value of 2.05 Å to 2.40 Å; by contrast, the remaining three Cu-N bonds shorten from a mean value of 1.99 Å to 1.91 Å |320|.

It is interesting to mention proof has been offered that solution- and solid-state structures are similar |321|.

Finally, a rather similar structural reorganization occurs in the $|Cu(L_{61})|^{2+}/|Cu(L_{61})|^+$ redox change |322|.

In this connection, Figure 92 shows the structures of the redox pair.

352

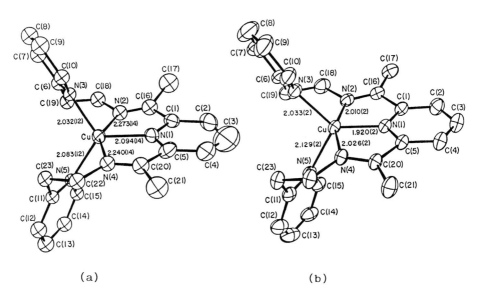

(a)   (b)

Fig. 92. Perspective views, with bond lengths, of: (a) $|Cu(L_{61})|^+$;
(b) $|Cu(L_{61})|^{2+}$ (from Ref. 322).

Even if a detailed structural comparison with $|Cu(L_{56})|^{2+}/$ $|Cu(L_{56})|^+$ reveals some minor but significant differences in interplanar angles, likely arising from the different pendant groups, in this case also the copper(I) complex may be best described as trigonal-bipyramidal (with the three pyridine nitrogens defining the trigonal plane), while the copper(II) complex may be best described as square-pyramidal (with the axial site occupied by the nitrogen of a pendant pyridine).

Here too, one-electron reduction not only induces elongation of the Cu-N bonds (from an average value of 2.02 Å to 2.14 Å), but the two copper-imino nitrogen bonds are significantly weakened (from a mean value of 2.02 Å to 2.26Å). At variance with L_{56} complexes, in this case the three remaining Cu-N bonds slightly increase (from a mean value of 2.03 Å to 2.07 Å)|322|.

9. CONCLUSIONS

We have presented evidence to the theoretical statement that in mononuclear copper complexes the slowest process is the stereochemical change accompanying a redox process, the easiest and fastest is the relevant electrochemical step.

In the heavily represented four-coordinate complexes, the commonest planar copper(II)-to-tetrahedral-copper(I) conversion is rather unfavourable, whereas the planar copper(II)-to-planar copper(III) change is the most readily accessible. However, if ligand architecture forces a non-planar, or even better a pseudotetrahedral geometry in the copper(II) coordination, the copper(II)/copper(I) rearrangement becomes more and more facilitated; concomitantly, the preparation of the copper(III) species is made more difficult.

Perhaps less well-known is the fact that, not only are five-coordinate copper(II) complexes with N_5, N_3S_2, N_2S_3 donor sets stable, but also stable are their corresponding copper(I) complexes – with geometrical coordinations ranging from square-pyramidal to trigonal-bipyramidal. In contrast, in strained macrocyclic CuS_5 complexes, one-electron reduction induces the square-pyramidal copper(II) geometry to convert to a tetrahedral CuS_4 asssembly, leaving one sulfur-atom uncoordinated.

Even less known is the fact that there is evidence that the common octahedral CuN_6 coordination of copper(II) complexes can be assumed also by copper(III) species. This geometry seems, on the contrary, totally forbidden for copper(I) derivatives, even if some suspicion of its existence in the CuN_4S_2 coordination exists. On the other hand, once again octahedral copper(II)-S_6 macrocyclic assemblies reduce to tetrahedral copper(I)-S_4 species, leaving two sulfur atoms uncoordinated.

Considerably less clear are the structure-redox relationships in truly tricoordinate copper complexes, particularly because of the lack of appropriate X-ray structural informations.

Finally, we must recall that the standard potential of a redox change is conditioned not only by stereochemical aspects, but also by electronic effects from the ligand. In fact electron withdrawing moieties favour the addition of electrons, while electron donating groups favour the removal of electrons. This means that the redox potential is a measure of both the stereochemical and electronic effects, which may operate jointly or oppositely, and many times these contributions are difficult to separate. We include in electronic effects, not only those deriving from side-arm groups of the ligand, but also from the π-accepting ability of the donor atom sets, as expressed by their softness. In this connection, it is useful to recall that the soft acid copper(I) is stabilized by ligation with heteroatoms in the order $S > N > O$; this means that the reduction of copper(II) complexes is the more favoured the higher the number of sulfur atoms present in the ligand. Conversely, the borderline acid copper(II) is stabilized by ligation with borderline nitrogen donor atoms.

10. ABBREVIATIONS FOR SOLVENTS

ACAC acetylacetone
ACAN acetic anhydride
CLN 1-chloronaphthalene
DMF N,N-dimethylformamide
DMSO dimethylsulfoxide
NB nitrobenzene
PC propylene carbonate
Py pyridine

ACKNOWLEDGEMENTS

The below mentioned publishers are gratefully acknowledged for permission to reproduce material from their publications.

ACADEMIC PRESS (J. Mol. Biol.);

ADENINE PRESS (Biological and Inorganic Copper Chemistry, Vol.II.);

AMERICAN CHEMICAL SOCIETY (Advances in Chemistry Series 201; Inorg. Chem.; J. Am. Chem. Soc.);

AMERICAN INSTITUTE OF PHYSICS (J. Chem. Phys.);

CHAPMAN AND HALL (Transition Met. Chem.);

CHEMICAL SOCIETY OF JAPAN (Bull. Chem. Soc. Jpn.);

ELSEVIER SCIENCE PUBLISHING COMPANY (J. Inorg. Biochem.);

ELSEVIER SEQUOIA S.A. (Inorg. Chim. Acta; J. Electroanal. Chem. Interfacial Electrochem.);

JAPAN ANTIBIOTICS READER ASSOCIATION (J. Antibiot.);

NATIONAL RESEARCH COUNCIL OF CANADA (Can. J. Chem.);

PERGAMON PRESS (Polyhedron);

THE ROYAL SOCIETY OF CHEMISTRY (J. Chem. Soc., Chem. Commun.; J. Chem. Soc., Dalton Trans. ; J. Chem. Res.).

REFERENCES

1 P.Zanello, S.Tamburini, P.A.Vigato and G.A.Mazzocchin, Coord.Chem.Rev., 77 (1987) 165.
2 Copper proteins; T.G.Spiro, Ed.; Wiley Interscience; New York, 1981.
3 Copper Coordination Chemistry: Biochemical and Inorganic Perspectives; K.D.Karlin and J.Zubieta, Eds.; Adenine Press: Guilderland, N.Y., 1983.
4 Biological and Inorganic Copper Chemistry; K.D.Karlin and J.Zubieta, Eds.; Adenine Press: Guilderland, N.Y., 1986.
5 B.Reinhammar and B.Malmström, in Ref. 2, p.109.
6 J.A.Fee and P.E.Di Corleto, Biochemistry, 12 (1973) 4893.
7 M.J.Ettinger and D.J.Kosman, in Ref. 2, p.219.
8 J.L.Anderson, T.Kuwana and C.R.Hartzell, Biochemistry, 15 (1976) 3847.
9 L.Heyrovsky and J.Kuta, in "Principles of Polarography", Academic Press, New York, 1966, p.156.
10 W.E.Geiger, Progr.Inorg.Chem., 33 (1985) 275.
11 A.J.Bard and L.R.Faulkner, Electrochemical Methods. Fundamentals and Applications. John Wiley. New York, 1980.
12 H.B.Gray and E.I.Solomon, in Ref. 2, pp.1-39.
13 M.Calvin and R.H.Bailes, J.Am.Chem.Soc., 68 (1946) 949.
14 H.F.Holtzclaw, Jr., K.W.R.Johnson and F.W.Hengeveld, J.Am.Chem.Soc., 74 (1952) 3776.
15 H.F.Holtzclaw, Jr., A.H.Carlson and J.P.Collman, J.Am.Chem.Soc., 78 (1956) 1838.
16 J.N.Burnett, L.K.Hiller, Jr. and R.W.Murray, J.Electrochem.Soc., 117 (1970) 1028.
17 R.L.Lintvedt, H.D.Russel and H.F.Holtzclaw, Jr., Inorg.Chem., 5 (1966) 1603.
18 R.L.Lintvedt, G.Ranger and B.A.Schoenfelner, Inorg.Chem., 23 (1984) 688.
19 C.W.Anderson, K.R.Lung and T.A.Nile, Inorg.Chim.Acta, 85 (1984) 33.
20 T.S.Piper and R.L.Belford, Mol.Phys., 5 (1962) 169.
21 J.W.Carmichael, Jr., L.K.Steinrauf and L.Belford, J.Chem.Phys., 43 (1965) 3959.
22 C.Lecomte, D.Bayeul, N.Senglet and R.Guilard, Polyhedron, 7 (1988) 303.
23 S.Harmalker, S.E.Jones and D.T.Sawyer, Inorg.Chem., 22 (1983) 2790.
24 F.Rohrscheid, A.L.Balch and R.H.Holm, Inorg.Chem., 5 (1966) 1542.
25 P.R.Sharp and A.J.Bard, Inorg.Chem., 22 (1983) 3462.
26 U.Casellato, P.A.Vigato, D.E.Fenton and M.Vidali, Chem.Soc.Rev., 8 (1979) 199.
27 R.L.Lintvedt, M.D.Glick, B.K.Tomlonovic, D.P.Gavel and J.M.Kuszaj, Inorg.Chem., 15 (1976) 1633.
28 P.Zanello, P.A.Vigato and G.A.Mazzocchin, Transition Met.Chem., 7 (1982) 291.

29 E.R.Dockal, T.E.Jones, W.F.Sokol, R.J.Engerer, D.B.Rorabacher and L.A.Ochrymowycz, J.Am.Chem.Soc., 98 (1976) 4322.

30 D.R.Rorabacher, M.J.Martin, M.J.Koenigbauer, M.Malik, R.R.Schroeder, J.Endicott and L.A.Ochrymowycz, in Ref. 3, p.167.

31 S.C.Rawle, G.A.Admans and S.R.Cooper, J.Chem.Soc., Dalton Trans., (1988) 93.

32 W.Rosen and D.H.Bush, Inorg.Chem., 9 (1970) 262.

33 W.N.Setzer, C.A.Ogle, G.S.Wilson and R.S.Glass, Inorg.Chem., 22 (1983) 266.

34 K.Wieghardt, H.-J. Küppers and J.Weiss, Inorg.Chem., 24 (1985) 3067.

35 A.J.Blake, A.J.Holder, T.I.Hyde and M.Schröder, J.Chem.Soc., Chem.Commun., (1987) 987.

36 A.J.Blake, R.O.Gould, A.J.Holder, T.I.Hyde, A.J.Lavery, M.O.Odulate and M.Schröder, J.Chem.Soc., Chem.Commun., (1987) 118.

37 S.C.Rawle, T.J.Sewell and S.R.Cooper, Inorg.Chem., 26 (1987) 3769.

38 H.-J. Küppers, K.Wieghardt, Y.-H.Tsay, C.Krüger, B.Nuber and J.Weiss, Angew.Chem.Int.Ed.Engl., 26 (1987) 575.

39 J.A.McCleverty, Progr.Inorg.Chem., 10 (1970) 49.

40 D.Coucouvanis and J.P.Fackler, Jr., J.Am.Chem.Soc., 89 (1967) 1346.

41 L.Persaud and C.H.Langford, Inorg.Chem., 24 (1985) 3562.

42 D.T.Sawyer, G.S.Srivatsa, M.E.Bodini, W.P.Schaefer and R.M.Wing, J.Am. Chem.Soc., 108 (1986) 936.

43 K.W.Plumlee, B.M.Hoffman, J.A.Ibers and Z.G.Soos, J.Chem.Phys., 63 (1975) 1926.

44 J.D.Forrester, A.Zalkin and D.H.Templeton, Inorg.Chem., 3 (1964) 1507.

45 S.Alvarez, R.Vicente and R.Hoffmann, J.Am.Chem.Soc., 107 (1985) 6253.

46 D.Snaathorst, H.M.Doesburg, J.A.A.J.Perenboom and C.P.Keijzers, Inorg. Chem., 20 (1981) 2526.

47 R.Williams, E.Billig, J.H.Waters and H.B.Gray, J.Am.Chem.Soc., 88 (1966) 43.

48 S.I.Shupack, E.Billig, R.J.H.Clark, R.Williams and H.B.Gray, J.Am.Chem.Soc., 86 (1964) 4594.

49 S.Zàlis and A.A.Vlcek, Inorg.Chim.Acta, 58 (1982) 89.

50 B.Werden, E.Billig and H.B.Gray, Inorg.Chem., 5 (1966) 78.

51 J.P.Fackler, Jr. and D.Coucouvanis, J.Am.Chem.Soc., 88 (1966) 3913.

52 S.Boyde, C.D.Garner and W.Clegg, J.Chem.Soc., Dalton Trans., (1987) 1083.

53 D.Coucouvanis, J.Am.Chem.Soc., 93 (1971) 1786.

54 T.Imamura, M.Ryan, G.Gordon and D.Coucouvanis, J.Am.Chem.Soc., 106 (1984) 984.

55 M.G.Kanatzidis, N.C.Baenziger and D.Coucouvanis, Inorg.Chem. 24 (1985) 2680.

56 G.Matsubayashi, K.Takahashi and T.Tanaka, J.Chem.Soc., Dalton Trans., (1988) 967.

57 R.Eisenberg, Progr.Inorg.Chem., 12 (1972) 295.

358

58 F.J.Hollander, M.L.Caffery and D.Coucouvanis, J.Am.Chem.Soc., 96
   (1974) 4682.
59 D.Coucouvanis, F.J.Hollander and M.L.Caffery, Inorg.Chem., 15
   (1976) 1853.
60 G.Cauquis and D.Lachenal, J.Electroanal.Chem., 46 (1973) 41.
61 A.R.Hendrickson, R.L.Martin and N.M.Rohde, Inorg.Chem., 15
   (1976) 2115.
62 J.G.M.van der Linden and H.G.J.van de Roer, Inorg.Chim.Acta, 5
   (1972) 254.
63 H.Jwasaki and K.Kobayashi, Acta Cryst., B36 (1980) 1655.
64 B.H.O'Connor and E.N.Maslen, Acta Cryst., 21 (1966) 828.
65 D.Coucouvanis, Progr.Inorg.Chem. 11 (1970) 233.
66 J.G.Wijnhoven, T.E.M.van den Hark and P.T.Beurskens,
   J.Cryst.Mol. Struct., 2 (1972) 189.
67 V.Sakaguchi and A.W.Addison, J.Chem.Soc., Dalton Trans., (1979)
   600.

68 M.D.Glick, D.P.Gavel, L.L.Diaddario and D.B.Rorabacher,
   Inorg.Chem., 15 (1976) 1190.
69 V.P.Pett, L.L.Diaddario, Jr., E.R.Dockal, P.W.Corfield,
   C.Ceccarelli, M.D.Glick, L.A.Ochrymowycz and D.B.Rorabacher,
   Inorg.Chem., 22 (1983) 3661.
70 L.L.Diaddario, Jr., E.R.Dockal, M.D.Glick, L.A.Ochymowycz and
   D.B. Rorabacher, Inorg.Chem., 24 (1985) 356.
71 E.R.Dockal, L.L.Diaddario, M.D.Glick and D.B.Rorabacher,
   J.Am.Chem.Soc., 99 (1977) 4530.
72 M.J.Martin, J.F.Endicott, L.A.Ochrymowycz and D.B.Rorabacher,
   Inorg. Chem., 26 (1987) 3012.
73 E.W.Ainscough, A.M.Brodie and K.C.Palmer, J.Chem.Soc., Dalton
   Trans., (1976) 2375.
74 E.N.Baker and G.E.Norris, J.Chem.Soc., Dalton Trans., (1977)
   877.
75 M.M.Olmstead, W.K.Musker and R.M.Kessler, Inorg.Chem., 20 (1981)
   151.
76 W.K.Musker, M.M.Olmstead, R.M.Kessler, M.B.Murphey, C.H.Neagley,
   P.B. Roush, N.L.Hill, T.L.Wolford, H.Hope, G.Delker, K.Swanson
   and B.V. Gorewit, J.Am.Chem.Soc., 102 (1980) 1225.
77 E.Weber and F.Wögtle, Liebigs Ann.Chem., (1976) 891.
78 C.R.Lucas, L.Shuang, M.J.Newlands, J.-P.Charland and E.J.Gabe,
   Can.J. Chem., 66 (1988) 1506.
79 P.W.R.Corfield, C.Ceccarelli, M.D.Glick, I.W.-Y.Moy,
   L.A.Ochymowycz and D.B.Rorabacher, J.Am.Chem.Soc., 107 (1985)
   2399.
80 J.A.R.Hartman and S.R.Cooper, J.Am.Chem.Soc., 108 (1986) 1202.
81 E.R.Brown and J.R.Sandifer, in "Physiscal Methods of Chemistry.
   Electrochemical Methods". B.W.Rossiter and J.F.Hamilton,
   Eds.J.Wiley, New York. Vol.2, 1986, Chapt.4.
82 L.J.Zompa, Inorg.Chem., 17 (1978) 2531.
83 A.D.Beveridge, A.J.Lavery, M.D.Walkinshaw and M.Schröder,
   J.Chem.Soc., Dalton Trans., (1987) 373.
84 P.Zanello and P.Leoni, Can.J.Chem., 63 (1985) 922.

85 H.Gampp, M.M.Roberts and S.J.Lippard, Inorg.Chem., 23 (1984) 2794.
86 M.Kodama and E.Kimura, J.Chem.Soc., Dalton Trans., (1977) 1473.
87 W.L.Kwik, K.P.Ang and A.W.N.Tay, Polyhedron, 7 (1988) 695.
88 I.M.Kolthoff and J.J.Lingane, in "Polarography", Interscience Publisher, New York, (1952), Vol.1, p.223.
89 J.J.Driscoll and D.J.Kosman, in Ref. 3, p. 75.
90 A.H.Lewin, R.J.Michl, P.Ganis, U.Lepore and G.Avitabile, J.Chem.Soc., Chem.Commun., (1971) 1400.
91 K.Nilsson and A.Oskarsson, Acta Chem.Scand., A 36 (1982) 605.
92 J.S.Haynes, S.J.Rettig, J.R.Sams, J.Trotter and R.C.Thompson, Inorg. Chem., 27 (1988) 1237.
93 H.A.Laitinen, E.I.Onstott, J.C.Bailar, Jr., and S.Swann, Jr., J.Am.Chem. Soc., 71 (1949) 1550.
94 D.Banerjea and P.Banerjee, J.Indian Chem.Soc., 49 (1972) 16.
95 P.Stevens, S.M.Waldek, J.Strohl and R.Nakon, J.Am.Chem.Soc., 100 (1978) 3632.
96 L.Fabbrizzi, A.Poggi and P.Zanello, J.Chem.Soc., Dalton Trans., (1984) 1495.
97 M.Kodama and E.Kimura, J.Chem.Soc., Dalton Trans., (1981) 694.
98 E.Kimura, T.Koike, R.Machida, R.Nagai and M.Kodama, Inorg.Chem., 23 (1984) 4181.
99 H.C.Freeman, J.E.W.L.Smith and J.C.Taylor, Acta Cryst., 14 (1961) 407.
100 A.Levitzki and M.Anbar, Chem.Commun., (1968) 403.
101 J.J.Bour, P.J.M.W.L.Birker and J.J.Steggerda, Inorg.Chem., 10 (1971) 1202.
102 P.J.M.W.L.Birker, Inorg.Chem., 16 (1977) 2478.
103 P.J.M.W.L.Birker, J.Chem.Soc. Chem.Commun., (1977) 444.
104 F.Akhtar, D.M.L.Goodgame, M.Goodgame, G.W.Rayner-Canhan and A.C.Skapski, Chem.Commun., (1968) 1389.
105 G.Ivarsson, Acta Chem.Scand., 27 (1973) 3523.
106 N.C.Li, J.M.White and E.Doody, J.Am.Chem.Soc., 76 (1954) 6219.
107 S.Knapp, T.P.Keenan, X.Zhang, R.Fikar, J.A.Potenza and H.J.Shugar, J.Am.Chem.Soc., 109 (1987) 1882.
108 S.Knapp and H.J.Schugar, private communication.
109 A.D.Mighell, C.W.Reimann and F.A.Mauer, Acta Cryst., B 25 (1969) 60.
110 M.Munakata, S.Kitagawa, A.Asahara and H.Masuda, Bull.Chem.Soc.Jpn., 60 (1987) 1927.
111 K.M.Davies and B.Byers, Inorg.Chem., 26 (1987) 3823.
112 B.J.Hathaway, Coord.Chem.Rev., 52 (1983) 87.
113 J.Foley, S.Tyagi and B.J.Hathaway, J.Chem.Soc.,Dalton Trans., (1984) 1.
114 P.J.Burke, D.R.McMillin and W.R.Robinson, Inorg.Chem., 19 (1980) 1211.
115 P.J.Burke, K.Henrick and D.R.McMillin, Inorg.Chem., 21 (1982) 1881.
116 J.R.Hall, M.R.Litzow and R.A.Plowman, Aust.J.Chem., 18 (1965) 1331.
117 E.Müller, C.Piguet, G.Bernardinelli and A.F.Williams, Inorg.Chem., 27 (1988) 849.

360

118 C.-W.Lee and F.C.Anson, Inorg.Chem., 23 (1984) 837.
119 A.G.Lappin, M.P.Youngblood and D.W.Margerum, Inorg.Chem., 19 (1980) 407.
120 G.S.Patterson and R.H.Holm, Bioinorg.Chem., 4 (1975) 257.
121 I.V.Nelson and R.T.Iwamoto, Anal.Chem., 35 (1963) 867.
122 N.Al-Shatti, A.G.Lappin and A.G.Sykes, Inorg.Chem., 20 (1981) 1466.
123 H.Nakai and Y.Noda, Bull.Chem.Soc.Jpn., 51 (1978) 1386.
124 P.C.Healy, L.M.Engelhardt, V.A.Patrick and A.H.White, J.Chem.Soc., Dalton Trans., (1985) 2541.
125 S.K.Hoffmann, P.J.Corvan, P.Singh, C.N.Sethulekshmi, R.M.Metzger and W.E.Hatfield, J.Am.Chem.Soc., 105 (1983) 4608.
126 R.Hämäläinen, M.Ahlgrén, U.Turpeinen and T.Raikas, Cryst.Struct.Comm., 8 (1979) 75.
127 J.F.Dobson, B.E.Green, P.C.Healy, C.H.L.Kennard, C.Pakawatchai and A.H.White, Aust.J.Chem., 37 (1984) 649.
128 G.Bidan, B.Divisia-Blohorn, J.-M.Kern and J.-P.Sauvage, J.Chem.Soc., Chem.Commun., (1988) 723.
129 H.Yokoi and A.W.Addison, Inorg.Chem., 16 (1977) 1341.
130 R.C.Elder, E.A.Blubaugh, Jr., W.R.Heineman, P.J.Burke and D.R.McMillin, Inorg.Chem., 22 (1983) 2777.
131 Y.Nishida, M.Takeuchi, N.Oishi and S.Kida, Inorg.Chim.Acta, 96 (1985) 81.
132 A.W.Addison and J.H.Stenhouse, Inorg.Chem., 17 (1978) 2161.
133 M.Elder and B.R.Penfold, J.Chem.Soc., (A) (1969) 2556.
134 R.Tewari and R.C.Srivastava, Acta Cryst., B 27 (1971) 1644.
135 C.H.Wei, Inorg.Chem., 11 (1972) 2315.
136 S.G.McGeachin, Can.J.Chem., 46 (1968) 1903.
137 D.Datta and A.Chakravorty, Inorg.Chem., 22 (1983) 1085.
138 S.K.Mandal, L.K.Thompson, E.J.Gabe, F.L.Lee and J.-P.Charland, Inorg.Chem., 26 (1987) 2384.
139 D.E.Nikles, N.J.Powers and F.L.Urbach, Inorg.Chem., 22 (1983) 3210.
140 M.Linss and U.Weser, Inorg.Chim.Acta, 138 (1987) 163.
141 A.W.Addison, T.N.Rao and E.Sinn, Inorg.Chem., 23 (1984) 1957.
142 M.Linss and U.Weser, Inorg.Chim.Acta, 125 (1986) 117.
143 A.W.Addison, M.Palaniandavar, W.L.Driessen, F.Paap and J.Reedijk, Inorg.Chim.Acta, 142 (1988) 95.
144 D.Hall, T.N.Waters and P.E.Wright, J.Chem.Soc., Dalton Trans., (1973) 1508.
145 T.N.Sorrel and D.L.Jameson, Inorg.Chem., 21 (1982) 1014.
146 K.D.Karlin, J.C.Hayes, J.Hutchinson, J.R.Hyde and J.Zubieta, Inorg. Chim.Acta, 64 (1982) L 219.
147 A.W.Addison, H.M.J.Hendriks, J.Reedijk and L.K.Thompson, Inorg.Chem., 20 (1981) 103.
148 T.Sakurai, H.Oi and A.Nakahara, Inorg.Chim.Acta, 92 (1984) 131.
149 Y.Nakao, M.Onoda, T.Sakurai, A.Nakahara, I.Kinoshita and S.Ooi, Inorg.Chim.Acta, 151 (1988) 55.
150 F.P.Bossu, K.L.Chellappa and D.W.Margerum, J.Am.Chem.Soc., 99 (1977) 2195.
151 H.C.Freeman and M.R.Taylor, Acta Cryst., 18 (1965) 939.

152 L.L.Diaddario, W.R.Robinson and D.W.Margerum, Inorg.Chem., 22 (1983) 1021.

153 J.M.Anast, A.W.Hamburg and D.W.Margerum, Inorg.Chem., 22 (1983) 2139.

154 P.Worthington, P.Hambright, R.F.X.Williams, J.Reid, C.Burnham, A.Shamim, J.Turay, D.M.Bell, R.Kirkland, R.G.Little, N.Datta-Gupta and U.Eisner, J.Inorg.Biochem., 12 (1980) 281.

155 A.Girardeau, H.J.Callot and M.Gross, Inorg.Chem., 18 (1979) 201.

156 K.M.Kadish and M.M.Morrison, J.Am.Chem.Soc., 98 (1976) 3326.

157 E.B.Fleischer, C.K.Miller and L.E.Webb, J.Am.Chem.Soc., 86 (1964) 2342.

158 R.H.Felton and H.Linschitz, J.Am.Chem.Soc., 88 (1966) 1113.

159 A.Wolberg and J.Manassen, J.Am.Chem.Soc., 92 (1970) 2982.

160 H.J.Callot, A.Girardeau and M.Gross, J.Chem.Soc., Perkin II, (1975) 1321.

161 A.Girardeau, H.J.Callot, J.Jordan, I.Ezhar and M.Gross, J.Am.Chem.Soc., 101 (1979) 3857.

162 D.W.Clack and N.S.Hush, J.Am.Chem.Soc., 87 (1965) 4238.

163 G.P.-Heiling and G.S.Wilson, Anal.Chem., 43 (1971) 550.

164 D.Lexa and M.Reix, J.Chim.Phys. (Paris), 71 (1974) 511.

165 A.S.Hinman, B.J.Pavelich and K.McGarty, Can.J.Chem., 66 (1988) 1589.

166 A.Girardeau, A.Louati, M.Gross, H.J.Callot, L.K.Hanson, R.K.Rhodes and K.M.Kadish, Inorg.Chem., 21 (1982) 1581.

167 A.Antipas, D.Dolphin, M.Gouterman and E.C.Johnson, J.Am.Chem.Soc., 100 (1978) 7705.

168 D.Kuila, A.B.Kopelove and D.K.Lavallee, Inorg.Chem., 24 (1985) 1443.

169 D.K.Lavallee, A.B.Kopelove and O.P.Anderson, J.Am.Chem.Soc., 100 (1978) 3025.

170 D.A.Case and M.Karplus, J.Am.Chem.Soc., 99 (1977) 6182.

171 P.F.Richardson, C.K.Chang, L.D.Spaulding and J.Fajer, J.Am.Chem.Soc, 101 (1979) 7736.

172 P.F.Richardson, C.K.Chang, L.K.Hanson, L.D.Spaulding and J.Fajer, J.Phys.Chem., 83 (1979) 3420.

173 J.-H.Fuhrhop, K.M.Kadish and D.G.Davis, J.Am.Chem.Soc., 95 (1973) 5140.

174 A.M.Stolzenberg, L.O.Spreer and R.H.Holm, J.Am.Chem.Soc., 102 (1980) 364.

175 A.M.Stolzenberg and M.T.Stershic, Inorg.Chem., 26 (1987) 1970.

176 A.M.Stolzenberg, P.A.Glazer and B.M.Foxman, Inorg.Chem., 25 (1986) 983.

177 T.Kakutani, S.Totsuka and M.Senda, Bull.Chem.Soc.Jpn., 46 (1973) 3652.

178 C.J.Brown, J.Chem.Soc. (A), (1968) 2488.

179 D.W.Clack, N.S.Hush and I.S.Woolsey, Inorg.Chim.Acta, 19 (1976) 129.

180 R.O.Loutfy and Y.C.Cheng, J.Chem.Phys., 73 (1980) 2902.

181 A.Wolberg and J.Manassen, J.Am.Chem.Soc., 92 (1970) 2982.

182 Y.Orihashi, M.Nishikawa, H.Ohmo, E.Tsuchida, H.Matsuda, H.Nakanishi and M.Kato, Bull.Chem.Soc.Jpn., 60 (1987) 3731.

362

183 L.D.Rollmann and R.T.Iwamoto, J.Am.Chem.Soc., 90 (1968) 1455.
184 A.B.P.Lever, S.Licoccia, K.Magnell, P.C.Minor and B.C.Ramaswamy, A.C.S. Symp.Ser., 201 (1982) 237.
185 A.Louati, M.El Meray, J.J.André, J.Simon, K.M.Kadish, M.Gross and A.Giraudeau, Inorg.Chem., 24 (1985) 1175.
186 K.Miyoshi, H.Tanaka, E.Kimura, S.Tsuboyama, S.Murata, H.Shimizu and K.Ishizu, Inorg.Chim.Acta, 78 (1983) 23.
187 M.Kodama and E.Kimura, J.Chem.Soc., Dalton Trans., (1976) 116.
188 E.Kimura, T.Koike, K.Uenishi, M.Hediger, M.Kuramoto, S.Joko, Y.Arai, M.Kodama and Y.Iitaka, Inorg.Chem., 26 (1987) 2975.
189 L.Fabbrizzi, A.Poggi and P.Zanello, J.Chem.Soc., Dalton Trans., (1983) 2191.
190 E.Zeigerson, G.Ginzburg, D.Meyerstein and L.J.Kirshenbaum, J.Chem.Soc., Dalton Trans., (1980) 1243.
191 R.L.Deming, A.L.Allred, A.R.Dahl, A.W.Herlinger and M.O.Kestner, J.Am. Chem.Soc., 98 (1976) 4132.
192 P.Zanello, R.Seeber, A.Cinquantini, G.A.Mazzocchin and L.Fabbrizzi, J.Chem.Soc., Dalton Trans., (1982) 893.
193 L.F.Lindoy, N.E.Tokel, L.B.Anderson and D.H.Bush, J.Coord.Chem., 1 (1971) 7.
194 C.B.Castellani, L.Fabbrizzi, M.Licchelli, A.Perotti and A.Poggi, J. Chem.Soc., Chem.Commun., (1984) 806.
195 J.M.Palmer, E.Papaconstantinou and J.F.Endicott, Inorg.Chem., 8 (1969) 1516.
196 D.C.Olson and J.Vasilevskis, Inorg.Chem., 10 (1971) 463.
197 D.P.Rillema, J.F.Endicott and E.Papaconstantinou, Inorg.Chem., 10 (1971) 1739.
198 M.Nappa, J.S.Valentine, A.R.Miksztal, H.J.Shugar and S.S.Isied, J.Am.Chem.Soc., 101 (1979) 7744.
199 S.Muralidharan and G.Ferradi, Inorg.Chem., 20 (1981) 2306.
200 M.Kodama and E.Kimura, J.Chem.Soc., Dalton Trans., (1976) 2341.
201 L.Fabbrizzi, Comments Inorg.Chem., 4 (1985) 33.
202 P.A.Tasker and L.Sklar, J.Cryst.Mol.Struct., 5 (1975) 329.
203 L.Fabbrizzi, C.Mealli and P.Paoletti, J.Chem.Res., (S), (1979) 170.
204 A.W.Addison and E.Sinn, Inorg.Chem., 22 (1983) 1225.
205 L.Fabbrizzi, M.Micheloni and P.Paoletti, Inorg.Chem., 19 (1980) 535.
206 B.S.Nakani, J.J.B.Welsh and R.D.Hancock, Inorg.Chem., 22 (1983) 2956, and references therein.
207 R.Buxtorf and T.A.Kaden, Helv.Chim.Acta, 57 (1974) 1035.
208 R.E.De Simone, E.L.Blinn and K.F.Mucker, Inorg.Nucl.Chem.Lett., 16 (1980) 23.
209 T.Sakurai, K.Kobayashi, A.Hasegawa, S.Tsuboyama and K.Tsuboyama, Acta Cryst., B 38 (1982) 107.
210 L.Fabbrizzi, M.Licchelli, A.Perotti, A.Poggi and S.Soresi, Isr.J.Chem., 25 (1985) 112.
211 L.Fabbrizzi and A.Poggi, J.Chem.Soc., Chem.Commun., (1980) 646.
212 L.Fabbrizzi, A.Perotti and A.Poggi, Inorg.Chem., 22 (1983) 1411.
213 L.Fabbrizzi, F.Forlini, A.Perotti and B.Seghi, Inorg.Chem., 23 (1984) 807.

214 E.Kimura, M.Shionoya, M.Okamoto and H.Nada, J.Am.Chem.Soc., 110 (1988) 3679.
215 L.Fabbrizzi, T.A.Kaden, A.Perotti, B.Seghi and L.Siegfried, Inorg.Chem., 25 (1986) 321.
216 J.S.Rybka and D.W.Margerum, Inorg.Chem., 20 (1981) 1453.
217 T.J.Truex and R.H.Holm, J.Am.Chem.Soc., 94 (1972) 4529.
218 C.L.Bailey, R.D.Bereman and D.P.Rillema, Inorg.Chem., 25 (1986) 3149.
219 C.L.Bailey, R.D.Bereman, D.P.Rillema and R.Nowak, Inorg.Chem., 23 (1984) 3956.
220 L.Fabbrizzi, A.Lari, A.Poggi and B.Seghi, Inorg.Chem., 21 (1982) 2083.
221 S.P.Roe, J.O.Hill and R.J.Magee, Monatsh.Chem., 118 (1987) 463.
222 R.R.Gagné, J.L.Allison and D.M.Ingle, Inorg.Chem., 18 (1979) 2767.
223 M.J.Maroney and M.Wicholas, Inorg.Chim.Acta, 77 (1983) L 237.
224 N.Ooi, G.-E.Matsubayashi and T.Tanaka, Inorg.Chim.Acta, 114 (1986) 25.
225 O.P.Anderson, C.M.Perkins and K.K.Brito, Inorg.Chem., 22 (1983) 1267.
226 G.Ferraudi, P.Barrera, J.Granifo, J.H.Ham IV, D.P.Rillema and R.Nowak, Inorg.Chem., 24 (1985) 281.
227 N.Takvoryan, K.Farmery, V.Katovic, F.V.Lovecchio, E.S.Gore, L.B. Anderson and D.H.Bush, J.Am.Chem.Soc., 96 (1974) 731.
228 J.Labuda, V.Plaskòn and V.V.Pavlishchuk, Inorg.Chim.Acta, 146 (1988) 13.
229 T.H.Lu, T.J.Lee, B.F.Liang and C.S.Chung, J.Inorg.Nucl.Chem., 41 (1981) 2333.
230 A.E.Elia, E.C.Lingafelter and V.Schomaker, Croat.Chem.Acta, 57 (1984) 653.
231 E.J.Pulliam and D.R.McMillin, Inorg.Chem., 23 (1984) 1172.
232 A.J.Jircitano, R.I.Scheldon and K.B.Mertes, J.Am.Chem.Soc., 105 (1983) 3022.
233 M.R.Caira, L.R.Nassimbeni and P.R.Wooley, Acta Cryst., B 31 (1975) 1334.
234 R.R.Gagné, J.L.Allison, R.S.Gall and C.A.Koval, J.Am.Chem.Soc., 99 (1977) 7170.
235 A.W.Addison, M.Carpenter, L.K.-M.Lau and M.Wicholas, Inorg.Chem., 17 (1978) 1545.
236 R.R.Gagné, J.L.Allison and G.C.Lisensky, Inorg.Chem., 17 (1978) 3563.
237 O.P.Anderson and A.B.Packard, Inorg.Chem., 18 (1979) 1940.
238 J.A.Bertrand, J.H.Smith and D.G.VanDerveer, Inorg.Chem., 16 (1977) 1484.
239 C.O.Dietrich-Buchecker, J.P.Sauvage and J.M.Kern, J.Am.Chem.Soc., 106 (1984) 3043.
240 C.O.Dietrich-Buchecker and J.P.Sauvage, Chem.Rev., 87 (1987) 795.
241 M.Cesario, C.O.Dietrich-Bucheker, J.Guilhem, C.Pascard and J.P.Sauvage, J.Chem.Soc., Chem.Commun., (1985) 244.
242 M.G.Simmons, C.L.Merrill, L.J.Wilson, L.A.Bottomley and K.M.Kadish, J.Chem.Soc., Dalton Trans., (1980) 1827.

243 C.L.Merrill, L.J.Wilson, T.J.Thamann, T.M.Loehr, N.S.Ferris and W.H.Woodruff, J.Chem.Soc., Dalton Trans., (1984) 2207.

244 J.A.Goodwin, D.M.Stanbury, L.J.Wilson and R.A.Scott, in Ref. 4 Vol.II, pp. 11-25.

245 J.D.Korp, I.Bernal, C.L.Merrill and L.J.Wilson, J.Chem.Soc., Dalton Trans., (1981) 1951.

246 M.Pasquali, C.Floriani, A.Gaetani-Manfredotti and A.Chiesi-Villa, J.Am.Chem.Soc., 100 (1978) 4918.

247 M.G.B.Drew, C.Cairns, S.G.McFall and S.M.Nelson, J.Chem.Soc., Dalton Trans., (1980) 2020.

248 H.Umezawa, T.Takita, Y.Sugiura, M.Otsuka, S.Kobayashi and M.Ohno, Tetrahedron, 40 (1984) 501, and references therein.

249 S.J.Brown, P.K.Masharak and D.W.Stephan, J.Am.Chem.Soc., 110 (1988) 1996, and references therein.

250 Y.Iitaka, H.Nakamura, T.Nakatani, Y.Muraoka, A.Fujiü, T.Takita and H.Umezawa, J.Antibiot., 31 (1978) 1070.

251 N.J.Oppenheimer, C.Chang, L.O.Rodriguez and S.M.Hecht, J.Biol.Chem., 256 (1981) 1514.

252 J.C.Dabrowiak and F.S.Santillo, J.Electrochem.Soc., 126 (1979) 2091.

253 J.Zubieta, K.D.Karlin and J.C.Hayes, in Ref. 3, pp.97-108.

254 K.D.Karlin, J.C.Hayes, S.Juen, J.P.Hutchinson and J.Zubieta, Inorg. Chem., 21 (1982) 4106.

255 M.Kodama and E.Kimura, J.Chem.Soc., Dalton Trans., (1978) 104.

256 E.Kimura, A.Sakonaka and R.Machida, J.Am.Chem.Soc., 104 (1982) 4255.

257 I.Bertini, D.Gatteschi and A.Scozzafava, Inorg.Chem., 16 (1977) 1973.

258 J.H.Ammeter, H.B.Bürgi, E.Gamp, V.Meyer-Sandrin and W.P.Jensen, Inorg. Chem., 18 (1979) 733.

259 B.J.Hathaway, Coord.Chem.Rev., 52 (1983) 87.

260 P.Chauduri, K.Oder, K.Wieghardt, J.Weiss, J.Reedijk, W.Hinrichs, J.Wood, A.Ozarowski, H.Stratemaier and D.Reinen, Inorg.Chem., 25 (1986) 2951.

261 O.P.Anderson, J.Chem.Soc., Dalton Trans., (1972) 2597.

262 E.I.Onstott and H.A.Laitinen, J.Am.Chem.Soc., 72 (1950) 4724.

263 K.Tanaka and N.Nakano, Bull.Chem.Soc.Jpn., 47 (1974) 2222.

264 R.R.Ruminski, Inorg.Chim.Acta, 103 (1985) 159.

265 M.P.Suh and S.-G.Kang, Inorg.Chem., 27 (1988) 2544.

266 J.Becher, D.J.Brockway, K.S.Murray, P.J.Newman and H.Toftlund, Inorg.Chem., 21 (1982) 1791.

267 J.M.Latour, G.A.Leonard, D.Limosin, D.C.Povey and S.S.Tandom, in Ref. 4, pp.221-230.

268 L.Stamp and H.Tom Dieck, Inorg.Chim.Acta, 147 (1988) 199.

269 D.F.Rohrbach, W.R.Heineman and E.Deutsch, Inorg.Chem., 18 (1979) 2536.

270 P.Zanello and A.Cinquantini, Transition Met.Chem., 10 (1985) 370.

271 R.R.Gagné, C.A.Koval, T.J.Smith and M.C.Cimolino, J.Am.Chem.Soc., 101 (1979) 4571.

272 B.C.Whitmore and R.Eisenberg, Inorg.Chem., 22 (1983) 1.

273 H.Doine, F.F.Stephens and R.D.Cannon, Inorg.Chim.Acta, 75 (1983) 155.

274 A.W.Addison, Inorg.Nucl.Chem.Lett., 12 (1976) 899.

275 T.Izumitani, M.Nakamura, H.Okawa and S.Kida, Bull.Chem.Soc.Jpn., 55 (1982) 2122.

276 M.Nakamura, H.Okawa and S.Kida, Inorg.Chim.Acta, 62 (1982) 201.

277 G.E.Gurr, Acta Cryst., B 24 (1968) 1511.

278 E.N.Baker, D.Hall and T.N.Waters, J.Chem.Soc.(A), (1970) 406, and references therein.

279 R.C.Elder and M.C.Hill, Inorg.Chem., 18 (1979) 729.

280 T.P.Cheeseman, D.Hall and T.N.Waters, J.Chem.Soc.(A), (1966) 1396.

281 Y.Nakao, N.Nonagase and A.Nakahara, Bull.Chem.Soc.Jpn., 42 (1969) 452.

282 A.Cinquantini, R.Seeber, R.Cini and P.Zanello, J.Electroanal.Chem., 134 (1982) 65.

283 Y.Nishida, M.Takeuchi, N.Oishi and S.Kida, Inorg.Chim.Acta, 96 (1985) 81.

284 Y.Nishida, N.Oishi, H.Kuramoto and S.Kida, Inorg.Chim.Acta, 57 (1982) 253.

285 E.N.Baker, D.Hall and T.N.Waters, J.Chem.Soc.,(A), (1970) 396, and references therein.

286 P.Zanello, S.Tamburini, P.A.Vigato and G.A.Mazzocchin, Transition Met. Chem., 9 (1984) 176.

287 B.Adhikary, S.K.Mandal and K.Nag, J.Chem.Soc., Dalton Trans., (1988) 935.

288 R.Graziani, M.Vidali, U.Casellato and P.A.Vigato, Transition Met.Chem., 3 (1978) 138.

289 C.Bisi Castellani, O.Carugo and A.Coda, Acta Cryst., C 44 (1988) 267.

290 C.Bisi Castellani, A.Buttafava, O.Carugo and A.Poggi, J.Chem.Soc., Dalton Trans., (1988) 1497.

291 P.J.Burke and D.R.McMillin, J.Chem.Soc., Dalton Trans., (1980) 1794.

292 J.Ellis, G.M.Mockler and E.Sinn, Inorg.Chem., 20 (1981) 1206, and references therein.

293 P.Zanello, R.Cini and A.Cinquantini, Inorg.Chim.Acta, 74 (1983) 89.

294 R.Cini, P.Zanello, A.Cinquantini, A.Colligiani, C.Pinzino and G.Valentini, Inorg.Chim.Acta, 88 (1984) 105.

295 U.Casellato, D.Fregona, S.Sitran, S.Tamburini, P.A.Vigato and P.Zanello, Inorg.Chim.Acta, 95 (1984) 309.

296 P.Zanello, A.Cinquantini, P.Guerriero, S.Tamburini and P.A.Vigato, Inorg.Chim.Acta, 117 (1986) 91.

297 L.Sacconi and I.Bertini, J.Am.Chem.Soc., 88 (1966) 5180.

298 P.C.Healy, G.M.Mockler, D.P.Freyberg and E.Sinn, J.Chem.Soc., Dalton Trans., (1975) 691.

299 E.D.McKenzie and S.J.Selvey, Inorg.Chim.Acta, 18 (1976) L 1.

300 J.P.Gisselbrecht and M.Gross, A.C.S. Advances in Chemistry Series, N. 201, (1982) 109.

301 M.Herceg and R.Weiss, Acta Cryst. B 29 (1973) 542.

366

302 E.Kimura, T.Koike, Y.Shimizu and M.Kodama, Inorg.Chem., <u>25</u> (1986) 2242.

303 J.Lisowski, M.Grzeszczuk and L.Latos-Grazynski, Recl.Trav.Chim. Pays-Bas, <u>106</u> (1987) 319.

304 L.Latos-Grazynski, J.Lisowski, M.M.Olmstead and A.L.Balch, J.Am.Chem. Soc., <u>109</u> (1987) 4428.

305 K.D.Karlin and S.E.Sherman, Inorg.Chim.Acta, <u>65</u> (1982) L 39.

306 J.M.Guss, P.R.Harrowell, M.Murata, V.A.Norris and H.C.Freeman, J.Mol.Biol., <u>192</u> (1986) 361.

307 P.Zanello, Comments Inorg.Chem., <u>8</u> (1988) 45.

308 V.V.Pavlishchuk, P.E.Strizhak, K.B.Yatsimirskii and J.Labuda, Inorg.Chim.Acta, <u>151</u> (1988) 133.

309 M.Gullotti, L.Casella, A.Pintar, E.Suardi, P.Zanello, and S.Mangani, J.Chem.Soc., Dalton Trans., in press.

310 L.Casella, Inorg.Chem., <u>23</u> (1984) 2781.

311 L.Casella, M.Gullotti, A.Pintar, F.Pinciroli, R.Viganò and P.Zanello, J.Chem.Soc., Dalton Trans., in press.

312 M.G.B.Drew, C.Cairns, S.M.Nelson and J.Nelson, J.Chem.Soc., Dalton Trans., (1981) 942.

313 E.John, P.K.Bharadway, K.K.-Jespersen, J.A.Potenza and H.Schugar, J.Am.Chem.Soc., <u>108</u> (1986) 5015, and references therein.

314 F.T.Greenaway, L.M.Brown, J.C.Dabrowiak, M.R.Thompson and V.M.Day, J.Am.Chem.Soc., <u>102</u> (1980) 7782.

315 A.H.Alberts, R.Annunziata and J.-M.Lehn, J.Am.Chem.Soc., <u>99</u> (1977) 8502.

316 A.Bencini and F.Mani, Inorg.Chim.Acta, <u>154</u> (1988) 215.

317 P.Comba, N.F.Curtis, G.A.Lawrance, M.A.O'Leary, B.W.Skelton, and A.H.White, J.Chem.Soc., Dalton Trans., (1988) 2145.

318 P.Comba, N.F.Curtis, G.A.Lawrance, M.A.O'Leary, B.W.Skelton, and A.H.White, J.Chem.Soc., Dalton Trans., (1988) 497

319 P.J.Hochgesang and R.D.Bereman, Inorg.Chim.Acta, <u>156</u> (1989) 213.

320 J.A.Goodwin, G.A.Bodager, L.J.Wilson, D.M.Stanbury, and W.R.Scheidt, Inorg. Chem., <u>28</u> (1989) 35.

321 J.A.Goodwin, L.J.Wilson, D.M.Stanbury, and R.A.Scott, Inorg.Chem., <u>28</u> (1989) 42.

322 J.A.Goodwin, D.M.Stanbury, L.J.Wilson, C.W.Eigenbrot, and W.R.Scheidt, J.Am.Chem.Soc., <u>109</u> (1987) 2979.

SUBJECT INDEX

(grouped by chapter)

Chapter 1(pp. 3-92)

Chapter 2(pp. 95-177)

Chapter 3(pp. 181-355)